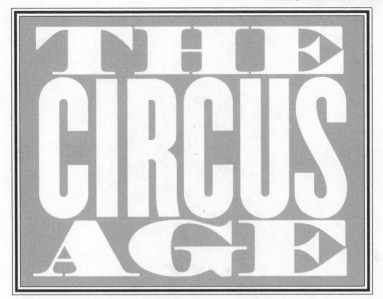

THE CIRCUS AGE

CULTURE
& SOCIETY
UNDER THE
AMERICAN
BIG TOP

JANET M. DAVIS

THE UNIVERSITY OF
NORTH CAROLINA PRESS
CHAPEL HILL & LONDON

THE CIRCUS AGE

© 2002
The University of North Carolina Press
All rights reserved
Designed by Richard Hendel
Set in Monotype Bell, Champion, and Madrone types by Tseng Information Systems, Inc.
Manufactured in the United States of America

Published with the aid of a University Cooperative Society Subvention Grant
awarded by the University of Texas at Austin.

Library of Congress Cataloging-in-Publication Data
Davis, Janet M.
The circus age: culture and society under the American big top / Janet M. Davis.
p. cm.
Includes bibliographical references and index.
ISBN-13: 978-0-8078-2724-6 (cloth : alk. paper)
ISBN-10: 0-8078-2724-X (cloth : alk. paper)
ISBN-13: 978-0-8078-5399-3 (pbk. : alk. paper)
ISBN-10: 0-8078-5399-2 (pbk. : alk. paper)
1. Circus—Social aspects—United States. 2. Circus—United States—History—
20th century. I. Title.
GV1803 .D38 2002
791.3'0973—dc21
2002000863

cloth   06   05   04   03   02   5   4   3   2   1
paper   10   09   08   07   06   7   6   5   4   3

To Jeff

and in memory of my mother,

Jean B. Davis,

January 5, 1931–April 25, 2002

CONTENTS

ILLUSTRATIONS

A section of color plates follows page 140.

PREFACE

I discovered the topic of this book at Chicago's Museum of Science and Industry on a brisk March day in 1990. While I wandered amid pickled slices of the human body, a giant, ceaselessly swinging pendulum, a dimly lit Main Street in Chicago (circa 1910), a walk-in submarine, and airplanes through the ages, I came upon a photographic display of turn-of-the-twentieth-century American circus parades. It mesmerized me. I was a new graduate student in modern South Asian history at the time and keenly interested in the ways that the British had used indigenous art, architecture, and pageantry to consolidate their colonial authority in India. Strongly influenced by Edward Said's landmark treatise, *Orientalism*, I wanted to study the relationship between British Indian culture (specifically popular culture) and politics.[1] Yet here, in these grainy black-and-white photographs of provincial American main streets—complete with clapboard houses and tidy picket fences—were gigantic crowds gazing with giddy pleasure at camels, caged tigers, and elaborately caparisoned elephants topped with fake South Asian *mahouts* and *howdahs*. In short, these pictures of American circus parades contained a strikingly similar web of orientalist images that I had seen in British representations of South Asia. What did this seemingly apropos circus orientalism reveal about turn-of-the-century American culture? What did the circus tell its audiences about empire in a nation where empire—ostensibly—did not exist? Why was this circus display at the Museum of Science and Industry of all places? Over the next two years, these persistent questions led me into a graduate program in U.S. history where I (much) later completed a doctoral dissertation that became the basis for this book.

Before I encountered the circus at the Museum of Science and Industry, I had had little contact with this cultural form. During childhood, I had seen a few shows, and even played "Fearless Fanny, the Lion Tamer" as a high school student in a local children's theater production. But that was it. I simply regarded the circus as a beguiling childhood pleasure, rendered complete with pink lemonade and peanuts.[2] Ernest Hemingway once wrote that the circus "is the only ageless delight that you can buy for money."[3] The circus scholar Marcello Truzzi defines the circus as a "traveling and organized display of animals and skilled performers within one or more circular stages known as 'rings' before an audience encircling these activities."[4] Circus traditions existed in ancient Rome (where the yawning circular "circus maximus" satisfied spectators with bloody gladiator brawls and chariot races), India, China, Mexico, Russia, Eastern Europe, and Southeast Asia. Within its self-contained universe of rings, the circus offers metaphysical entertainment. It showcases the interconnectedness of the human and animal, and playfully tests one's potential to transfigure physical laws. The circus contains innumerable bodies—lithe and muscular, fat, hairy, shockingly thin, flexible, glass- and fire-eating, legless, armless. I soon realized that the circus fit right into the Museum of Science and Industry's profusion of preserved innards, massive bodily replicas, and technologies that have enabled humans to fly and breathe under water.

I also quickly learned that the physical structure of the circus was far from ahistorical, even if some of its offerings in the ring were age-old. Instead, I recognized that the circus is a dazzling mirror of larger historical processes. In the United States, the circus's growth and development closely chronicled that of the nation, because the circus—a traveling amusement— was dependent on the same transportation networks that helped facilitate U.S. expansion. The enormous three-ring railroad circuses that I saw pictured in the photographic display at the Museum of Science and Industry were the product of a newly consolidated nation-state comprising transcontinental railroads, new communications technologies, and a new overseas empire.

Furthermore, a unique set of primary source materials drew me inside the dynamic aggregate of workers, animals, show owners, and audiences who created the circus's representational power. Each large railroad circus kept a route book, a meticulous daily diary of each performance stop that chronicled the size of the audience and its ethnic composition, in addition to unusual incidents like births, accidents, fights, storms, animal rampages, and thefts. Published and unpublished manuscripts by show owners and performers, newspaper editorials, circus fiction, show programs, music, trade periodicals, state laws, photographs, lithographs, and film led me deeper into the rich, convoluted world of the railroad circus. Influenced by the interdisciplinary methodology of American studies and its attention to culture and identity formation, the politics of inclusion and exclusion, and the interplay between the local, national, and global, I have attempted to use the turn-of-the-twentieth-century railroad circus as a way to explore then-prevailing attitudes about gender, race, labor, sexuality, monopoly formation, nationalism, and empire.[5]

At its heart, this book looks to the circus as a way to understand ideological processes. Defining ideology as a system of ideas, I use the term on several levels, among individuals (impresarios, for example), classes (in analyzing why middle-class purity reformers did *not* target the circus as an object of censure), and at the broadest national level (in considering turn-of-the-century nationalism and empire). The sociologist Karl Mannheim contends that ideology must be located in actual social practices, among ordinary people, as well as elite power brokers. He has written that ideology and utopia are dialectical social constructions—the former dedicated to preserving the status quo while the latter tries to overthrow it.[6] "Once the individual has grasped the method of orienting himself in the world, he is inevitably driven beyond the narrow horizon of his own town and learns to understand himself as part of a national, and later of a world, situation."[7] Consequently, this book explores (among other things) the unsettling ideological power of witnessing an elephant passing by one's front yard in, say, Keokuk, Iowa, the intricate network of international machinations that brought the elephant to the United States through the global circus trade, and the various racial and sexual stereotypes that the elephant symbolized about its country of origin. At a time before the proliferation of mass media technologies, the railroad circus presented to its audiences a global sensory blitz—immediate, live images that mirrored the nation's position in the modern world.

ACKNOWLEDGMENTS

The process of writing this book has been somewhat of a circus, a logistical spectacle made possible only through the collective generosity of colleagues, friends, and family. It is a pleasure to thank those who have helped me over the past decade.

My first debt is to Linda Gordon, who supervised the dissertation on which this book is based. In addition to Linda's unflagging faith that I would finish, she helped immeasurably in the project's conceptualization and organization, paying close attention to stylistic matters and syntax, while always pushing me to think more fully about the big picture. I am very lucky to have had such an extraordinary mentor. I was equally fortunate to work with Paul Boyer, whose good humor and excellent suggestions have improved the project considerably. This book first came to life as a seminar paper for Tom McCormick. Throughout the years, Tom has offered constant support, superb criticism, and friendship. Kirin Narayan also made invaluable observations and provided a delightful combination of mentoring and irreverence during my years at UW-Madison. This book has also benefited from Bill Cronon's incisive observations. Thanks to David Zonderman for his careful reading of an earlier version of this project. Judy Cochran, graduate advisor for the UW history department, deserves special thanks. Even as she struggled with lung cancer, Judy was incredibly generous with her time and expertise. She is deeply missed.

My work has been nurtured by scholarly conversations at professional meetings. The following individuals have given me much helpful feedback: Ronald Inden, Claire Potter, Katherine Morrisey, Bethel Saler, Alison Kibler, Robert Rydell, Neil Harris, Bluford Adams, Doug Mischler, Alexander Saxton, Carolyn Strange, Tina Loo, Bonnie Huskins, Karen Dubinsky, Fred Dahlinger, Fred Pfening Jr., Fred Pfening III, Stuart Thayer, Richard Reynolds III, Bill Slout, Steve Gossard, Jay Cook, Martha Burns, Brett Mizelle, Clay McShane, Kasey Grier, Jennifer Price, Andrew Isenberg, Nigel Rothfels, Molly Mullin, Jennifer Ham, Alex Missal, Gail Huesch, Anneke Leenhouts, Kay Schaffer, Jan Todd, and Patricia Vertinsky. Other discussions with Laurie Hovell-McMillin, Andrew Neather, Mike Sappol, Cynthia Enloe, David Roediger, and Jeff Hyson have also clarified my thinking.

Thanks to the following individuals who have provided references, clippings, and other research materials: Stuart Thayer, Fred Dahlinger, Richard Reynolds III, Bill Slout, Steve Gossard, Monica Drane, Laura Beausire, Jyotsna Uppal, Estelle Young, Tony Harkins, Alison Kibler, Linda Frost, Molly Mullin, Shelley Fisher Fishkin, Jeff Meikle, Bill Goetzmann, Kathie Tovo, Bernth Lindfors, Vicki Howard, Denise Spellberg, John Haddad, Sally Clarke, Jan Todd, Kim Hewitt, Susan Traverso, Tommie Meyers, Lisa Tetrault, Heidi Harkins, Kathy Messerich, Jeff Osborne, Andrea Osborne, Zachary Osborne, and Jean Davis.

A giant thanks to my friends and colleagues who have read parts or all of the project at various stages. Members of the Women's History Dissertator Group sharpened my ideas on many occasions. In addition, Laura McEnaney, Alison Kibler, Jyotsna Uppal, and Eleanor Zelliot offered close readings of individual chapters at significant moments in the book's development. A huge thanks to Fred Dahlinger for his careful reading of an earlier draft of the entire manuscript. His comments were absolutely invaluable. I am also indebted to Cindy Aron's terrific suggestions for improving the whole manuscript. A heartfelt thanks to Jeff Osborne, who performed a crucial eleventh-hour reading of the work in its entirety. I am grateful to Paul Hockings's comments on an early portion of this work that appeared in *Visual Anthropology* (1993); the editors there have graciously granted me permission to reprint a revised version of that essay in Chapter 6. I greatly appreciate the insightful suggestions of Robert Rydell and an anonymous reviewer who read the manuscript for the University of North Carolina Press.

This project would not have been possible without the tremendous support of archivists and librarians. The staff at Circus World Museum's Robert L. Parkinson Library and Research Center has been outstanding. The library's director, Fred Dahlinger, is an amazing font of circus knowledge. Unsparing with his time and resources, Fred has offered his support at every stage of this project. Every researcher should be so lucky as to have such wonderful guidance. Thanks also to Meg Allen, who kindly processed all of my photographic requests and double-checked various references. Debbie Walk at the John and Mabel Ringling Museum of Art in Sarasota, Florida, generously allowed me to stay at the archives past business hours. Likewise, Heidi Taylor has been very helpful with my photo requests. Thanks to Mary Ann Jensen at the Princeton University Library in Princeton, New Jersey, for guiding me through the Joseph T. McCaddon Circus Collection. Meg Sherry Rich kindly expedited the permissions process at Princeton.

Extended quotations are published with permission of the Princeton University Library. Joan Barborek and other staff members at the Hertzberg Circus Collection and Library in San Antonio, Texas, have aided my quest for materials. (As this book goes to press, I am sad to report that the City of San Antonio has decided to close this invaluable research facility.) The staff at the Harold McCracken Research Library at the incredibly beautiful Buffalo Bill Historical Center in Cody, Wyoming, have been very helpful. In particular, thanks to Francis Clymer for archival assistance, and Elizabeth Holmes and Ann Marie Donoghue for help with photographic requests. Thanks to Molly Long at Feld Entertainment, Inc., for granting me permission to use Ringling Bros. and Barnum & Bailey poster images in this book. Lastly, a word of appreciation for Jamie Creadle at the McFadden-Ward House and Museum in Beaumont, Texas, who provided wonderful research materials and terrific hospitality during my stay there.

This project has been greatly enhanced by the generous support of various institutions. A dissertation fellowship from the American Association of University Women provided access to far-flung archival material and provided precious, uninterrupted writing time while I was in graduate school. A National Endowment for the Humanities Summer Stipend, a Summer Research Assignment from the University of Texas, and a University of Texas Research Grant have also given me crucial support. Thanks to Allison Perlman, whose outstanding performance as my research assistant for a different project inevitably influenced sections of this book. A University Cooperative Society Subvention Grant awarded by the University of Texas at Austin enabled me to include color photographs in this book — thus giving a taste of the profuse vivid poster images that turn-of-the-century audiences experienced in advance of the circus.

In the fall of 1998, I landed at the University of Texas at Austin as a nervous, brand-new assistant professor. My colleagues in the American Studies Department and in my other home in the History Department quickly made me comfortable, and have continued to provide essential intellectual kinship and a congenial environment for teaching and research.

A group of extraordinary teachers at Carleton College influenced my career path. I am pleased to thank Diethelm Prowe, Bardwell Smith, Bob Bonner, and Eleanor Zelliot for their wise counsel. Bob Bonner, in particular, steered me away from law school at a critical moment. Eleanor Zelliot continues to be a mentor, a model of generosity and great integrity.

It has been a pleasure to work with the staff at the University of North Carolina Press. My editor, Sian Hunter, has provided sharp insights, tremendous patience, and fine humor at every turn. Assistant managing editor Pamela Upton has been a delight, answering my murky queries with constant clarity and good cheer. Fred Kameny's copyediting has been superb. A gigantic thanks to everyone at the Press.

Throughout this project, my friends and family have sustained me. Thanks to old friends Estelle Young, Laura Beausire, and Monica Drane for their circus (and other) insights. Thanks also to Tony Harkins and Tracy Harkins, and new friends here in Austin—Denise Spellberg, Caroline Castiglione, and Sally Clarke. Steve Hoelscher, Kris Nilsson, Erika Hoelscher, Janice Bradley, Mark Smith, and Georgia Xydes have made life in Austin particularly enjoyable. And finally, a huge thanks to my family. My sisters, Kathy Messerich and Betsy Moran, and my brother, Steve Davis have been a constant source of affection and amusement. My mother, Jean Davis, provided an enormous wealth of research information, and even went to bat for me at the archives at a dire moment. Throughout my life, she has provided love and enthusiastic support for my various adventures. She and Karen Osborne both helped hasten the project's completion by taking care of my children at various moments. My father, Hugh Davis, and Heidi Harkins also provided vital support. I owe my greatest debt to my partner in life, Jeff Osborne, and to our children, Andrea and Zachary. The kids have been remarkably patient with my circus musings, and have provided welcome relief with cat songs, animal noises, escalator rides, geography games, golf ball hunts, and lounging sessions on the couch. Jeff has provided steadfast love, intellectual support, and an abiding sense of silliness. He has also taken care of the business of daily life—watching the children, feeding the pets, washing the dishes, paying the bills, and the like—which is the main reason why I have finally been able to finish this book.

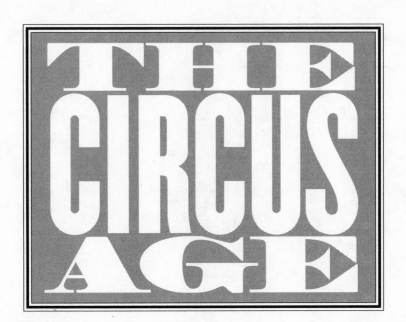

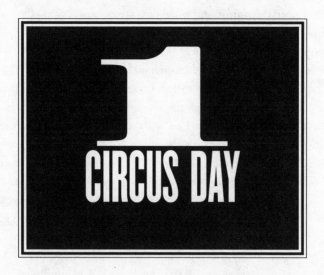

CIRCUS DAY

On June 11, 1999, the Ringling Bros. and Barnum & Bailey circus crept into Austin, Texas, at dusk. Arriving at a downtown rail yard on that still, sultry evening, the circus quietly conveyed its animal stock to the nearby show site without any announcement to the public, in order to avoid traffic, insurance hassles, and most important, confrontations with animal rights activists.[1] The circus had been advertised in the local newspaper and on television, but the media paid little attention to its actual presence during its two-day stint. The *Austin American-Statesman* contained only one blurb about the circus, sandwiched next to a notice about a local traffic death: "Elephant dung for the taking: Bring your own shovel and a bucket today if you want to scoop up manure from the elephants owned by the Ringling Bros. and Barnum & Bailey Circus."[2] The circus performed four times at the Erwin Center, an expansive, air-conditioned indoor arena at the University of Texas. In the blinding heat and sunshine, parents with small children streamed to the show from adjacent parking lots, grateful to enter the climate-controlled cool surrounding the circus. Meanwhile, the hustle and bustle of community life and commerce continued outside uninterrupted.

Yet a hundred years earlier, a large railroad circus shut a town down. Months before, people knew that it was coming: scores of "advance men" and billposters had already plastered all over dull barns, storefronts, and saloons thousands of vivid lithographs of wild animals and scantily clad performers emblazoned in splashes of peacock blue, orange, molten red, yellow, grass green, plum, and gold to advertise the upcoming show. In

1892 Adam Forepaugh's circus, for one, announced its impending presence in Philadelphia by mummifying an eight-story building with 4,938 lithographs, in addition to pasting thousands of other posters around the city.[3] In detail, local newspapers eagerly chronicled the circus's movement, along with complete information about its arrival time.

On "Circus Day" (as it was called in newspapers, memoirs, and show programs across the nation), shops closed their doors, schools canceled classes, and factories shut down. In 1907 the Board of Education in Bridgeport, Connecticut, voted to close the schools on Circus Day, and children in Paterson, New Jersey, successfully lobbied school authorities to dismiss classes.[4] When the Adam Forepaugh circus arrived in South Bend, Indiana, that same year, the Studebaker Wagon Works locked its doors so that its seven thousand employees could see the program.[5] Special trains offering discounted "excursion" fares transported rural circus-goers living within a fifty-mile radius of the show grounds. Roads became thick with people, horses, and wagons. A resident of Clifton, Arizona, remembered that when Buffalo Bill's Wild West came to town in 1913, some local farmers sold part of their hay and grain supply in order to take their entire families to the show.[6] Farmers traveled by horse and wagon twenty to forty miles and spent scant cash on novelty items like popcorn, cotton candy, and pink lemonade.[7] Known as "rubber necks" to circus workers, rural residents craned constantly to take it all in. Sherwood Anderson was mesmerized by Circus Day as a boy in Clyde, Ohio: "When a circus came to the town where Tar [Anderson] lived he got up early and went down to the grounds and saw everything, right from the start, saw the tent go up, the animals fed, everything."[8] In 1904 a newspaper in the mill town of Ashland, Wisconsin, near the shores of Lake Superior, noted the circus's impact: "All the roads brought in large train loads of people who came here to attend the circus and many people arrived last evening. All the mills on this side of the bay stopped work today noon and almost all business is at a standstill and everyone is taking the circus."[9]

The railroad circus overwhelmed large cities as well. When Barnum & Bailey opened its annual season in New York City in 1905, the route book reported that both the matinee and evening programs on March 24 at Madison Square Garden (where the self-styled "Greatest Show on Earth" traditionally opened each year) were "big," "packed." Many others were turned away. The next day, there was an "immense crush" at the doors when huge crowds were refused entry at the already overflowing arena.[10] The Ringling Bros. circus virtually shut down New Orleans in 1898. According to the

Daily Picayune, "Last night the Ringling Bros.' Circus came near depopulating the city. It looked as if everybody had gone to the big show. If you wanted to see anybody you had only to look through the crowd, for they were all there."[11]

On Circus Day, thousands of spectators spilled into the streets to watch the free parade (fig. 1). Barnum & Bailey's New York City parade in 1891 had 400 horses, 16 elephants, 1,000 circus performers, and copious animals from the menagerie. This living sensory mass of color, sound, and odor proceeded slowly down Fifth Avenue, weaving through congested Manhattan until it reached Madison Square Garden.[12] The scene was equally grand in provincial towns. In 1904 a filmmaker captured brief, grainy images of Barnum & Bailey's parade in Waterloo, Iowa, on celluloid: thick crowds, jiggling dromedaries, zebra herds, a forty-horse hitch, a military band, intricate, gilded "cage" wagons, each housing panting feline predators, smiling, waving women dressed in gauzy, kimonolike gowns atop the elephants, and a calliope at the rear of this moving expanse.[13] Knowing that throngs of people watched the parade from second- and third-story windows, the John Robinson circus built fancy tin roofs on its wagons (called "cottage cages") with brightly painted designs that could be viewed from above.[14]

Long, winding lines at the ticket wagon greeted audience members who had not purchased their tickets in advance. Warren S. Patrick, treasurer of the Walter L. Main circus, remarked that selling 8,000 to 9,000 tickets in forty minutes (approximately 1,000 others had been sold in advance) was tough on his hands. "My mental calculation is invariably right; but now and then my fingers, after a severe strain, may drop one or two [quarters], too many or too little."[15] Inside the show grounds, crowds wandered around gawking at the enormous tented city that could stretch across ten acres (fig. 2). Along the noisy midway, candy "butchers" (vendors) sold lemonade, palm frond fans, sausages, and roasted peanuts. Remembering Circus Day in his hometown of Galesburg, Illinois, Carl Sandburg vividly recalled the midway men who beckoned audiences with "oily tongue" to play games of chance for cheap prizes: "Only ten cents for a ring and the cane you ring is the cane you get."[16] An hour before each big-top production, masses of people gathered at the sideshow tent lined with colorful banners depicting the Fat Lady, the Skeleton Man, the Dog-Faced Boy, and the others inside. A velvety-voiced spieler (or talker) lured patrons to part with a dime and come inside during the "blow off," a tantalizing outdoor display of seminude women flexing their muscles, a "living picture gallery" tattoo artist, or perhaps a rousing rendition of skin snapping by the Elastic Skin Man. During

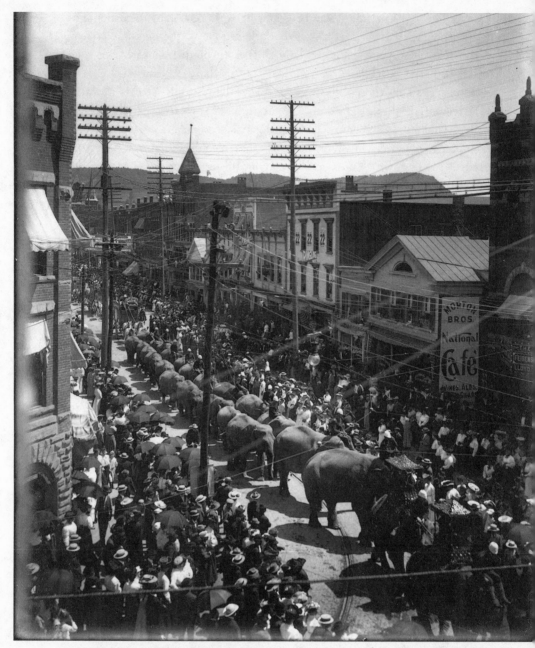

*Figure 1. Ringling Bros. circus parade, Oneonta, N.Y., July 22, 1905. On "Circus Day,"
people took to the streets and upper-story windows to watch herds of elephants (twenty-
four here) and other parts of the free parade wind ponderously through town.
(Photograph courtesy of Circus World Museum, Baraboo, Wis., RB-N81-05-35)*

Figure 2. Hillside view of show grounds at Red Wing, Minn., Ringling Bros., 1915.
Spread out across approximately ten acres, the circus was a vast, temporary canvas city.
(Photograph courtesy of Circus World Museum, Baraboo, Wis., RB-N81-15-4-1)

the "blow off," some spielers even quietly intimated that audiences *might* see nude women at the adjacent "Gentlemen Only" "cooch" show.

Once inside the menagerie tent attached to the big top, spectators saw big cats and bears lounge, eat chunks of meat, and pace in their cages, while llamas, giraffes, educated pigs, horses, chimpanzees, and peacocks fidgeted nearby. The lively strains of the brass circus band—including operatic selections, marches, and plantation melodies—told the milling audience members that it was time to head inside the big top for the main program. Candy butchers shouted and scurried around the cavernous big top, a massive canvas space propped aloft by huge poles and ropes that could hold over 10,000 people (fig. 3). A grand, paradelike entry processional of animals and performers marked the start of the main program. Approximately twenty to twenty-five other acts followed. An international constellation of players worked simultaneously on three rings and two stages. At a typical Ringling Bros. show, performers heralded from twenty-two countries, including Persia, Japan, and Italy; fifty clowns cavorted around the serious acts in vignettes of intentional chaos.[17] The athletic prowess of these sleek,

[*Circus Day*]

5

Figure 3. Barnum & Bailey, big-top interior, from advertisement in route book for tent maker, 1904. Note the presence of the sloping thrill act apparatus to the center left. Measuring approximately 200 feet wide by 460 feet long, the huge big top made its human inhabitants, shown center right, look tiny. (Interior photograph courtesy of Circus World Museum, Baraboo, Wis., B+B-N45-04-6)

muscular bodies was startling. As a boy in rural Iowa, the writer Hamlin Garland observed that "the stark majesty of the acrobats subdued us into silent worship."[18] Mark Twain's Huck Finn echoed this sentiment as he solemnly watched big-top feats in a small Arkansas community: "It was a powerful fine sight; I never see anything so lovely . . . the men looking ever so tall and airy and straight . . . and every lady's rose-leafy dress flapping soft and silky around her hips, and she looking like the most loveliest parasol."[19] The big-top program ended with a series of rousing horse races on the arena's outer hippodrome track.

The mammoth circus audience was also part of the spectacle, as thousands of "strangers" from around a county streamed into town. Big cities overflowed. Provincial communities became temporary cities, complete with anonymous, pushing crowds. Fred Roys of New York compared the religious revivals of his youth to Circus Day. "Them religious revivals they used to have . . . they was great doin's. When I was a kid we used to look forward to 'em like we did the circus. Sometimes they was as good as a circus."[20] Newspapers focused on the crowd as a defining element of Circus Day. In 1890 one journalist described the "show" of nearly ten thousand people from around a county filing into Barnum & Bailey's big top: "It was the biggest crowd of people ever in one tent in the city. A great sea of faces stretched out in every direction, representing all of the country thirty miles around. To see so many people was the best part of the 'performance.'"[21] Newspapers provided detailed lists of trains bringing specific numbers of people from outlying communities to the circus.[22]

Yet the sheer physical presence of a circus and its swirling masses was often bewildering. When Ringling Bros. played at Mount Pleasant, Iowa, on a steamy summer day in 1894, the huge throng became confused: "Pandemonium reigned and it seemed as if everybody was panic stricken. Families were parted, children screamingly hunted for parents, and parents distractedly hunted for children. Almost everyone was drenched to the skin and many a toilet was hopelessly ruined. Fortunately no one was hurt and the damage to the property little or nothing."[23]

Furthermore, the thousands of patrons tightly packed under the canvas tents were vulnerable in bad weather. The "blow down," or severe storm, was common. At Adam Forepaugh's 1893 date in Sioux Falls, South Dakota, audience members were trapped under the heavy big top after a gale force wind collapsed the tent.[24] That same summer, in River Falls, Wisconsin, seven people were killed after lighting struck one of the center poles at the Ringling Bros. circus.[25] When a windstorm "swayed and rocked" the big top at the Ringling Bros.' stint in Sherman, Texas, in 1900, spectators were so jittery that many of the more than 10,000 there "made a wild rush to get out."[26] Tornadoes, hail, wind storms, torrential rain, and knee-deep mud were some of the weather hazards that could abruptly end the program. But these attendant weather-related dangers were also part of the jarring excitement on Circus Day.

This diverse, elephantine community disruption otherwise known as Circus Day reached its peak at a turbulent historical moment. In 1903 ninety-eight circuses and menageries—the highest number in U.S. history—traveled the nation. At least thirty-eight of these rumbled by rail, and several journeyed coast to coast in a single season.[27] The historian Robert Wiebe characterizes this era as a time when a provincial "nation of loosely connected islands" was giving way to an anonymous, modern, urban, industrial society. Speculative investments in land and capital, the formation of large corporations, and accelerated industrialization defined the burgeoning post–Civil War economy. The proliferation of national railroad networks, the spread of the telegraph and telephone, the rise of the unscrupulous Gilded Age "robber baron," and the stirrings of the nascent automobile industry all helped destabilize an older, provincial way of life. Consequently, Wiebe argues that a "search for order" animated the Progressive project of corporate regulation and urban reform.[28]

The face of the nation also changed rapidly during this era. From 1890 to 1924, about 23 million immigrants poured into the United States hoping to find prosperity in the nation's expanding industrial economy.[29] The federal census of 1890 declared that the frontier had officially "closed," for there was no longer a clear line between settlement and wilderness in the trans-Mississippi West. By 1920 the U.S. census revealed that 51 percent of the population lived in cities with more than 2,500 residents. In the early twentieth century, the manufacturing output of the United States now exceeded that of Great Britain, Germany, and France combined.[30] Some Americans, like Edward Bellamy (*Looking Backward*, 1888), held great faith in the eventual utopian promise of industrial society. Henry Ford envisioned the automobile as a democratic symbol for the "great multitude," produced by efficient, well-mannered workers on an assembly line, whose productivity and thrift would also make them good car buyers.[31]

Yet this enthusiasm was hardly universal. Sherwood Anderson despaired that industrial development was strangling the nation's spiritual life: "The land was filled with gods but they were new gods and their images, standing on every street of every town and city, were cast in iron and steel. The factory had become America's church and duplicates of it stood everywhere, on almost every street of every city belching black incense into the sky."[32] While observing the New York skyline in 1904, Henry Adams noted pessimistically that New York City was becoming a powder keg of change: "The city had an air and movement of hysteria, and the citizens were crying, in every accent of anger. . . . Prosperity never before imagined, power never yet wielded by man, speed never reached by anything but a meteor, had made the world irritable nervous, querulous, unreasonable and afraid."[33]

Periodic panics and depressions in 1873, 1893, and 1907 magnified this "irritable, nervous, querulous" milieu. From 1881 to 1905, approximately 7 million workers participated in 37,000 strikes to protest low wages and dangerous conditions across the nation, from Homestead, Pennsylvania (1892), to Coeur d'Alene, Idaho (1899).[34] In *How the Other Half Lives* (1890), Jacob Riis's stark photographs documented devastating scenes of urban poverty and grinding factory work. Scattered across the Great Plains and the South, Populists formed the People's Party in 1892, ran candidates for national office, and lobbied for federal price supports, standardized shipping charges, and the free coinage of silver as a way to protect the small farmer gouged by railroad companies and grain elevator operators favoring corporate agricultural producers. Even seven years after the Panic of 1893 amid relative prosperity, national unemployment levels stood at a sobering 12 percent.[35]

At the turn of the century, five transcontinental railroads now criss-crossed the country; mass-produced "safety" bicycles were omnipresent, and crawling gasoline-powered cars dotted the nation; electric street cars clanged their way through congested cities, newly illuminated with glowing electric lights.[36] (In Cleveland in 1886, the circus owner Walter L. Main and his father were able to buy twenty horses at the rock-bottom price of $200 for their new circus[37] because the city was replacing its horse-drawn street-cars with electric trolleys.)[38] In an era of accelerated overseas immigration and rural migration to U.S. cities, a polyglot urban culture took shape in which growing numbers of women worked outside the home and partici-pated in a shared work and leisure culture with men.[39] Between 1870 and 1910 the number of women working for wages doubled from 4 to 8 million (a rise from 13 to 23 percent of the total workforce).[40] Women also entered public life through their participation in Progressive Era reform move-ments, which challenged nineteenth-century Victorian notions of "private" and "public" spheres. As a sign of the times, the suffrage leader Charlotte Perkins Gilman in her polemic *Women and Economics* (1898) advocated a sys-tem of state-supported childcare so that mothers could work outside the home. Some men, however, feared that the new urban industrial economy was rendering their bodies, brains, and authority useless. As an architect of a therapeutic "strenuous life," Theodore Roosevelt advocated vigorous exercise and "extreme" experiences in the wilderness.

The rise of the American overseas empire also defined this period of up-heaval. The 1890s heralded the arrival of what the historian Emily Rosen-berg has termed the "promotional state," when the government, in con-junction with private industry, aggressively sought new overseas markets for American surplus goods. Policymakers and business leaders viewed this "crisis of overproduction" as the cause of depression and labor strife in the 1890s and argued that new overseas markets would be a safety valve for domestic ills.[41] As part of its stated mission of promoting democratic self-determination in Cuba and the Philippines, the United States vanquished the decrepit Spanish empire in the four-month Spanish-American War in 1898, thereby gaining new overseas possessions previously belonging to Spain—including Puerto Rico, Guam, the Philippines, and Cuba.

Racism, particularly in the form of social Darwinism, was an integral ideological component of empire building. The new overseas empire grew in tandem with the rise of Jim Crow segregation, disenfranchisement, and lynching at home. Race riots charred the urban landscape in Wilmington, North Carolina, in 1898 and Atlanta in 1906, among other places. Critics of overseas expansion noted the inconsistencies of the Republican Party's position as the architect of both Reconstruction and U.S. imperialism. The *Boston Evening Transcript* observed that Southern race policy was paradoxically "now the policy of the Administration of the very party which carried the country into and through a civil war to free the slave."[42]

Race-thinking shaped contemporary interpretations of domestic demographic trends as well. Although Euroamerican fertility rates had fallen steadily since the late eighteenth century, this demographic reality did not become a "crisis" until the turn of the century, when the flood of immigrant "others" reached record numbers. In 1901 the sociologist Edward Ross coined the term "race suicide," which quickly became a popular expression of native-born anxiety. Euroamericans often saw these newly arrived millions as fecund aliens who threatened to turn the native-born into a racial minority, potentially stripped of their political and social power.

The railroad circus provides a vivid cultural window into this era's complex and volatile web of historical changes. This book argues that the turn-of-the-century railroad circus was a powerful cultural icon of a new, modern nation-state. This vast, cosmopolitan cultural form was the product of the same economic and social forces that were transforming other areas of American life. That is to say, the railroad circus was a cultural artifact of what Alan Trachtenberg has aptly called "incorporation."[43] Its immensity, pervasiveness, and live immediacy transformed diversity—indeed history—into spectacle, and helped consolidate the nation's identity as a modern industrial society and world power. The railroad circus represented a "human menagerie" (a term popularized by P. T. Barnum) of racial diversity, gender difference, bodily variety, animalized human beings, and humanized animals that audiences were unlikely to see anywhere else.

But the circus's celebration of diversity was often illusionary, because the circus used normative ideologies of gender, racial hierarchy, and individual mobility to explain social transformations and human difference. At first glance, this is a problematic claim because the nomadic circus traveled on the fringes of community life—in fact, as subsequent chapters will demonstrate, its workers consciously felt that they were a breed apart from the rest of society. Indeed, performers themselves embraced cultural diversity within this international, multiracial "traveling town." Still, the circus clearly promulgated the major social currents of the day. As the semiotician Paul Bouissac has written: "[The circus] is a kind of mirror in which the culture is reflected, condensed and at the same time transcended; perhaps the circus seems to stand outside the culture only because it is at its very center."[44]

This book will focus on the largest railroad circuses because these speedy, if ungainly, three-ring shows had much wider cultural exposure than small, plodding, horse-drawn wagon shows. Little wagon circuses traveled regionally, primarily in rural areas, while the biggest railroad outfits (possessing over fifty railroad cars) such as Barnum & Bailey, the Ringling Bros., and Adam Forepaugh & Sells Brothers bridged rural and urban, roaring across the entire nation in a single season. These railroad circuses frequently employed over 1,000 people and hundreds of animals. This book will include railroad Wild West shows as part of its analysis of the circus because both amusements took place in an arena surrounded by an audience, were financed by the same investors (James A. Bailey, for one), had a similar division of labor, and overlapped considerably in their content at the turn of the century: some Wild West shows had a sideshow, and many circuses featured Wild West acts with "cowboys" and Native Americans (plate 1). In addition, Wild West shows had trick riding acts that strongly resembled circus stunts, and contained an international conglomeration of talent, including acrobats.

But there were important differences between the circus and Wild West shows. Unlike the circus, Wild West shows generally took place in an open air arena (usually a baseball field, a racetrack, or a driving park) because an errant spray of lead from the shooting acts could shred a circus big top. Only the grandstand was covered by canvas. And, as the historian Joy Kasson contends, William F. "Buffalo Bill" Cody was always obsessed with realism in his efforts to create an "authentic" popular portrait of the nation's frontier past and present, even though he did so through myth and melodrama.[45] Circus impresarios, on the other hand, aimed to amuse, tantalize, educate, and perplex their audiences with a jarring mix of the real—"genuine" exotic human and animal acts—and the pointedly unnatural—educated dogs, boxing elephants, or human "iron jaw" acts in which performers dangled from the heights by their teeth. Because this book is primarily concerned with the circus, its treatment of the Wild West is, by necessity, limited to the ways that Wild West shows intersect structurally and ideologically with the circus. Other scholars have given the Wild West much fuller treatment on its own terms.[46]

Tracing the circus's development from its arrival in the United States in the early 1790s to its maturation in the late nineteenth century, chapter 2 will suggest that the evolution of the circus during the nineteenth century is a cultural metonym for national expansion and infrastructural development. At the height of its physical maturation at the dawn of the twentieth century, the railroad circus provides a complex case of cultural and ideological production because multiple groups participated in crafting its contradictory meanings: owners, managers, laborers, performers, animals, and audience members.

Chapter 3 demonstrates that the railroad circus's physical production processes were an important part of its cultural power. Circus exhibitions of labor—from advertising the show months in advance to the physical set-up and disassembly of the tented city—were just as potent to audiences as the incredible bodily stunts on the scripted program. The railroad circus functioned as a "traveling company town." Its intricate social system and bonds of solidarity among circus employees enhanced the outfit's distance from the outside world. In general, workers abided by strict rules of conduct that maximized production and helped the circus adhere to tight railroad schedules, thus giving its far-flung audiences an intimate look at the logistics and ideology of the new industrial order.

Chapter 4 explores the contradictions embedded in the labor of circus women. The growing visibility of the female performer mirrored the rise of women's visibility in public life, as suffrage activists, consumers, temperance reformers, and participants in sports and leisure. Yet showmen were uneasy about the radical potential of these "New Women." Consequently, they deployed a number of strategies to reconfigure strong Euroamerican circus women into dainty, domestic ladies, and women of color into educational artifacts. But impresarios also promoted female propriety with a wink, subtly reminding audiences that the circus was an excellent place to see seminude women.

Chapter 5 elucidates how exhibitions of male gender could both reinforce and subvert social norms. Although circus press agents advertised Euroamerican male big-top players as ideal modern men, actual performances were sites of gender play that could provide audiences with liberating alternatives to disciplined lives of manly capital accumulation. Animal acts, often scripted as male, helped define human constructions of male gender. Male audiences also engaged in transgressive masculine fanfare by using Circus Day as an opportunity for drinking and ritualized violence.

The railroad circus was at its most au courant in its celebration of America's emerging role as a global power. Chapter 6 analyzes how circus and Wild West spectacles (dramatizations of allegories, or reenactments of contemporary and historical events) helped naturalize for American consumers the entanglements of the United States in remote countries. During the Spanish-American War era, small-town newspapers breathlessly covered war, revolution, and America's military presence in faraway locales. The circus brought these distant episodes home. It also participated in foreign relations on a different plane: as an international business buying animals and hiring people from overseas. These logistics were animated by the same jingoistic *Weltanschauung* that marked the vast staged spectacles under canvas.

No other amusement saturated consumers like the circus at the turn of the century. Neither vaudeville, movies, amusement parks, nor dance halls equaled the circus's immediate physical presence—that is to say, towns did not shut down in their midst. These popular forms were integrated into local economies and local systems of surveillance, while the railroad circus was an ephemeral community ritual invading from without. Contemporary international expositions capitalized on the public's fascination with distant cultures through ethnological village displays along the midway, but one had to travel to a large city such as Chicago, Atlanta, Omaha, Buffalo, or St. Louis in order to experience a world's fair. The traveling circus, in contrast, came to one's doorstep. Disconnected from daily life, the nomadic circus had a distance from community ties that enhanced its ability to serve as a national and even international popular form, because American railroad shows traveled overseas. Adeline Blakeley, an ex-slave, identified the railroad circus as a national popular form while telling her life story to an interviewer for the Works Progress Administration's Federal Writers' Project: "I remember once Barnum & Bailey were coming to Fort Smith [Arkansas]. We were going down . . . but Bud [her employer's child] got sick and we couldn't go. When Helen [her employer at the turn of the century] and I went to California, we all saw the same circus together. . . . There we were . . . seeing the show we had planned to see way back in Arkansas."[47]

The peripatetic fin-de-siècle circus reached virtually all Americans. It educated and challenged people, irrespective of their ability to read or their distance from the metropolis. Its live visual presence made it a popular forum on science, race-thinking, gender ideologies, U.S. foreign relations, and national identity. Hamlin Garland remarked:

[The circus] was our brief season of imaginative life. In one day—in a part of one day—we gained a thousand new conceptions of the world and of human nature. It was an embodiment of all that was skillful and beautiful in manly action. It was a compendium of biologic research but more important still, it brought to our ears the latest band pieces and taught us the most popular songs. It furnished us with jokes. It relieved our dullness. It gave us something to talk about. . . . We always went home wearied with excitement, and dusty and fretful—but content. We had seen it. We had grasped as much of it as anybody and could remember it as well as the best. Next day as we resumed work in the field the memory of its splendors went with us like a golden cloud.[48]

Like vaudeville, the chain store, the "cheap nickel dump," and the amusement park, the circus helped consolidate a shared national leisure culture at the turn of the century. But in contrast to these mostly urban forms of entertainment, the circus was ubiquitous in all regions of the nation, small towns and urban centers alike: from New York City to Modesto, California, to Greenville, Texas, to New Orleans, to Butte, Montana, to Mazomanie, Wisconsin . . . and on and on. Circus Day disrupted daily life thoroughly, normalized abnormality, and destabilized the familiar right at home, day after day, town after town.

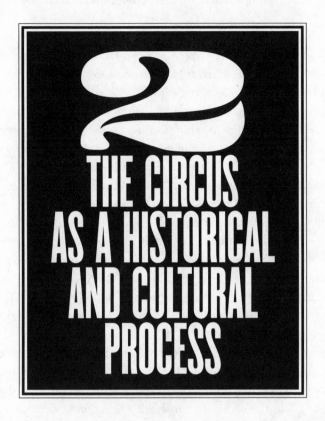

2

THE CIRCUS AS A HISTORICAL AND CULTURAL PROCESS

The mammoth, three-ring railroad circus rattling across America at the turn of the century was a monstrous version of its former self. A hundred years earlier U.S. circuses contained only a smattering of acrobats, clowns, and trained animals. Playing in drafty wooden arenas in population centers along the eastern seaboard, these outfits had no parade, menagerie, grand entry, spectacle, sideshow, aftershow concert, or Wild West show. For that matter, there was no circus of multi-act arena performances at all in colonial North America. Before the Revolutionary era, individual clowns, animal trainers, jugglers, and acrobats wandered from town to town, demonstrating their talents in theaters and tavern yards and on street corners. In 1774 the Continental Congress banned traveling shows (along with cockfighting and horse racing and "every species of extravagance and dissipation") to foster republican virtue among the nascent citizenry of a nation on the verge of independence.[1] After the Revolution, these restrictions (already ignored by many citizens, especially Virginia gentlemen who loved horseracing) were lifted at the national level. Shortly thereafter, the circus came to the United States.

Like other quintessentially "American" cultural icons such as "Yankee Doodle" and the tune for the "Star-Spangled Banner," the circus was a British import. In 1793 the English trick rider John Bill Ricketts staged the first multi-act exhibition of riders, trick horses, clowns, acrobats, jugglers, and rope walkers in a circular arena in Philadelphia. His distinguished audience included President George Washington. To the delight of his patrons, Ricketts threw an orange into the air and caught it on the tip of his sword while standing atop a galloping horse. In England, Rickets had been a student of the retired dragoon (cavalryman) Philip Astley, who started an open-air riding school in London in 1768. Curious crowds gathered to watch these equestrian acrobatics, and the enterprising Astley began to charge admission. Ten years later, Astley created the world's first circus amphitheater near Westminster Bridge, including multi-act displays of acrobatic riding, aerial stunts, clowning, and sleight of hand.[2] Astley's protégé arrived in the United States in 1792 and opened a riding studio in Philadelphia, at that time the young nation's largest city. From there, Ricketts's American circus career flourished.

In its formative years in the United States, the circus did not travel great distances. At each designated site, Ricketts and his competitors constructed in advance large wooden arenas (to accommodate spectators and the forty-two-foot ring length needed for a frolicking horse and rider).[3] Because construction costs were high, early circuses generally limited their tours to large urban areas—Boston, New York, and Philadelphia—where they could be guaranteed a big audience to cover building expenses. (Ricketts did venture, nonetheless, into Maine and Canada in 1797.)[4] Most of these arenas were probably roofless, and were either sold (often as lumber) after a production had finished or used as the same site for later dates. Fire was an enormous liability that drove many circus proprietors out of business. As a result, they strictly forbade audiences to smoke at the arena.[5] Ricketts left the U.S. circus market in 1800 to search for a more lucrative entertainment sites in the West Indies, where he was lost at sea.[6]

In the early nineteenth century people referred to the nascent circus as the "rolling show."[7] But these circuses "rolled" ponderously by horseback, wagon, and boat because they were often mired in mud, jammed in ice, or buried in snow. In 1825 Joshuah Purdy Brown (1802?–1834) of New York completely transformed the circus business when he began showing in a canvas tent. The circus historian Stuart Thayer observes that this revolutionary development "led to the establishment of the rituals of itinerancy."[8] Its constant movement soon made the American circus unique in comparison to its relatively settled European counterpart. Circus proprietors were no longer dependent upon urban population centers because they no longer had to invest significant capital into making the arena. Consequently, after 1825 the circus market expanded swiftly, now reaching into previously isolated rural areas. Because the canvas tent created an increasingly nomadic work environment, a more complex division of labor emerged. Circus owners now needed workers to ride ahead, so that they could advertise the upcoming production. Showmen also began to conduct a morning parade on the day of the circus to attract attention. Additional employees helped facilitate movement—hauling the tent, ring fence, and circus performers.[9] This new labor system was a prototype for the giant railroad circuses later in the century.

Once the circus began to travel by wagon and exhibit under canvas, it merged with another popular form of public entertainment, the animal menagerie, when Joshuah Purdy Brown combined the two in 1828. Circus owners were eager to join the animal menagerie business because many Protestant clergymen denounced the circus for its seminude athletes and the practice of gambling on the show grounds. Based on protests from the clergy, Vermont and Connecticut state laws banned the circus in the antebellum era.[10]

Colonial animal menageries provided audiences with glimpses of faraway places long before the exhibition of foreign people became a standard part of the circus. Before the advent of the exotic animal trade in the middle of the nineteenth century, speculative sea captains often purchased or traded wild animals in Africa and Asia, and sold them to fledgling U.S. menageries.[11] The "Lyon of Barbary" (from the Barbary Coast in Africa) arrived in the British North American colonies in 1716; in 1721 came the first camel, in 1733 the first polar bear, and in 1768 the first leopard.[12] The mainstay of the circus, the elephant, first landed at New York City in 1796 on Captain Jacob Crowninshield's ship, *America*, from Bengal. The first tigers arrived from Surat, India, at Salem, Massachusetts, in 1806 on the ship *Henry*.[13] The animal menageries were exhibited in barns, inn-yards and stables, or any other public venue where the animals could be hidden from curious, nonpaying spectators.[14] The Crowninshield elephant, for one, attracted thousands of customers, including President John Adams, "to see the elephant." In fact, "to see the elephant" soon became an important part of the American lexicon, a powerful synonym for experiencing battle used commonly during the Mexican War and the Civil War.[15]

The menagerie business proliferated in the nineteenth century. Although proprietors always marketed their exhibits as educational, many were hastily assembled. Henry David Thoreau visited a menagerie in Plymouth, Massachusetts, in 1851 and was frustrated that the proprietors, a "few stupid and ignorant fellows," knew nothing about their animals: "The absurdity of importing the behemoth & then instead of somebody appearing to tell which it is — to have to while away the time — though your curiosity is growing desperate — to learn one fact about the creature — to have [Dandy] Jack [a riding monkey] and the poney introduced!!!"[16] Although some menageries produced informational pamphlets on their animals as early as the 1840s, their nods to pedagogy paled next to the turn-of-the- century railroad showmen, who consistently published colorful, lavishly illustrated programs carefully describing the origins and habitat of each animal in the menagerie.

Throughout the history of the American circus, its growth was tied to the physical expansion of the nation. As the population grew, the circus moved to take advantage of new markets. Because it traveled, the circus was dependent upon internal improvements and new inventions like the steamboat and the railroad. The circus traveled through the Appalachian mountains on the Cumberland Gap Road (or National Road), topped with gravel at federal expense, after Daniel Boone marked the area in the 1790s. Even though the federal government refused to sponsor canal and road construction during the antebellum era (a precedent set in 1817 when President James Madison vetoed the internal improvements bill of Senator John C. Calhoun of South Carolina), state and local governments used public funds to build a flurry of roads and canals. With the completion in 1825 of the Erie Canal, the "big ditch" sponsored by Governor DeWitt Clinton of New York that extended for 364 miles, the circus — previously limited to eastern and southern states — now had access to the old Northwest Territories and beyond. As the populations of Ohio, Indiana, Illinois, Michigan, and Wisconsin boomed after the Civil War, these states (home to fertile, inexpensive farmland and located at the crossroads of rivers, roads, and railroads) became circus centers. Wisconsin, for one, was home to Dan Castello and W. C. Coup, who were partners with P. T. Barnum in his 1871 circus debut.[17] The Ringling brothers, who became the biggest circus proprietors in the nation at the turn of the century, also hailed from Wisconsin.

The steamboat also hastened the circus's movement and growth. After Robert Fulton invented this new technology in 1807, several circuses created giant river palaces. The discordant calliope was designed to be heard along the river banks, announcing the arrival of the circus with its ghostly, fluted timbre. During the 1850s, at the height of the river boat circus's popularity, Gilbert R. Spalding's and Charles J. Rogers's circus created an expansive "Floating Palace" that played dramatic spectacles like *Hamlet* and the contemporary temperance hit, *Ten Nights in a Bar Room* for over a thousand spectators at a time.[18] But in the Civil War years, river boat circuses were edged out of the water by the movement of Union and Confederate troops.

Wagon circuses trod cautiously during the Civil War. Working as an apprentice for the Robinson and Lake circus in 1863, the sixteen-year-old future impresario James A. Bailey wrote in his diary about close encounters with Confederate soldiers. "Sunday, July 26, 1863: "The first place that the people commenced to make preparations to receive the Rebels was at Madison. The town was full of drunking Soldiers, we left there Saturday, and went to Versailles when we got there the Rebels were at Osgood we got away from Versailles at 7 O'Clock am and the Rebels came in at 10 O'Clock."[19]

From the 1830s to the 1860s, the first railroad circuses were smaller than contemporary overland wagon shows. The railroad eventually enabled the circus to become a transcontinental entertainment, but early railroad circuses had no menagerie, sideshow, or street parade because constant rail travel was difficult and expensive: although there were 30,500 miles of track in place nationwide by 1860, railroads could not accommodate the circus's frequent movements because the rail system was a patchwork of different track gauge and no circus had yet devised an efficient system of loading and unloading its rented system (i.e. company issue) railway cars. The circus historian Fred Dahlinger writes that from the 1850s to the 1870s, many railroad circuses were also "gilley" productions because manual laborers transported the stock from the railway depot to the grounds—a dangerous and tedious process. Some circuses stopped only at towns whose show site was right next to the railroad tracks.[20] Dahlinger and Thayer add that railroad travel was financially risky as well: the equipment was expensive, and all railroad-related expenses had to be paid up front. Wagon shows, by contrast, had far fewer advance expenditures—only the license and lot rental had to be paid before ticket revenues were received. The train crew itself, moreover, represented yet another expense for railroad showmen.[21]

Many audiences felt cheated by the early railroad circus and character-ized it as an abbreviated amusement that still charged the same price as the bigger overland wagon show (usually twenty-five to fifty cents).[22] Some cir-cus employees, too, were resistant to this change. Jules Turnour, a clown with Ringling Bros., felt initially uncomfortable during the show's inaugu-ral season on rails in 1890: "Somehow I didn't like the change at first. I had become so accustomed to the wagon traveling at night, to the wild, free, clean abandon of the life, that I did not fancy the idea of sleeping on a stuffy train, with smoke and cinders to bother me. . . . The wagon life may have been hard traveling, but it was in the open."[23] Yet rail travel was, on the whole, easier for circus workers, enabling them to rest more soundly be-tween stops than in the bumpy, jarring overland wagon. In time, railroad showmen contemptuously referred to overland circuses as "mud shows."[24] By the 1860s greater numbers of circuses traveled at least part of the sea-son by rail. Using eight railway cars, and moving overland part of the time, Dan Castello's Circus and Menagerie made the first transcontinental tour in American circus history in 1869. Castello's trip occurred just weeks after the California railroad magnate Leland Stanford tapped (and missed) the golden spike marking the riotous completion of the Transcontinental Rail-road in May 1869 at Promontory Point, Utah.[25]

Soon thereafter, the railroad circus flourished. Although Barnum, Coup, and Castello debuted with a wagon show in 1871, they quickly moved to the railroad in 1872. In mid-1872 Barnum and Coup solved the logistical problems of loading and unloading when they bought their own specially designed railroad cars, including flat cars, sleeping cars crammed with extra bunks, well-ventilated stock cars, and palace cars (designed with elabo-rate partitions and special feeding areas for valuable ring stock — animals working under the big top). Consequently, Barnum, Coup and Castello's Great Traveling World's Fair easily surpassed the biggest overland wagon shows, complete with a parade, menagerie, and sideshow. Long a fixture at Barnum's Museum in New York City (1841–68), the sideshow of "freaks" became a major attraction at Barnum's expanding railroad circus. Some long-term entertainers at the American Museum, such as William Henry Johnson, joined the Barnum, Coup and Castello sideshow. In 1872 Barnum, Coup, and Castello adopted two big-top rings to accommodate their grow-ing audience under the increasingly crowded canvas tent. The inaugural railroad season in 1872 was a financial success, and the railroad quickly be-came the standard method of transport for large circuses. Just six years after its start as a regional wagon show, Ringling Bros. became a railroad outfit in 1890, starting with eighteen cars during its inaugural season.

The railroad circus of Barnum, Coup and Castello in 1872 was a blueprint for the gargantuan railroad exhibitions at the turn of the century. But in terms of sheer size, theirs paled in comparison to later railroad circuses. When the 1872 season started (and before they purchased their own cars later in the season), Barnum and Coup used a shabby group of leased "system" cars that were twenty to thirty feet long. The combined car length was about 1,200 feet. By 1897 Barnum & Bailey's Greatest Show on Earth used its own sixty-foot cars, which totaled 3,600 feet in length.[26] In 1903 the Ringling Bros. had sixty-five cars.[27] The size of the big top tent also grew dramatically. In the 1840s the Mabie Brothers' one-ring tent measured eighty-five feet in diameter; by 1890 the Barnum & Bailey three-ring tent was approximately 460 feet long. At the turn of the century, scores of sixty-foot railway cars simultaneously carried several hundred and sometimes over a thousand circus workers, performers, animals, tents, food, and props with greater speed than over land. By 1910, when over thirty circuses traveled by rail, the Ringling Bros. show and the Barnum & Bailey circus each used eighty-four railway cars to transport their productions nationwide, and Buffalo Bill's Wild West Show traveled with fifty-nine railway cars.[28] Large circuses owned their railroad cars, and contracted with car makers to design palace cars and special elongated flat cars to accommodate their massive gilded wagons. Yet most railroad circuses were still small by comparison, ranging from two to twenty cars, which held dining and sleeping areas, baggage, animals, cage wagons, and seats.

Once the railroad circus expanded, its content became increasingly complex. After Barnum merged with the veteran circus owner James Bailey in 1880, the show adopted three rings the following year. In 1895 one Barnum & Bailey press release declared: "A Single Animal has Given Place to Herds and Droves."[29] The enlarged, spacious big top changed the nature of circus acts, giving players more room to maneuver and consequently making stunts increasingly elaborate. Flying trapeze artists, for example, increased their troupe size and added aerial somersaults to their acts in the 1880s.[30] Lena Jordan is credited with executing the first successful triple somersault on the flying trapeze in 1896. In the early 1900s Ernie Clark, a member of the "Flying Clarkonians" family, awkwardly attempted the quadruple somersault in practice sessions, but consistently pulled off the triple in front of the audience.[31]

The enormous transcontinental railroad network completed after the Civil War transformed the circus into a frenetic three-ring, two-stage, cross-country extravaganza. Now able to travel on a network of uniform railroad gauge, the circus's rising ubiquity was a symbol of national expansion and consolidation during the Gilded Age. In 1892 a poster described Barnum & Bailey's spectacle "Columbus and the Discovery of America" as an inexpensive whirlwind world tour: "Special Cheap Rate Excursions from Everywhere by All Lines of Travel, Wonderland Itself Laid Bare" (fig. 4). (Of course, wonderland was laid "bare" in other ways, too, because the production boasted hundreds of barely dressed "oriental" ballet girls.) In the context of a growing leisure culture at the turn of the century, the circus advanced itself as a convenient means of taking a global tour without having to leave home. David Nasaw notes that between 1870 and 1900, real income for nonagricultural workers rose by over 50 percent; concurrently, the cost of living dropped 50 percent. Moreover, the average manufacturing worker labored three and a half hours less in 1910 than in 1890; the decline in the number of hours of the white-collar workweek was even greater.[32]

The Wild West show was a product of the same technological and cultural currents that enabled the circus to expand. Colonel William F. "Buffalo Bill" Cody, a Civil War veteran, hunter, actor, businessman, and politician, created the first American Wild West show in 1883, a large railroad outfit that was able to blanket the nation by adopting the railroad circus's division of labor and mode of transportation. Buffalo Bill's Wild West had an elaborate program: cowboys, American Indians, horses, buffalo, Indian raids on settler's cabins and wagon trains, ersatz prairie fires and cyclones, bison hunts, military drills, shooting acts, races, and dramatic reenactments from the Indian Wars and of overseas battles at the turn of the century. Using the technological medium that helped hasten the frontier's actual disappearance, Cody's railroad outfit produced national narratives of "civilization," "progress," and nostalgia for preindustrial American Indian cultures and "wild" spaces, like the circus.

Although the railroad allowed the circus to travel great distances quickly and broaden its routes, it also diminished visits to the smallest rural villages that were too tiny to be profitable. Similar to their predecessors in wooden arenas, these giant railroad shows primarily played cities and towns where ticket sales could exceed their huge capital investments. Yet as noted in chapter 1, rural residents still widely attended the railroad circus, because large show owners worked with railroad companies to offer discounted "ex-

Figure 4. "Columbus and the Discovery of America," Barnum & Bailey, 1892. Featuring
a cast of 1,200 people, 400 horses, and scores of other animals, this spectacle promised
a quick global tour without straying far from home. (Lithograph courtesy of
Circus World Museum, Baraboo, Wis., with permission from Ringling Bros.
and Barnum & Bailey,® The Greatest Show on Earth,® B+B-NL44-92-1U-1)

cursion" fares for rural residents living along railroad lines within a fifty-mile radius of a show stand. In Cuero, Texas, a local newspaper advertised the times and places in various hamlets where an excursion train (charging round-trip fares of $1 for adults and fifty cents for children)[33] would pick up passengers for the Ringling Bros.' performance in Cuero on November 10, 1898.[34] In 1888 Louis E. Cooke, Barnum & Bailey's general railroad contractor and excursion agent, contracted with railroad companies for 568 excursions which brought an additional 419,026 patrons to the circus during its 192-day season.[35]

Although the large railroad circuses covered the most territory during the season, the majority of circuses at the turn of the century were still small, "dog and pony" "mud" shows. These circuses traveled primarily by wagon and played in isolated villages. Some circus audiences became bored with these small productions, at which the now ordinary horse dominated the animal holdings. In 1899 a newspaper writer in Stoughton, Wisconsin, complained about the lack of variety at the Gollmar Brothers' recent exhibition. "To tell the truth, the show might have been better and there might have been a little more to the parade besides horses and colored wagons. . . . It was almost as much as could be expected of any twenty-five cent circus, and we can remember having seen even poorer exhibitions than the Gollmar's."[36]

Nevertheless, others felt that the large, modern railroad productions "glutted rather than fed." William Dean Howells waxed nostalgic about his experiences at an old-fashioned dog-and-pony show in 1902, comparing it favorably to the small wagon circuses of his youth in the "old days": "I felt the old thrill of excitement, the vain hope of something preternatural and impossible. . . . There was, in fact, an air of pleasing domesticity diffused over the whole circus. This was, perhaps, partly an effect from our extreme proximity to its performances."[37] The modern three-ring railroad circus, by contrast, was overwhelming, "too big to see at once" with its huge canvas enclosure of rings and stages, a distinctly American cultural form whose scripted chaos and singular indigestibility departed sharply from its intimate one-ring European antecedents.

The visually oriented three-ring circus flourished in tandem with multiple visual forms at the turn of the century: department stores filled with mirrors and reflective glassy surfaces, early motion picture actualities seen at saloons, railway stations, circuses, and world's fairs, and splashy new newspaper formats with big photo-filled sports pages; the three-ring circus was symbolic of an emergent "hieroglyphic civilization" which the historian Warren Susman has characterized as "a significant break for a culture that had taken form under Bible and dictionary."[38] William Dean Howells thought that the modern circus was "an abuse and an outrage. . . . [The circus] has become too much of a good thing. . . . I'm still very fond of it, but I come away defeated and defrauded. . . . [I] have been given more than I was able to grasp."[39] Historians have concurred with this assessment. Neil Harris posits that turn-of-the-century audiences became "glutted" and reduced to "passive bedazzlement" in this overwhelming visual feast.[40] Indeed, the antebellum one-ring circus was intimate by comparison, an entertainment whose "talking clowns" and ring masters integrated witty, gossipy commentary about local politics into the program. But turn-of-the-century audiences and circus workers alike still used the colossal three-ring circus as a site for imaginative play, violence, and economic opportunity.

CULTURAL SPECTACLE

How did these multiple groups—often with conflicting interests—participate in the physical and ideological making of the railroad circus? This study treats the circus on its own terms—instead of solely symbolic terms—as a diverse conglomeration of workers and audiences who actively produced its ideological content.[41] Impresarios, many of whom were McKinley Republicans supporting overseas expansion, big business, and Progressive reforms, grandly (but always with a wink) proclaimed that the circus was a magnificent exemplar of national progress. They consciously framed their exhibitions of the world with normative tropes about labor, racial inequality, separate spheres, and U.S. hegemony that often contradicted the lived experiences of the multicultural members of the traveling circus community. As a whole, the railroad circus was the product of rich members of the "culture industry," but it also was (in the words of Stuart Hall and other Marxian cultural studies theorists) "contested terrain."[42] Scholars use this term to locate class conflict in the seemingly apolitical realm of popular culture; but one can take this oft-used phrase a step further to explore the conflicted relationship between popular culture and historical constructions of gender, race, and sexuality. With its competing visions of normality and subversion, the ubiquitous turn-of-the-twentieth-century railroad circus represents a potent case study of contradictory cultural production. (Readers who wish to avoid the following discussion on cultural theory should skip ahead to chapter 3.)

In an era of accelerated European immigration, circus acts codified ethnic difference as racial difference. This book uses "white" and "whiteness" with some hesitation because skin color was not unilaterally a conclusive marker of "racial" identity at the turn of the century. Ethnicity still defined one's race, be it Yankee, Italian, Irish, German, or Russian. As recent works on the historical construction of "whiteness" have suggested, white racial identity was (and still is) interconnected with the changing status of African Americans in American society and the arrival of various immigrant groups over time. Consequently, whiteness is not just about skin color but is part of a complex matrix of power relations.[43]

Turn-of-the-century circus acts articulated the instability of white racial identity through clownish caricatures of ethnic difference. When the bagpipe player William Shearer solicited the Ringling Bros. in 1903, he stressed his ability to play ethnic stereotypes as a selling point: "I take the liberty of writing to you to ask if you can use me to play with your Circus this coming season. I am a first-rate performer on the Highland Bag Pipe. . . . I do a novelty musical act which always takes well in the Side Show. I play on an Irish potato on a common wooden potatoe masher, on a German beer steine, on a tin coffee pot and finish with a good lively Strathspey and Reel on my pipes, as well which is very unique and always pleases the audience."[44]

At the same time, the circus helped consolidate a shared sense of white racial privilege among its diverse, white ethnic audiences; Euroamerican spectators came, in part, to laugh at what they ostensibly were not: preindustrial, slow, bumbling, naive, or "savage." The circus played a double function because it codified European ethnicity as racial difference, while simultaneously promoting a uniform "white" American racial identity.

Despite the presence of oppressive racial representations, circus people —many of whom were social outsiders—often found a refuge of sorts in this nomadic community of oddballs. In fact, the circus often provided a better income than was available elsewhere. (Female stars, for one, made just as much as their male counterparts or more, and a few women, such as Mollie Bailey and Nellie Dutton, became successful circus owners. Bailey was sole owner of a small circus, "a Texas show for Texas people," at the turn of the century.)[45] Lottie Barber, a fat lady at the sideshow also known as "Jolly Dolly Dimples," remarked: "My fat is my kingdom, my riches. You can tell your thin ladies . . . that my big bulk has kept the wolf from my door for thirty-five years. I've never been broke since I struck the show business."[46] Although such commercial exhibits of physical difference may seem offensive by today's norms, Lottie Barber and her comrades at the sideshow were unfailingly pragmatic about their unusual bodily capital, viewing their own physical limitations as an opportunity to make a living in a society that might otherwise shun them.

These nomadic circus strangers helped subvert contemporary norms about gender and the body. Dressed in sleek leotards and wearing closely cropped hair, circus men and women often looked indistinguishable from each other, particularly as they exhibited equally difficult feats of agility. Judith Butler suggests that in privileging gender as a social construction, feminist theorists have unwittingly transformed "sex" into an unchanging, indisputable material reality. Instead, Butler argues that "sex" is an unstable, discursive formation that is largely defined by rigid, heterosexual ideals.[47] With their blurring of male and female bodies, circus acts flattened sexual differences, and went so far as to challenge the distinction between human and animal. Trapeze artists and acrobats became birds and butterflies, while the "Learned Pig" solved simple math problems, and elephants, tigers, and bears danced upright. The circus encompassed an array of remarkably transgressive bodies: women grew long beards, armless ladies sewed with their feet, hairy people worked as "missing links," and midgets and giants played cowboys, royalty, and military figures.[48] Circus people also made light of the body's threshold for discomfort by engaging in seemingly agonizing activities as they swallowed swords or ate fire (both of which caused no pain if done correctly).

The railroad circus was an interactive cultural arena for workers, owners, and audiences; as such, the circus complicates scholarly ideas about representations of self and Other. On one level, the circus's spectacular pageant of the Other—a profusion of people of color working as "missing links," "savages," and "ape girls"—make it a popular counterpart to high cultural analyses like Edward Said's *Orientalism*, which use literary texts, paintings, magazines, and European travel writing to investigate how European (and American) imperialists have depicted—and dominated—the rest of the world.[49] Although extremely useful to this study, these approaches run the risk of compounding the stereotype of the Other as mute.[50] Several scholars have demonstrated that in live performance, the relationship between self and Other is constantly in flux—even when the performance reinforces racist norms—because as an entertainer, one returns the audience's gaze with one's own, thus undermining the controlling function of the gaze.[51]

The circus disturbed the seemingly safe staged distance between self and Other because it was interactive: the entertainer-as-Other talked back to audiences, teased them, and fooled them. Duping was a central part of the circus; consequently audiences were always vulnerable as they unwittingly became part of the "show." The Ringling Bros.' program in 1894, for instance, treated its unsuspecting audience members gathering for the big-top show to a "fight" between a foppish "city dude" and an "innocent-looking German countryman." The German became increasingly angry as the city dude strutted around trying to impress female audience members with his "eye-glass and cane, high collar and general extravagance of dress." Finally, the frightened dude charged around the hippodrome track while the German bombarded him with chunks of bread, sausage links, and a pail of beer. As the big-top program was about to start, the two retreated to the men's dressing room; only then did the audience learn that the dude and the bumpkin were "in the play."[52]

The circus crowd itself was part of the "human menagerie" at the circus. For audience members, the presence of huge masses generated the same kind of excitement as the extraordinary human and animal athletes on the program under canvas. Emily Dickinson witnessed how Circus Day transformed ordinary neighbors into virtual strangers: "The show is not the show, / But they that go. / Menagerie to me / My neighbor be. / Fair play— / Both went to see."[53] Circus workers duly observed and recorded these "performances," thus making everyone part of the production. The spectacle of these crowds became especially exciting when people fought, became drunk, gambled, or panicked in the face of a storm or rampaging animal.

The giant railroad circus, then, was a dialogical cultural process because its "show" was multifaceted, a spectacular conversation of sorts between performers, workers, animals, the elements, and the audience. In analyzing the literature of François Rabelais, Mikhail Bakhtin demonstrates that carnivals, Church feasts, agricultural feasts, and civil ceremonies in early modern France were occasions for shared laughter, at which participants poked fun at authority figures and celebrated the grotesque body. These special events suspended time and dissolved social hierarchies. Bakhtin contends that the "carnival spirit" was more than a simple safety valve: "This carnival spirit offers the chance to have a new outlook on the world, to realize the relative nature of all that exists, and to enter a completely new order of things."[54] But several scholars such as Robert Allen, Terry Eagleton, Susan Davis, Peter Stallybrass, and Allon White have taken Bakhtin to task for his uniformly utopian and radical vision of the "folk." Terry Eagleton, in particular, argues that elites sanctioned carnival as a "contained popular blow-off" that unwittingly reinforced the social order.[55] Residual parts of carnival (both conservative and potentially radical) like masking, the grotesque, mobbing, and theatrical inversions of social hierarchies heightened the volatile mood on Circus Day. Time was in abeyance as towns shut down. People feasted. Dressed in an array of grotesque costumes, clowns twisted their bodies and laughed at the existing order. Humans and animals aped each other, calling into question what it meant to be male, female, indeed even human.

Echoing the scenes of constant gorging at carnival captured on a Breugel canvas, Circus Day audiences spent their money freely. In Herman, Texas, for one, Miss Pauline Janes's shop offered special "Circus Day Bargains!": "Remember these prices prevail on 'circus day' only. My store is the best place from which to view the parade. Come and see me."[56] On Circus Day, the sociologist William Graham Sumner's iconic savings bank depositor was hardly the "hero of American civilization." Because consumers were wont to spend on Circus Day, many communities accused the circus of indirect thievery. Local residents complained that the nomadic circus was a "drain" on the local economy, as townspeople spent their scarce cash there instead of buying goods within the community. In 1900 the Georgia legislature virtually taxed the circus out of the state: on the grounds that it made a "big pile of money out of the community," it was required to pay from $300 to $1,000 a day[57] (depending on the size of the town) for the privilege of exhibiting there.[58]

Although Circus Day was a carnivalesque occasion for community consolidation, it was also, paradoxically, a time of community fragmentation. Established bonds of intimacy within watchful communities temporarily dissolved into anonymity, which gave people license to engage in illicit activities. Robert Allen reminds us that "carnivals can become riots."[59] A group of boys at a circus date in Appleton, Wisconsin, in 1910, for instance, loitered, smoked cigarettes, and set five sheds on fire near the railway depot at 2 A.M. while waiting for the Ringling Bros. train to arrive.[60] Newspapers frequently published veritable catalogues of criminal activities committed while the circus was in town — who was robbed, and the value of the goods stolen. In 1905 the *Clinton* (Iowa) *Daily Herald* noted that Barnett's millinery lost $25 in a robbery.[61] The robbers entered through the rear door while the unwary clerks stood outside watching the free Ringling Bros. circus parade. In addition, two diamond rings were stolen at a house in nearby Sterling, Illinois. The article concluded that both crimes were "doubtless committed by some of the thieves who follow the circus about."[62] After the Ringling Bros. circus blew through town, the *Sherman* (Tex.) *Weekly Democrat* reported, "It wouldn't be a complete circus day without a horse theft and

Tuesday's circus was no exception to the rule."[63] On the eve of Buffalo Bill's Wild West in 1898, the *Arkansas Democrat* of Little Rock cautioned local residents to "Be Careful Tomorrow: Crooks Will Abound and Stores and Dwellings Should Be Watched."[64] A newspaper in Mount Pleasant, Iowa, warned its readers to "look out for bums": "The *News* begs to inform the people that this is Circus Day, and to warn them that it would be wise to make doors and windows doubly sure. About every show, no matter how well regulated of itself, a horde of bums, thieves and confidence men have been drifting into town until now it is safe to say that fifty are in town looking for a chance to commit some depredation. At the Pork House and stock pens a crowd of them can be seen plotting together. Tonight especially should caution be observed."[65]

Although large railroad circus owners hired Pinkerton detectives to maintain order on the show grounds, audiences and circus workers alike often flagrantly disregarded the presence of law enforcement. In 1908 the acrobat Aristodemo Frediani observed that Barnum & Bailey's Pinkerton detectives ignored fights, short-change "artistry," gambling, and the workingmen who sneaked under the bleachers to steal umbrellas and canes from unsuspecting audience members.[66] Community fears about crime and disorder suggest that the circus's overwhelming presence (complete with its horde of "sneak-thieves") could have solidified provincial xenophobia, especially when local newspaper articles occasionally called circus players "gypsies," a characterization that transmogrified this entire itinerant community into a liminal racial Other. But the historical evidence suggests otherwise. Virtually no town banned the circus outright at the turn of the century, whereas several states had done so in antebellum America. Residents rightly expressed their anxieties about illicit activities on Circus Day, but few (if any) moved to abolish it wholesale.

The specter of community fragmentation continued on Circus Day as audiences occasionally responded to the show's beckoning vision of a big, exciting world by "running away." The act of "running away" involved breaking away from one's community for the imagined economic opportunities and unfettered life with the circus. A manager for Buffalo Bill's Wild West wrote to a frantic sister whose brother had seemingly disappeared, to reassure her that he was alive and well and working in the dining tent.[67] From California, Gail O. Downing wrote to his pregnant wife Orilla Downing in Cody, Wyoming, about an unexpected turn of events: "I am now dropping you a little surprise. I am headed for Cheyenne Wyo. to join C. B. Irwins Wild West Show."[68] When Barnum & Bailey played at North Adams,

Massachusetts, in 1903, a local newspaper reported, "About a half a score of young men in the city who are out of employment joined the Barnum & Bailey show and will leave the city with the aggregation this evening. The men will act as canvasmen [workers who erected the canvas tents]. They are promised good wages and board, but the work is hard."[69] Because turnover rates were high among workingmen, circus managers constantly hired people throughout the season, disrupting community bonds along the way. "Running away" also represented a potential escape from the shackles of gendered and racial conformity that limited ordinary community life. Despite the lure of mobility at the circus, proprietors used these same normative stereotypes to market their shows.

The presence of the circus further assaulted community ties when children slipped away to join the show. After a Barnum & Bailey date in Lynn, Massachusetts, in 1907, the local police searched two days for two girls, thirteen-year-old Hazel Kimball and fifteen-year-old Clara Appleton, who disappeared after the circus departed. Finally, Hazel's frantic mother traveled to Salem, where she found her daughter washing dishes in the circus cook tent. Meanwhile, Clara hid under a wagon to avoid being taken home, because she wanted to become a snake charmer. The girls had joined the circus simply by hiding in a circus railroad car when the outfit departed Lynn.[70] Amid thousands of strangers, railroad cars, animals, wagons, and tents, children could easily disappear with the circus.

In this crowded, carnivalesque environment of thieves, idle teen-agers, vanishing children, and "profligate" consumers, the specter of violence was omnipresent. After the evening program had finished, drunks and "toughs" leered at disrobing female artists in the women's dressing tent and picked fights with circus workers or fellow spectators. Despite the potential leveling of social hierarchies that the circus's carnivalesque presence promised, frequent altercations on the show grounds exposed deep racial tensions. Nonetheless, several writers idealized Circus Day as a racial safety valve, a moment of unmitigated merriment for weary African American farmers in particular. The *Arkansas Democrat* reported that "cotton-hoeing and cotton-picking the year round would make life a heavy burden to the colored brother if unrelieved by the annual circus."[71] Yet such tranquil characterizations belied the presence of very real racial violence—most often directed at people of color. In Cuero, Texas, the Ringling Bros. 1892 route book noted: "During a street brawl here to-day among the natives, one Mexican was stabbed to the heart, another all cut up and a white man had his ears bitten off."[72] As various theorists, media scholars, and historians have written, "low" groups often used carnivalesque celebrations like the circus as opportunities for "displaced abjection" to empower themselves by demonizing people even "lower" on the social hierarchy than themselves.[73]

The potential for violence on Circus Day speaks, in part, to its overwhelming presence and its diverse audience base. As a result, the circus occupied an ambivalent position in the spectrum of turn-of-the-century popular entertainment. Wild West show, vaudeville, and other allied amusements claimed to be decorous and orderly, each quickly noting that it was "not a circus."[74] Such commentary acknowledges the circus's potentially precarious position in a broader and much debated cultural hierarchy in fin-de-siècle America.[75] But in contrast to the all-male world of burlesque and concert saloon, *everyone* went to the circus: from President Theodore Roosevelt, who received a personal invitation scrolled on satin from James A. Bailey in 1903, to hundreds of inmates from local insane asylums across the country who were brought to the circus by their wardens.[76] In many ways, the composition of the turn-of-the-century circus audience mirrored those of the mixed Jacksonian-era theater, because women, men, and children of different social class and ethnicity sat together under the same canvas big top tent. In 1898 the *Galveston Daily News* noticed the diverse crowd attending the Ringling Bros. circus: "Men, women and children from all walks of life and all avenues of trade and profession and wards and precincts were there and as one big family."[77]

Yet seating arrangements at the largest railroad circuses also reflected contemporary social hierarchies based on class and race (fig. 5). Wealthy and middle-class Euroamerican spectators sat in the comfortable and expensive (usually $1 to $2)[78] "starbacks" or reserved box seats—the best seats in the big top, located along the center ring.[79] Depending on the show, working-class patrons paid twenty-five or fifty cents to sit at either end of the big top on unreserved bleachers (so called because of their resemblance to long bleaching boards), also known as "blues" (the practice of painting bleacher seats blue started in the mid-nineteenth century for unknown reasons).[80] Recent immigrants, Native Americans, many working-class circusgoers, and stray children paid a "blues" price to sit in the gallery or "straw house," an open "pit" area between the hippodrome track and the seating area on which straw was placed to accommodate a few thousand more spectators.

The racial geography of the circus audience reflected the proliferation of Jim Crow segregation at the turn of the century. Until 1900 southern segregation laws had applied primarily to passenger trains; thereafter, these

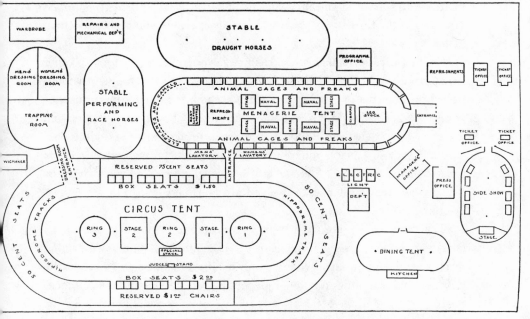

Figure 5. Barnum & Bailey program, sketch of show grounds, 1903. Although virtually everyone attended the circus, seating arrangements under the big top were generally stratified on the basis of social class and, in the Jim Crow South, uniformly racially segregated. (Interior sketch courtesy of Circus World Museum, Baraboo, Wis., B+B-N45-03-10)

laws (both de facto and de jure) extended to virtually every aspect of public life: separate toilets, water fountains, waiting rooms, orphanages, schools for the blind and deaf, Bibles for court testimony, parks, swimming pools, restaurants, streetcars, and steamboats.[81] In the South, black circus-goers rode in separate "Jim Crow" railroad cars. Under the big top, black patrons generally sat segregated from other spectators in the gallery.[82] Outside the tents, black and white audiences bought their concessions at separate snack stands. The Louisiana state legislature passed a law in 1914 mandating racially segregated entrances, exits, and ticket windows at circuses and other tent shows; the law also specified that ticket sellers remain a minimum distance of twenty-five feet from each other.[83] Racial segregation had a long history at the circus in the southern United States: throughout the nineteenth century, newspapers noted separate points of entry and segregated seating areas, in the pit [standing room] or gallery, for black circus audiences. Newspapers also mentioned that African Americans were supposed to attend the circus at specified times and dates.[84] However, segregationist practices at the turn-of-the-century circus were more comprehensive than in earlier years.

The historical evidence is less clear regarding the segregation of other racial groups at the circus. Route books and press releases frequently mentioned the presence of Native American, Chicano, and Asian American audiences, but generally do not specify where they sat, only that they sat en masse.[85] One article from 1903 did note that several Chinese attending Barnum & Bailey's Madison Square Garden date paid $1.50 for expensive box seats.[86] One spectator, Li Kung Chang, stated that he would never sit in the gallery, "where the representatives of Italy, Germany and Ireland are most prominent."[87]

Ticket prices, ranging from twenty-five cents to $2, made Circus Day a fairly expensive amusement for its day.[88] (It should be noted, however, that many dog-and-pony outfits charged only a dime for admission.)[89] Vaudeville tickets sold for a dime to a dollar, depending on the theater and the location of one's seat in the orchestra or gallery; "cheap nickel dumps" and dime museums cost what their names suggest; burlesque halls, concert saloons, and "ten-twenty-thirty" theaters, which featured "blood and thunder" melodramas, ranged in price from a dime to thirty cents.[90] Such amusements were part of a spectrum of ordinary, mostly urban leisure activities, whereas a large railroad circus or Wild West show might come to town only once or twice a year; as a result, residents could save in advance so that they might spend on Circus Day.

Like vaudeville, amusement parks, world's fairs, and the nascent movie industry, the railroad circus was an essential component of a burgeoning mass culture.[91] In the new urbanizing society at the turn of the century, immigrants and the native-born from all social classes increasingly participated in shared forms of popular entertainment.[92] David Nasaw explains that the new mass culture was "a by-product of the enormous expansion of cities."[93] Collectively, these popular forms helped bring about the development of twentieth-century mass culture forms like radio, television, and Disney's empire that capitalized on middle-class notions of propriety to produce virtuous entertainment for all classes. Unlike these amusements, though, the circus did not experience a development exclusively tied to the growth of cities; instead its evolution, as suggested earlier in this chapter, depended upon continental expansion and internal improvements.

The diversity of the audience at the turn-of-the-century circus was amplified by the presence of children. This development was especially striking

[*A Historical and Cultural Process*]

because the antebellum circus had been primarily an adult entertainment. In the 1880s P. T. Barnum called himself "the Children's Friend" and welcomed "children of all ages." Barnum and many social purity reformers argued that the circus offered all Americans—especially impressionable children—great moral lessons about courage, discipline, and bodily fortitude. Large railroad showmen frequently sponsored Orphans' Day productions in which local orphans were able to attend the circus free of charge. On April 12, 1894, orphanages in New York City collectively sent 4,491 children to Barnum & Bailey's circus.[94] Impresarios also sent sick children to the circus, where their health was reportedly restored: "Patients [once sick with hydrophobia] Now Cured," blared one story.[95] In 1902 the National Biscuit Company introduced Barnum's Animals, crackers encased in a vivid "take-along" package covered with pictures of animals. The popular new women's magazine Ladies' Home Journal (1883) had pages and pages of colorful circus cutouts for children: female bareback riders clad in tutus, bare-chested Native American men, pipe-smoking seals, floppy clowns, boxing kangaroos, erect ringmasters, educated pigs.[96]

In a popular setting, the circus complemented the ideas of contemporary intellectuals like Ellen Key, John Dewey, and G. Stanley Hall, who argued that play was an important part of childhood development. Rudyard Kipling's Jungle Book (1894) and other children's books portrayed a child-centered world in which animals talked and children had exciting adventures in far-flung locations. Gary Cross has written that play began to replace work as a way for middle-class children to learn adult roles in an urban industrial society (where their labor was increasingly superfluous—although working-class youth still toiled to help support their families). Cross observes that in the context of a burgeoning consumer economy and changing attitudes about children's play, the toy business expanded rapidly at the turn of the century.[97] The Progressive leaders of the Playground Movement contended that the creation of urban play spaces could foster self-control through bodily conditioning.[98] In this social context, the circus ballyhooed itself as a site of uplift where children could watch superlative physical discipline in a fun setting.

The circus also inspired other aspects of the flowering children's consumer culture. Circus novels for children were common at the turn of the century, as were circus toys. Schoenhut's popular Humpty Dumpty Circus (1903) was a wooden, jointed play set of circus athletes and animals which could be twisted into myriad poses. Both toy manufacturers and circus proprietors used contemporary imperialism to create salable commodities. By

1910 the Humpty Dumpty Circus became "Humpty Dumpty in Africa," based on Theodore Roosevelt's African safari of 1909. The play set included a Roosevelt figure and a black guide, in addition to the usual stock of circus characters.[99] The modern child often first glimpsed the exotic Other through circuses and toys, a formative encounter that helped make colonial power relations part of the unconscious, "natural" world of child's play.

Many recent studies have broadened the parameters of diplomatic history to include topics like play sets and other facets of everyday life. Shaped by the new social history, itself a product of the social movements of the 1960s and early 1970s, these works consider how ordinary people (as well as elites) have participated in and shaped U.S. foreign relations. Using gender, race, and class as their analytic tools, practitioners of the new social history have demonstrated the interconnectedness of domestic culture — including sexuality, the division of labor, civil rights issues, and consumption patterns — and U.S. foreign policy.[100] Scholars of popular culture, influenced by the field of cultural studies and the work of Antonio Gramsci,[101] have also located power relations outside traditional political boundaries.[102] Robert Rydell, in particular, has pioneered this interdisciplinary approach by demonstrating that American international expositions at the turn of the century and in the 1930s promulgated U.S. domination overseas, Euroamerican racism, and the political and economic interests of "captains of industry." Building upon these studies, this book explores the powerful relationship between popular culture, ideology, national identity, and state formation.

3
SPECTACULAR
LABOR

When Barnum & Bailey's Greatest Show on Earth rolled into Kansas City, Missouri, in 1917, Emmett Kelly, a teen-aged industrial painter (and future hobo circus clown star) remembered the scene vividly. Transfixed by the size of the circus, Kelly counted a hundred railroad cars: "I could hardly believe the size of it. . . . The show traveled on four separate trains and looked like a big town. There was a blacksmith shop and big cook and dining tents and a barbershop tent and I could see a man delivering mail like a regular postman, and there were electric-light plants and water wagons—it was a sight I'll never forget."[1]

People awoke hours before dawn to catch the first glimpse of the mile-long configuration of circus trains pulling into town (fig. 6). Carl Sandburg recalled scrambling out of bed as a boy in Galesburg, Illinois, on Circus Day: "When the circus came to town we managed to shake out of sleep at four o'clock in the morning, grab a slice of bread and butter and make a fast walk to the Q. [Chicago, Burlington and Quincy Railroad] yards to watch the unloading in early daylight."[2] With amazement, countless spectators watched what circus folk called "the greatest *free* show on earth." They gazed at the dazzling gilded wagons rolling smoothly off railroad flatcars; they saw elephants assisting muscular men erecting voluminous canvas tents; they smelled huge vats of coffee that would produce two thousand cups for bleary-eyed workers, and sizzling bacon, sausage, eggs, and pan-

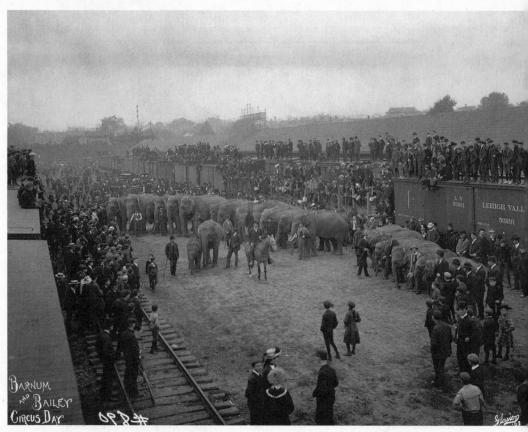

Figure 6. "Circus Day," Barnum & Bailey, Brockton, Mass., 1903. Unloading the circus train was part of the "show" for the masses who awoke before dawn to see this logistical spectacle. (Frederick Glasier Collection, neg. no. 890; black-and-white photograph, copy from glass plate negative, 10 × 8 in., museum purchase, Collection of the John and Mable Museum of Art Archives, Sarasota, Fla.)

cakes in the cook tent that would soon feed more than 1,000 employees. In 1890 the *Detroit Free Press* estimated that Barnum & Bailey employed an "army" of 800 men and 200 women.[3] In 1893 this circus also had 407 horses and ponies, 2 mules, 1 giant and 1 hairless horse, 12 elephants, 4 camels, and 8 dromedaries. There were 102 wagons, including 55 baggage wagons, 22 cages, 2 ticket wagons, 4 band wagons, and a clown cart. The ground crew of "roustabouts" (manual laborers) erected a total of 68,000 yards of canvas each working day and used 173,397 feet of rope as part of this process.[4] In 1908 the *Eau Claire* (Wis.) *Daily Telegram* observed that the Ringling Bros. circus employed 75 cooks and kitchen helpers who prepared 3,200 meals a

[*Spectacular Labor*]

day using 4,000 pounds of meat; for breakfast alone, the kitchen staff cooked 3,600 eggs and 800 pounds of mutton.[5] Spectators witnessed an efficient two-hour transformation of an empty lot into a vast, fragrant, nomadic city.

What did this gargantuan logistical display of human and animal labor tell its audiences about the Gilded Age? How did the railroad industrialize the circus? What did the circus's size and labor structure reveal about the ideologies of its owners and workers? This chapter shows that the railroad circus's great size, its participation in monopoly capitalism, its specialized division of labor, its ethos of efficiency, the individual "rags-to-riches" narratives of its owners, and its structural and ideological embodiment of a (traveling) company town gave its distant patrons an intimate look at the beliefs, values, and material practices of the new corporate order. Still, the railroad circus presented this industrial order in ways that evoked an older preindustrial world, where humans and animals were stronger than machines, and talented individuals could "rise" through hard work and self-discipline. Echoing Raymond Williams, who argues that culture is a "social material process," this chapter illuminates the thick, physical framework in which the circus produced its ideological content.[6]

THE CIRCUS AS BIG BUSINESS

At the turn of the century, the size and scale of the railroad circus mirrored the growth of big business and the expansion of the industrial workforce. Between 1895 and 1904 over 1,800 manufacturing companies merged into 157 horizontal combinations which dominated their respective markets.[7] By 1900 nearly 450 other companies employed over 1,000 people and more than 1,000 companies employed between 500 and 1,000. In the iron and steel industry, the workforce of the average firm, which stood at under 100 in 1870, had since quadrupled.[8] Other parts of the amusement industry ballooned as well. The Keith-Albee vaudeville circuit became a monopoly, a harbinger of the growth of big business in movies and radio, and later of such synergistic media conglomerates as Disney, Sony, and AOL Time Warner. Beginning in the 1920s, the "Big Five" motion picture companies (RKO, Metro-Goldwyn-Mayer, Paramount, Warner Brothers, and Twentieth Century Fox) virtually controlled American film production and distribution. The Big Five owned large movie theaters in lucrative markets and also practiced "block booking," selling only big blocks of movies (instead of individual films) to independent theaters.[9]

National media used the giant circus as a colorful trope for monopoly capitalism. Newspaper cartoons commonly depicted business trusts and industrialists as animal and human circus actors. In one cartoon, John D. Rockefeller and Averell Harriman, obese with bulging eyes, sat in plush box seats with bemused detachment as they watched a big-top performance in which terrified monkeylike citizens careened wildly atop galloping donkeys, labeled "trust extortion," which represented the oil, railroad, beef, sugar, and coal trusts.[10] Yet the relationship between the circus and big business was more than metaphorical, because the circus *was*—relatively speaking—big business at the turn of the century.

Although much smaller in scale than giant trusts like Standard Oil and U.S. Steel, a few giants created by several circus mergers in the late nineteenth century and the early twentieth controlled the railroad circus routes. P. T. Barnum and James A. Bailey merged their operations in 1880. By 1904 Bailey (Barnum had died in 1891) owned two of the nation's biggest shows—the Greatest Show on Earth and the Adam Forepaugh & Sells Brothers circus, itself the product of a merger of the Adam Forepaugh circus and the Sells Brothers circus in 1896. From 1895 Bailey also supplied capital, equipment, and managerial expertise for Buffalo Bill's Wild West.[11] The Ringling Bros. successfully captured the American circus market while Barnum & Bailey's circus toured Europe from 1897 to 1902. In the 1890s, and then again in 1904, these rivals agreed to separate their routing territory so that they would not overlap and draw business away from each other. After Bailey died in 1906, the Ringling Bros. bought his entire holdings during the depression of 1907 for $410,000, a bargain price.[12] Yet for purposes of comparison, U.S. Steel was worth $1 billion[13] in 1901 when it was created out of several horizontal combinations.[14] The Ringling Bros. operated the Barnum and Bailey circus as a separate unit until 1919, when they merged the two circuses into a single production, the Ringling Bros. and Barnum & Bailey's Combined Shows. In 1929 John Ringling, the last surviving brother, bought out the American Circus Corporation, a stock corporation composed of five circuses; the buyout, unfortunately, forced Ringling into debt. His financial resources were further eroded by the arrival of the Great Depression that same year. By 1932 Ringling, now ailing, was forced to relinquish control of the Ringling Bros. and Barnum & Bailey circus to unscrupulous creditors. Ringling remained as the show's president, but in name only. The firm New York Investors controlled his assets and forbade him to exercise any authority over the circus. Ringling's ex-friend Sam Gumpertz, a former Wild West rider, acrobat, Coney Island manager, and real estate magnate with close financial ties to the New York Investors, served as the general manager of the circus from 1932 until 1937, when members of the Ringling family regained control after John's death in the preceding year.[15]

In 1907 rival showmen balked when the Ringling Bros. purchased Barnum & Bailey. Although the federal courts never held that the circus business was a violator of the Sherman Anti-Trust Act (1890), other circus owners accused the Ringling Bros. of monopolistic practices. For example, E. Sherman Dandy, an agent for the Hagenbeck circus, charged that the Ringling "trust" paid railroad contractors for their competitors' routes. He also contended that they sabotaged smaller outfits by showering their routes with bills for upcoming trust shows. "[T]hey make . . . the country a chess board, and move their attractions from one point to another, canceling dates without regard for obligation to the public simply to put an established show ahead of one which is struggling for recognition."[16] Moreover, Dandy accused the Ringling Bros. and the railroads of price fixing: railroad companies agreed to haul the trust productions for one-fifth the regular rate in order to keep their business while forcing Ringling competitors to pay the regular rate.[17] Al Ringling justified such practices by stating that his family's circus operation was "nothing more than survival of the fittest." Other Gilded Age capitalists like Andrew Carnegie and John D. Rockefeller explained their success with such Spencerian language as well.[18]

Employing over a thousand workers and performers, the biggest railroad circuses like Barnum & Bailey and the Ringling Bros. shared other structural and ideological similarities with turn-of-the-century corporations. The circus's division of labor was highly specialized and was bound to the clock to meet fixed railroad schedules. As on the assembly line at Henry Ford's new automobile plant in Highland Park, Michigan, each railroad circus worker had a specific job that atomized and segmented his labor. And as on the alienating assembly line, railroad circus managers often treated workers as anonymous cogs in a vast production machine. At Barnum & Bailey, each workingman wore a numbered identity badge because numbers were easier to remember than hundreds of different names.[19] At small wagon circuses, or "mud shows," though, the division of labor remained "preindustrial" because these outfits ran independently of the railroad clock and the duties of the workforce overlapped considerably—performers and managers alike sold tickets, in addition to doing any other job that needed to be done.

Greater specialization and discipline at the large railroad circus emerged concurrently with other developments in the American industrial workplace. Industrialists extolled the financial rewards of "scientific management." An engineering executive and author, Frederick Winslow Taylor, used the stopwatch and the time-and-motion study to increase workers' output. Taylor analyzed various jobs, from iron forging to bricklaying, over a thirty-year period, beginning in the late 1870s, and concluded that shop managers could create a more congenial and productive workplace if they provided laborers with detailed instructions for each task and rewarded those who performed most efficiently. According to Taylor, "The task is always so regulated that the man who is well suited to his job will thrive while working at this rate during a long term of years and grow happier and more prosperous, instead of being overworked. Scientific management consists very largely in preparing for and carrying out these tasks."[20] Reflecting Taylor's objectives, railroad outfits meticulously planned and parceled out each part of the labor process. Impresarios carefully chose managers who designed the route a year ahead of the show date; they organized teams of workers who secured local contracts months in advance, and managers choreographed the activities of the workingmen who erected and tore down the tented city on Circus Day. The complicated act of simply advertising the production gave future audiences an intimate peek at the "industrialized" work spectacle they would witness on Circus Day.

PUTTING THE SHOW ON THE ROAD:
THE CIRCUS ADVERTISING MACHINE

Circus proprietors and routing agents wove a complex web of market research to plan their routes. The circus historians Fred Dahlinger and Stuart Thayer write that circus advertising became increasingly sophisticated from 1871 (when Barnum entered the circus business) to the turn of the century, when the Ringlings became undisputed "circus kings." While Barnum and his manager W. C. Coup focused on population, the Ringlings considered a range of marketing factors in determining where to perform most profitably. They analyzed seasonal patterns and regional weather conditions, specifically the incidence of drought and rain, crop reports, factory conditions, bank clearings, and the presence of summer resorts to determine an area's level of prosperity.[21] In October 1900 the *Greenville* (Tex.) *Evening Banner* proudly noted that six circuses were currently in Texas—a sure sign of a booming economy: "It is a well known fact that circuses always pick out the states where the people are most prosperous and the fact that so many are here now is a hint to outsiders that should not be overlooked."[22] Circus routing agents also studied how often competing circuses visited a specific region and the content of their rival productions. This was a particularly important consideration: when a showman's routing turf was seemingly intruded upon, he might destroy his rival's bills and plaster the remains over with his own posters ("sticker wars," in the words of Charles Ringling), or a brawl might erupt between competing outfits.[23] Proprietors and routing agents gathered this information through a flurry of correspondence with bankers, newspaper editors, railroad traffic managers, and postmasters. These men also worked in tandem with the railroad contractor, a circus manager who secured travel arrangements with railroad officials.[24]

Advertising the route was daunting. Months before an actual show date, teams of contractors, advertising agents, and billposters traveled to future markets in brightly painted railroad cars designed to advertise the circus. Long before Circus Day, these workers, known collectively as "advance men," provided townspeople with a prelude of the disciplined pageant of labor that was to come. The advance men secured various contracts for fuel, animal feed, water, fuel, eggs, milk, meat, and other perishables, and inundated future markets with colorful lithographs and handbills. Although the precise number of employees working ahead of the actual circus varied, Barnum & Bailey's circus in 1894 provides a window into the complicated advance system at the largest outfits and serves as the primary example for the following discussion. Barnum & Bailey had four advertising cars (confusingly called Cars Number One, Two, Four, and Six) which followed each other a week or two apart on the same route; each car cost $1,000 per week[25] to operate, and each typically had eight to eleven billposters, a boss-billposter, several lithographers, a manager, and on occasion an advance press agent who confirmed advertising arrangements with local newspapers.[26]

Car Number One, also known as the "skirmishing" or "opposition" car, was a trouble shooter. The advance men riding in it made certain that competing shows, collectively called the "opposition," did not steal a previously arranged date or sabotage earlier transportation contracts made soon after circus managers had determined the route for the upcoming season. The general contractor was the first circus worker to cover the route, and he made written arrangements for virtually everything that the outfit needed at each stop: licenses, exhibition grounds, billboards, liverymen (local drivers), animal feed, meat, hotels, and food for circus employees.[27] Often the general contractor and other advance men had legal training, in order to wade through complex contracts and local laws.

An advance agent, press agent, and several billposters traveled in Car

Number Two. In this car the agent in charge had a list of all contracts made in advance, as well as detailed information concerning the town's population and local roads; he also served as his crew's banker, accountant, and railway agent, carrying cash in a small safe from which he paid his team. He also kept track of every lithograph and sheet of paper received, where posted and by whom, and the number of complimentary tickets issued to people willing to have their buildings pasted. The contracting agent had already arranged where this crew would eat breakfast and the number of liverymen who would be assigned to take the billposters out on the country roads along the specified route. Using this information, the boss billposter divided his workers into groups which covered the town and rural routes. These groups moved quickly, because they only had one day to blanket each town and outlying areas. As Car Number Two rumbled to its destination, billposters stood on platforms with armloads of circus handbills, bombarding every small hamlet, farmhouse, or crossroads as they roared past.[28]

While en route in the train at night, billposters prepared the sticky flour-and-water-based paste which they used to put up the bills the next morning. Wearing "pasty suits," they boiled the paste in six big iron cans. On the road, they used two to three barrels of flour in towns and five or six in big cities. Each crew took several buckets of thick paste, to be diluted with water, and was required to keep track of how much paste was used and the number of bills posted. Car Number Two was outfitted with a shrill, piercing steam whistle which the advance team blasted as a way to announce its arrival in a new community.[29] After breakfast in town, each bill posting-crew and its liveryman rode into the countryside, searching for suitable pasting sites: barns, stables, or shops. But they did not (usually) paste a building without getting a written contract from its owner. In exchange for two to six complimentary tickets to the upcoming circus, the property owner allowed circus workers to cover his buildings with posters for a period of time specified by the circus.[30] Billposters often covered a spectacular breadth of territory; before the 1891 season, Forepaugh's agents Geoffrey Robinson and Whiting Allen posted circus bills atop Pike's Peak at its pinnacle, an elevation of 14,110 feet![31]

Billposters pasted approximately 5,000 lithographs per locality.[32] The Barnum & Bailey circus usually played for an entire month at Madison Square Garden in New York City and in nearby Brooklyn; in 1893 the circus plastered 27,110 sheets in New York City, including 9,525 on the railways, and an additional 8,186 in surrounding cities.[33] Charles Theodore Murray saw the advance men as magicians of sorts: "The circus bill-poster was a member of the Santa Claus family—coming from nowhere and vanishing into nothing, but leaving the glowing traces of his visit in highly colored pictorial illustrations that covered the dead walls in town and along the country roads. Sometimes it was done in the night when we were in bed; sometimes while we were at school. But I never succeeded in catching the circus man in the act." [34] A turn-of-the-century trade publication, *Billboard Advertising*, observed that the circus was the first business in the United States to master the use of the poster. Initially featuring just one color in its antebellum days, each fin-de-siècle railroad circus poster contained at least six or seven eye-catching colors. Some posters were designed as individual puzzlelike pieces that formed a single giant banner when pasted together; one banner, comprising thirty-two posters, was reportedly some 70 feet long. [35] In 1896 Ringling Bros. spent $128,000 [36] for posters alone. [37]

Months before Circus Day, the advance team transformed gray, weather-beaten barns and dull, brick stores into a colorful frenzy of clowns, tigers, semibare women, and elephants. Spectators knew far in advance that the circus was coming, so they could make transportation and work-release arrangements for the big day. In short, circus billposters marked the landscape, claimed it, and transformed it months before the actual onslaught of crowds, tents, and animals.

Subsequent cars made certain that the upcoming show remained visible. Cars Four and Six verified the arrangements made by the second car; retracing the routes of their fellow workers, these advance men and billposters checked to see if bills had been defaced, destroyed by rain, or covered by a rival or another advertiser. If a farmer or business owner violated the terms of the bill-posting contract, the circus rescinded his ticket privileges. Meanwhile, press agents (often ex-newspaper writers) confirmed previous advertising arrangements with local newspapers and obtained permission from the local drug or book store to sell tickets there on Circus Day. The press agent from Car Number Two made the initial contracts with the newspapers and submitted a different press release for each of the four or five dailies in a town. (After the show, press agents often submitted faux "reviews" of the show that were unflaggingly positive. These same "after-blast" reviews—ostensibly written by a local reporter—appeared verbatim at towns hundreds of miles apart.) [38] Press agents also gave local reporters, policemen, and politicians complimentary tickets ("comps") as a way to generate positive press and to provide additional surveillance on Circus Day.

Car Number Four publicized the upcoming production along the rural periphery, so that the most isolated audiences felt the circus's reach. Called the "excursion" car, Car Number Four traveled all railroad routes within a fifty-mile radius of the circus stop. The billposters on this car covered this area with bills advertising special train schedules and excursion rates. The manager of the excursion car verified the arrangements for special rail ticket prices and travel times that had been made earlier by the excursion agent, who worked directly with railroad officials to make special travel arrangements for circus audiences.[39] Within a week or two of the production date, Car Number Six finalized the arrangements and billing work done by the previous cars and quickly remedied any gaps in press work or bill posting.[40] After Car Number Six finished all remaining business, the circus was ready to come to town.

THE CANVAS CITY

The "army" of canvasmen who erected and tore down the billowing canvas tents was a crucial part of the total labor show. Audience members traveled in horse-drawn wagons over miles of bumpy dirt road in predawn darkness just to observe how circus workers (aided by horses and elephants) created a magical, movable city on an empty lot (fig. 7). At the turn of the century—an age of increased mechanization and de-skilling in the industrial workplace—human and animal labor still performed virtually all the on-site jobs, before gasoline-engine stake-drivers and other motorized machines began replacing some of the human labor in 1910–20. The creation of the tented city was a thrilling physical feat in which human labor functioned as a seemingly seamless, corporate body.

Immediately after the circus trains arrived at the show grounds, the boss canvasman, who directed the erection of the tents, scouted out the lot, which occupied ten acres or so. He first decided where the big top should stand, which determined the position of the other eleven tents. A group of canvasmen used iron rods to mark the positions of the five center poles, which would be the tent's center of gravity, and then mapped the perimeter of Barnum & Bailey's 1894 big top—440 feet by 180 feet—with more rods, topped with little flags, color-coded red or blue to identify each tent site. Next, the boss canvasman and his crew marked the placement of the menagerie tent, which was connected to the big top by a neck of canvas. The perimeter of the menagerie—360 feet by 160 feet—was identified by rods topped by white flags. This process was completed in just eight minutes from start to finish. The crew then quickly mapped out the placement of the

Figure 7. "Elephant Laborer," Barnum & Bailey, 1906. In addition to performing tricks under the big top, elephants helped set up and tear down the canvas city. (Photograph courtesy of Circus World Museum, Baraboo, Wis., B+B-N81-06-1-N)

dressing-room tent, two horse tents, the wardrobe tents, the sideshow tent, the freaks' dressing-room tent, and several smaller tents for the blacksmith shop, the repair shop, and so on. The canvasmen finished this whole job in half an hour.[41]

Next, the boss canvasman directed the unloading of the stake and chain wagons, the pole wagons, the canvas wagons, and other baggage wagons, each drawn by four or six horses. He divided the eighty-five muscular men who composed the "big top gang" into two groups, of which one laid the stakes into the ground and the other handled the sledges. Each stake was four to five feet long and two or three inches thick, and three-fourths of its length had to be hammered into the ground. Using sledges with three-foot handles and heads that weighed seventeen pounds, groups of about seven men stood in a circle and took turns hammering each stake into the ground, singing rhythmically as they worked. Each group had a leader who initially tapped the stake into position (fig. 8). Meanwhile, groups of pole riggers placed the tent's center poles into position. Within forty-five minutes a whistle signaled that the stake drivers and pole riggers had finished,

Figure 8. "Sledging Gang," Barnum & Bailey, Brockton, Mass., 1903.
With near military precision, stake drivers singing sea shanties and other songs
rhythmically pounded into the ground the heavy stakes used to secure the fifty-foot center
poles. (Frederick Glasier Collection, neg. no. 1319; black-and-white photograph,
copy from glass plate negative, 10 × 8 in., museum purchase, Collection of
the John and Mable Museum of Art Archives, Sarasota, Fla.)

and summoned additional groups of workingmen to help raise the poles. The center poles (as well as the linchpin "king" pole) were raised with heavy ropes attached to the stakes. In the middle of this process, other workers started joining sections of canvas that would form the tent's roof and side walls. They lifted the canvas using horses, pulley blocks, and a complex array of small side poles. By this time three huge cook wagons had arrived at the site of their tent, and butchers began chopping 500 pounds of meat into individual cutlets, while cooks prepared coffee and eggs. The working-men raised all twelve tents by six o'clock in the morning—just two and a half hours after they had typically arrived. At that moment, a loud bell or whistle alerted every circus worker that the dining tent was now open. Now all wagons were stationed at their proper places. The empty field had been transformed into a temporary canvas city.[42]

The complex disassembly process began immediately after the 8 P.M. performance began. First, workingmen took down the menagerie tent. All "cage stock" (animals not appearing in the big top) were loaded into cage wagons and taken to the train. Concurrently, workmen lowered the open-flame burners (called chandeliers) from the center poles and extinguished all but one of the naphtha (petroleum) jets.[43] Then, all players appearing in the "ethnological congress of strange and savage tribes" readied their trunks and boxes to be loaded into the baggage wagon. Quickly the canvas-men and pole-riggers disassembled and loaded the canvas side walls and scores of side poles in different wagons, and finally, using pulley ropes, lowered the five center poles and separated the 40,000 feet of billowing canvas into six units. Twenty workingmen were on hand to load the heavy canvas and center poles into specially designed wagons. This complicated process, involving dozens of workers, each performing a specific job, took approximately thirty-one minutes.[44] Meanwhile, upwards of 10,000 people sat under the big top, unaware that the tented city was being swiftly disas-sembled as they enjoyed the evening program.

The members of the transportation, or railroad, gang (also known as polers, or polemen) were busy soon after their evening meal. They loaded the wagons in the order in which they were to be used the following day, beginning with the bulky cook wagon. Working by lantern light under the direction of a boss transportation man, the railroad gang ran wagons up two inclined planes that were each thirty-five feet long, four inches thick, and sixteen inches wide, with four-inch guards on either side. From there, the gang rolled each wagon into position on a specific flatcar. Then another group of workers, the "razorbacks" (from "raise your backs!"), secured the wagons and cages into position atop the flatcars. The railroad gang and razorbacks quickly loaded the canvas, poles, stakes, cook tent apparatus, and 200 stock horses, so that the first section of the train could pull away by 12:15 A.M., carrying 300 sleeping cooks, tent polers, stake drivers, butchers, and others, who were the first circus workers to awake the next morning. Simultaneously, other transportation crews loaded the remaining two sec-tions of the train. The second section contained the ring stock, the wardrobe wagon, and all the seating apparatus for the big top. Scores of animal men and grooms slept in this section. The third section held the elephant cars and four sleepers, as well as the proprietor's private car, and the cars for the big-top artists, the freaks, and the ethnological congress. By 1 A.M., all three sections had departed for the next town. Three hours later, the working-men who were asleep on the first section would arise for another day with the circus.[45]

Audiences enjoyed the spectacle of human and animal labor so thoroughly that many were willing to pay for the opportunity to see it. As far back as the 1850s, impresarios capitalized on this fascination by instituting the after-show "concert," which immediately followed the evening performance. At the turn of the century, the size and scale of the circus's logistical operations made the concert a big draw. Impresarios charged spectators twenty-five cents, ostensibly to watch minstrel acts or Wild West stunts, but the real show was the workingmen bustling around, moving animals, equipment, and tearing down adjacent tents.

The train, or "iron horse," was an essential player in the circus labor performance. Its physical presence, the circus notwithstanding, was also a form of spectacle, from its earliest years in the 1820s to its explosive expansion in the Gilded Age. The earliest memory for Al Rosboro, a ninety-year-old ex-slave in the late 1930s, was of the construction of the railroad in White Oak, South Carolina, and the arrival of the town's first train: "When de fust engine come through, puffin' and tootin,' lak to scare 'most everybody to death. People got used to it but de mules and hosses of old marster seem lak they never did. A train of cars a movin' 'long is still de grandest sight to my eyes in de world. Excite me more now than greyhound busses, or airplanes in de sky."[46] Laura Ingalls Wilder recalled—with terror—her first encounter with a train as she and her family waited at the depot on the wide prairie of Tracy, Minnesota, in 1879:

> They could not talk very well, because all the time they were waiting, and listening for the train. At long last, Mary said she thought she heard it. Then Laura heard a faint, faraway hum. Her heart beat so fast that she could hardly listen to Ma. . . . The train was coming, louder. . . . The engine's round front window glared in the sunshine like a huge eye. The smokestack flared upward to a wide top, and black smoke rolled up from it. A sudden streak of white shot up through the smoke, then the whistle screamed a long wild scream. The roaring thing came rushing straight at them all, swelling bigger and bigger, enormous, shaking everything with noise. . . . Then the worst was over. It had not hit them; it was roaring by them on thick big wheels. Bumps and crashes ran along the freight cars

and flat cars and they stopped moving. The train was there, and they had to get into it.[47]

As the most far-flung reminder of the industrial society, the roaring train reordered the American landscape. Thousands of railroad workers tunneled out mountain passes with pick axes and dynamite, built towering bridges, and filled gorges with dirt to enable the train's movement with 165,000 miles of track by 1890.

In addition to its awesome physical presence, the train turned the pastoral circus into an enormous industrial amusement, with an elaborate division of labor and a disciplined, time-bound industrial work ethic. The historian E. P. Thompson has analyzed how the advent of machines in eighteenth-century England created a regimented, fiercely extractive workplace where time became a form of currency, something to be "spent," not "passed."[48] The train engendered a similar transformation at the circus workplace by essentially "speeding up" and specializing the labor process through a new dependency on railroad timetables. As mentioned earlier, trains enabled circus proprietors to expand their operations dramatically after the Civil War, once the railroad industry standardized its gauge and completed the first transcontinental railroad in 1869. Showmen made train travel more efficient for their circuses by building special flatcars from which they could easily load and unload wagons of ring stock, menagerie animals, and supplies.

Not only did the train discipline the circus, it also standardized the nation's sense of time. As travel by train became commoner after the Civil War, railroad managers became frustrated by the absence of uniform time zones across the nation. Each town kept its own clock: when the sun crawled directly overhead, church bells rang and townspeople adjusted their time-pieces accordingly to twelve o'clock, high noon. Railway companies were unable to enforce punctual arrival and departure times because people's measure of time differed from town to town. Consequently, in 1883, the rail-road business unilaterally—without seeking federal legislation—divided the nation into four time zones, thereby unifying the nation's perception of time, space, and place.[49]

Time-consciousness pervaded all aspects of the railroad circus. Even the peppy brass big-top band instilled labor discipline. Under the big top, each circus act was carefully scripted to music. Performers knew exactly when they were to enter the ring, based on precise musical cues. Similarly, laborers worked quickly to meet each day's grueling railroad schedule. Yet paradoxically, for circus audiences, Circus Day was all about the suspension of time, when daily routines came grinding to a halt. The colorful panoply of foreign animals and human performers seemingly compressed time and space when the entire world appeared on Main Street. Inside the crowded tents, time also seemed in abeyance as spectators tried to comprehend three rings, two stages, and an outer hippodrome track of constant, relentless activity.

"RAGGED BARNUM" AND OTHER CAPTAINS OF THE CIRCUS INDUSTRY

Like other turn-of-the-century capitalists, circus owners vigorously participated in perpetuating the popular American ideal of the self-made man. P. T. Barnum, James A. Bailey, William F. Cody, and the five Ringling brothers all came from modest means; most floated around a number of different occupations before entering the amusement business; and each man's success was the product of luck as well as pluck. These impresarios claimed that their own humble backgrounds augmented their productions' good character—even though their success was often predicated upon exaggeration and hoaxes.

This popular "rags to riches" mythology flowered during the Gilded Age. From the 1860s until his death in 1899, Horatio Alger wrote over a hundred popular novels featuring the exploits of poor, deserving boys who rose from unfortunate circumstances through hard work and help from respectable members of society. The steel magnate Andrew Carnegie interpreted his own rise from modest circumstances to great wealth as the result of prudence and perseverance. Yet paradoxically, long-term economic upheaval severely tested the ideal of the "self-made man" at the same historical moment when this myth became most visible in American culture. In the twenty years following the market crash of 1873, business failure rates in the United States reached around 95 percent.[50] Working conditions were brutal in the new industries: between 1890 and 1917, about 72,000 railroad workers were killed on the job and 2 million injured.[51] Industrial workers in the Knights of Labor and rural populists in the Farmers' Alliance and People's Party rejected the prevailing social Darwinism of Herbert Spencer and William Graham Sumner in favor of building "cooperative commonwealths" that would protect individual liberty through collective action. Their social movements advocated mutuality in an age when individual farmers and workers had become increasingly vulnerable to the impersonal vagaries of the industrial marketplace.

In contrast to this ethos of mutuality, P. T. Barnum stressed individual initiative throughout his career. Born in 1810 to a large, old family from Bethel, Connecticut, Barnum began working as a clerk in a country store at fifteen to support his family after his father died. Over the next two decades, Barnum ran several lotteries and gained notoriety in the amusement and museum business with hoaxes such as those of Joice Heth, an elderly former slave billed as the 161-year-old nurse of George Washington, and the Feejee Mermaid, a shriveled fish corpse with monkey parts attached to it.[52] From 1841 to 1865 Barnum created several American celebrities, specifically the midget couple Tom Thumb and Lavinia Warren, and the opera star Jenny Lind, the "Swedish Nightingale."[53] First published in 1854, Barnum's autobiography, *Life of P. T. Barnum*, was a hugely popular rags-to-riches story that began with the American arrival of Barnum's seventeenth-century ancestor, Thomas Barnum, an indentured servant. In tandem with the American Museum, the autobiography helped make Barnum rich enough to provide most of the funding needed (approximately $320,000)[54] to create his Great Traveling Museum, Menagerie, Caravan and Hippodrome (1871) with his partners Dan Castello and W. C. Coup, who provided the rest of the initial capitalization.[55] After becoming a circus owner, Barnum carried thousands of copies of his autobiography for sale at the show grounds, and when his new circus commenced rail travel in 1872 Barnum designated certain railroad cars just to carry piles of his book.[56] In his writings and speeches, Barnum highlighted his own background as ample proof that individuals could rise above unfortunate circumstances without state-sponsored assistance. Until the eve of the Civil War, Barnum was an enthusiastic Jacksonian Democrat who supported open markets and briefly owned slaves before eventually supporting abolitionism.[57] Free competition had always been lucrative for Barnum, because his public squabbles with his rivals— notably the white elephant war with his competitor Adam Forepaugh in 1884—only increased his notoriety.[58]

Barnum was also a vigorous reformer. A devout Universalist and temperance advocate, Barnum subscribed to the notions of human perfectibility that bloomed during the Second Great Awakening. Press agents echoed this gospel of self-discipline and self-improvement: "This man started out without a cent . . . [and] worked himself up, step by step, leap by leap . . . until he stands . . . without a rival, one of the richest men in the country . . . [h]e is restless, earnest determined, industrious, zealous."[59] Barnum's emphasis on self-discipline drew him to the temperance cause, a social movement led in part by bourgeois business owners who were alarmed by sluggish, drunken behavior in the workplace, where employees took frequent "grog" breaks that sabotaged productivity. After witnessing drunken behavior in Saratoga in 1847, Barnum drained his liquor cabinet permanently. He explained his actions in his autobiography: "I felt that I had now a duty to perform—to save others, as I had been saved, and on the very morning when I signed the pledge [of temperance], I obtained over twenty signatures in Bridgeport. I talked temperance to all whom I met, and very soon commenced lecturing upon the subject in the adjacent towns and villages. I spent the entire winter and spring of 1851–1852 in lecturing free, through my native State, always traveling at my own expense, and I was glad to know that I aroused many hundreds, perhaps thousands, to the importance of the temperance reform."[60] As a museum proprietor, Barnum produced several temperance dramas, including "The Drunkard." He lobbied vigorously for the Maine Laws in 1850 and continued to press for temperance legislation when he was elected to public office in the 1860s and 1870s. Barnum's focus on temperance and human perfectibility carried over into the administration of his circus, for he allowed no drinking or gambling among his employees. Barnum's crusade for human perfectibility also influenced his promotion of the circus as a pedagogical entertainment that could improve its audiences.

The life of P. T. Barnum's circus partner, James A. Bailey, was a variation of the rags-to-riches narrative. Born on July 4, 1847, in Detroit as James A. McGinnis, Bailey lost both his parents by the time he was eight years old. Although his mother left approximately $20,000 for the care of her seven children,[61] "Jimmy" was constantly beaten by his eldest sister and guardian, Catharine, and at eleven he ran away. Later, he privately recalled his childhood to his brother-in-law Joseph McCaddon: "Instead of being treated as a ward for whom considerable provisions were made, I was made to work like a dog, and on the least provocation was whipped. My sister had boys of about my own age, and for their misdeeds I was punished. . . . I was worked so hard that I was always late at school, so I was continually being punished after school; and then for being late in getting home I was whipped again. I stood that treatment until I was about eleven years old."[62]

Jimmy's escape from his violent family eventually led him to the circus, a free space, where he felt safe. He came of age immersed in the outfit's daily operations. From the age of thirteen until his death forty-six years later, Bailey worked at a circus, first as a billposter on the advance advertising team with Robinson and Lake's circus. Jimmy A. McGinnis became James A. Bailey after he adopted the surname of Robinson and Lake's advance agent, Fred H. Bailey. James Bailey's diary chronicled how the bulk of the labor fell upon him as the junior employee. "Sunday, August 30, 1863: in Grayville [Illinois?] Stevens was so sick that he could not do any work I had to put up the bills all alone in Albion i put up the bills And in Olney stevens and Mr. Bailey went to the show and left me to do the [   ] I put up a part of the bills up that night and put the rest up the next Morning before breakfast. J.B. Monday, August 31: James Bailey put up the Bills alone. Tuesday, September 1: James Bailey put up the Bills alone. September 2: Mr. Stevens and Bailey gone to the show. i put up bills. James Bailey."[63]

Bailey's early work with the circus familiarized him with all angles of its operation. McCaddon observed, "In the beginning of his career, he had driven over the roads of all the midwest and southern states, year after year . . . in advance of the old-fashioned wagon shows . . . and he would familiarize himself with junctional points, distances, [and] the chief industries or products."[64] Armed with intimate knowledge of the circus, Bailey later acted unilaterally, meddling and unable to relinquish his authority. He frequently threatened workers with "instant dismissal" and bristled when challenged. McCaddon noted that employees had to "learn to obey orders. If he directs you to post a bill upside down, be sure you understand correctly and don't argue about it."[65] Bailey's tight-fisted, autocratic efficiency was a stark contrast to the Barnum's savvy for publicity. While Barnum attracted audiences through public spectacles as a politician, reformer, writer, and promoter of celebrities and hoaxes, Bailey remained invisible, choosing to absorb himself in single-handed management of his circuses—down to the smallest detail.

Press agents praised Bailey's self-discipline. Accordingly, Bailey "made himself great," possessing "tremendous energy," working since boyhood, "with the untiring tenacity and ambition which later characterized his entire career."[66] Other newspaper articles observed that "[Bailey] Likes His Work More than Anything Else."[67] Yet Bailey's bitter past with his biological family left him brittle and cold to his partners and employees, and it undoubtedly contributed to a nervous breakdown which took him away from the circus in 1886–87. Even his partner P. T. Barnum addressed him as "Mr. Bailey."[68] When an earlier circus partnership, Cooper and Bailey, lost money while performing in Argentina in 1878, Bailey decided to "redlight" (i.e. desert) his workers by stranding them in Buenos Aires without pay or transportation back to the United States. Among the workers he red-lighted were his brother-in-law and Cooper's young nephew![69]

Bailey disliked the masses that thronged into his circuses, even though programs ballyhooed Bailey's three-ring circus as a quintessentially "democratic" amusement that appealed to "all classes." Bailey yearned to produce an elite and "tasteful" one-ring, European-style circus that would attract "higher class" urban patrons. Before his death, Bailey planned with McCaddon to open such a one-ring "Big City Circus" that would celebrate individual artists, consequently avoiding the "mechanical sameness" of the contemporary three-ring railroad production. McCaddon later recalled the defining features of their proposed circus:

> The circus delux must be a place of beauty and thrills. . . . No more meaningless advertising street parades. . . . No more cheap side shows, or concerts, or peddling of toy balloons and other cheap articles to the annoyance of patrons. No more menagerie of drowsy animals in narrow cages, dimly lighted. . . . The New Circus, in lieu of the old style menagerie will have the first tent devoted to an exhibition of animals, all highly trained . . . that will later be seen in the arena. There may also be strange and curious living freaks, attractive illusions and other interesting exhibits. . . . The New Circus will be in smaller tents, water proof, more compact, comfortably seated. . . . Reserved numbered chairs and private box seats may be purchased by diagram from one to two weeks in advance, so patrons may avoid the pushing crowds.[70]

Bailey planned to eliminate mass audiences by making tickets prohibitively expensive for the "pushing crowds" who flocked to his three-ring outfit. Yet he died in 1906, before he could make his exclusive, urban circus a reality. In general, Bailey's abusive childhood and virtual lifetime immersion in the circus business shaped his rigid and taciturn style of management which helped create an efficient—yet occasionally resistant—workforce.

William F. "Buffalo Bill" Cody also crafted a narrative of individual initiative and self-improvement. According to Joy Kasson, "Cody was a Gilded Age businessman who loved to portray himself as a rags-to-riches hero."[71] Like Barnum, Cody created a profitable public image centered around hard work and colorful exploits. Cody also used the language of uplift to characterize his Wild West, but did so in ways that departed from Barnum's emphasis on individual propriety and proper domesticity. Buffalo Bill's "mission" emphasized American Indian cultural preservation and gender equality—both of which were strands of contemporary social reform. Cody marketed his own experiences on the trans-Mississippi West as an example of romantic heroism, adventure, and upward mobility on the democratic frontier. But one should also bear in mind that he epitomized the flip side of the western success story: even though his Wild West outfit made him rich, he experienced tremendous financial failure with speculative boondoggles in gold mines and mineral springs.

Cody loomed large as a central character in his production. Although poorly educated as a boy, he wrote four autobiographies and made several movies, such as "The Battle of Wounded Knee," produced by the Colonel W. F. Cody Historical Pictures Company. Born in 1846 on the eastern edge of the trans-Mississippi West in Iowa, the youthful Cody was a messenger for a freight company that eventually ran the pony express and later served in the 7th Kansas regiment, drove a stage coach, and hunted thousands of buffalo (hence his nickname) in order to feed the hungry crews laying the track for the Kansas Pacific railroad. In 1868 Cody was hired as chief of scouts for the 5th Cavalry. He became an actor in the 1870s, and participated in what Kasson calls "plains showmanship": buffalo hunts, feats of marksmanship, and horse races.[72] In 1883 he opened his Wild West. With long silver hair, a clipped goatee, a cowboy hat, and a leather coat, Cody had an appearance that reinforced his image as a "real" western icon. Moreover, Cody and his partner, the actor Nate Salsbury, hired hundreds of Native American players, including at least one famous chief each year beginning with Sitting Bull in 1885, and a stream of cowboys and cowgirls, notably Johnny Baker and Annie Oakley.[73]

Like other Gilded Age showmen, Cody credited his success to hard work. He was a virtual whirlwind of capitalist promotion. In letters to his family, "excuse haste" was a constant refrain, as he apologized for his perpetual busyness. He was involved in dozens of risky ventures—from gold mines in Arizona to a proposed scheme to turn the Grand Canyon into an exotic game park. Despite his volatile financial decisions, Cody maintained his Alger-like optimism that hard work would lead to financial success.

Cody mandated discipline and sobriety for all employees: "And I will have no one with me that's liable to let whiskey get away with him in this business a man *must* be perfectly reliable and *sober*" (emphasis in original).[74] Yet Cody battled the bottle throughout his career. In 1905 he wrote to his favorite sister, Julia, that he had rejected liquor through his faith in God: "And I realize how easy it is to abandon sin and Serve him. . . . Through this knowledge I have quit drinking entirely. And quit doing rash things simply by Controling my pasions."[75] Employees remembered that Cody constantly "downed one tumbler after another of his high potency mixture" on the show grounds, but that he never even appeared "tipsy."[76] Yet his alcoholism, coupled with bad financial investments, eventually undermined the stability of his Wild West—his one profitable venture. When Cody died in 1917, his Wild West show had changed ownership several times and he had only a nominal monetary interest in the outfit.[77]

Cody's actual relationship to the western frontier that he so successfully recreated in the ring was ambivalent. He advertised his production as a glimpse at a "vanishing" way of life and mourned the "disappearance" of the trans-Mississippi West and its chief human artifact, the American Indian, as casualties of Euroamerican expansion. Yet he helped hasten industrial development and settlement through his participation in the building of a transcontinental railroad, and he vigorously plugged settlement and development of Wyoming, his adopted home state. He garnered federal subsidies for the construction of an automobile and horse-stage line from Cody, Wyoming, to Yellowstone National Park, schemed to turn a local hot springs into a mineral spa (to "advertise the state"), built the Irma Hotel in Cody to promote tourism, ran a local newspaper, *Cody Enterprise*, and planned to build the Cody Military College, or International Academy of Rough Riders, in Wyoming. In a letter to a member of the State Land Board, Cody wrote, "I am working for Wyoming all the time."[78]

Cody's reformist beliefs shaped the content of his productions. Elected to the Nebraska state legislature in 1872, he advocated women's suffrage and the rights of working women. Cody's Wild West exhibited several female sharpshooters, most famously Annie Oakley, who worked there for seventeen years beginning in 1885. And he used show programs as a platform to express his political views on subjects ranging from western expansion to women's rights:

> You take a single woman earning her living in a city and the average man looks at her suspiciously if he hears that she lives alone. That makes me tired. A woman who is capable of financiering for herself is capable of taking care of her morals, and if she wants to take an apartment and live alone where she can do her work more quietly, or have things her own way when she comes from business, she has just as much right to do so as a bachelor. If a woman is a good woman she will remain good alone; if she is bad, being surrounded and overlooked, and watched and guarded, and chaperoned by a hundred old women in a boarding house won't make her good. This applies to society women as well as to working women.[79]

[*Spectacular Labor*]

Cody's relationship with Native American was conceptually inconsistent. He publicly supported the rights of American Indians, although he had helped decimate the Plains Indians' chief food source, the buffalo. Cody provided a good income for hundreds of Indian employees, particularly refugees from the massacre at Wounded Knee, South Dakota, in 1890, yet these actors were hired to play roles that reinforced stereotypes of Native American "savagery."[80] Cody attempted to use his Wild West to help preserve Native American cultures at a time when the Bureau of Indian Affairs and reformers were mandating wholesale assimilation. Cody argued that efforts by the bureau to prevent Native Americans from earning an "honest dollar" at Wild West shows constituted virtual "imprisonment" on their reservations: "the Indians are becomeing restless cooped up on their reservations. And if they are not allowed some liberty they will sooner or later give our frontier people trouble."[81] Cody also publicly exhorted all Native Americans to engage in agricultural cultivation as a way to prevent their "extinction": "But now, with the abundant acres of land that his white conquerors, with great justice, have allotted to him in the shape of reservations . . . [h]e now finds that fences are to be made, ground broken up, seed planted. . . . [He must] follow in fact, what he has often claimed in desire and spirit to follow, 'the white man's road.'"[82] As a showman and capitalist, Cody made contradictory claims, asserting that Native Americans could "uplift" themselves through assimilation, while at the same time protesting that they were "cooped up" on reservations.

The Ringling brothers also used the myth of individual mobility to explain their meteoric rise in the circus business. Their identity as a family operation shaped their management structure and their "wholesome," "Sunday School" reputation. Resonating with Gilded Age industrialists, the Ringlings freely attributed their self-made millions to hard work and individual volition. Alf T. Ringling, who managed the press department, fashioned the brothers' autobiography as a self-improvement tale, breathlessly describing the boys' meteoric rise from modest means to great wealth:[83]

> They had attained these results by years of patient labor, by many hours of thoughtful counsel, by careful conservative means, by dint of the greatest individual and collective exertion, and by the steadfast, unwavering determination to have the greatest tented exposition in America. During their early career they had passed through storms that threatened their hopes with destruction. They had gone through danger and had experienced all of the vicissitudes that befall those who embark in great

and enormous undertakings. . . . But when others were giving up the struggle under such adversities, the Ringling Bros. would summon up all their strength. Their motto was never to stop moving, to keep their show going.[84]

The Ringling brothers' actual background substantiated their claims. Al, Otto, Alf, Charles, and John Ringling were the sons of an itinerant immigrant German harness maker, August Rüngeling (who later Anglicized the family name). The father worked throughout the Midwest before settling in Baraboo, Wisconsin, in the 1870s with his wife, Salomé, and their eight children. Living in McGregor, Iowa, from 1862 to 1871, the five boys were inspired in 1869 to start a circus when they saw the John Stowe & Company circus and its unusual (for its day) Appaloosa horses. Andrew Gaffney, a performer with the circus, gave the Ringlings a complimentary family show pass after August repaired Gaffney's leather props free of charge because Gaffney was a local.[85] The Ringling boys quickly fashioned a "concert company" comprising panoramic (pictorial) comedy sketches and charged their youthful audiences "ten pins" (literally straight pins) instead of cash.[86] In 1879 Al worked part time as a juggler and acrobat, in addition to steadier employment as a carriage trimmer.[87] Three years later the brothers began performing a blend of blackface minstrelsy, comic skits, dance, songs, and juggling routines in hall shows around Wisconsin as the Ringling Brothers' Classic and Comic Concert Company. Facing blizzards, clems (fights), and sometimes no business at all, the brothers experienced rough times during these early years of touring. At one Wisconsin town, lead miners staged a wrestling match on the outskirts that drew away nearly the entire audience: only a handful of boys and the janitor attended the show that night.[88] Still, the Ringling brothers accumulated approximately $1,000 to $1,200 by 1884, enough to expand their operations.[89] They hired the veteran showman Yankee Robinson to form Yankee Robinson and Ringling Brothers Great Double Shows for their inaugural season under canvas. Robinson, sixty-six, died that same season, but the brothers plowed on.[90] In contrast to P. T. Barnum, who financed his own entrance into the circus industry, the Ringlings during the early years financed their circus with a series of small, promptly paid bank loans of as little as $20 at a time.[91] As their operations grew (particularly once they began buying other outfits), they borrowed more money, taking out a $77,000 loan[92] in 1905 from a local bank.[93] By 1907 the Ringling Bros. circus owned Barnum & Bailey's Greatest Show on Earth, in addition to several other circuses, making the Ringlings the most powerful circus showmen in the United States.

Throughout their careers, the Ringlings capitalized upon every aspect of their business as "proper," as proof of their own self-discipline. The brothers' autobiography solemnly professed: "It is said of the Ringling Bros. that in all their association with each other not one unkind word has ever passed between them. Certain it is that not one of their hundreds of employees can name an instance that would contradict the assertion, and none of the brothers remembers an occasion where friction has occurred between them."[94] Alf T. Ringling attributed the circus's moral code to the brothers' having run the show. As part of this familial ethos, the brothers did not sign official contracts with one another, relying instead on blood loyalty. They also split all profits equally among themselves.[95] In some ways, their family circus was evocative of an idealized pastoral social order, where families, not strangers, labored together. Otto handled the finances. John managed the routing and transportation arrangements. Alf T. ran the press department and made certain that the circus was prominently featured in local newspapers. Al was the equestrian director, choosing acts for the big top and sideshow, and deciding the order of the program. Lastly, Charles was in charge of advertising the circus. To the end of his life each brother ran a critical part of the overall operation, which magnified the circus's clean family name.[96] Yet by the 1930s, the brothers' survivors had disintegrated into permanent, bitter factions.[97]

Because the brothers prohibited graft, games of chance, and insobriety wholesale, they began calling themselves a "New School of American Showmen" in 1891. By 1894 press agents had adopted the more familiar "Sunday School" moniker.[98] As a "Sunday School" outfit, Ringling Bros. ensured that its contracts contained strict rules of conduct,[99] and it hired Pinkerton detective agents to enforce its upright environment.[100] E. E. MacGilvra, a rancher from Montana, knew the Ringlings as a child in Baraboo: "They were fine fellows. They carried their own police force, their own detective force. I can remember as a kid standing along side the ticket wagon and watching these farmers come in, you know, big families and all excited and everybody tugging at papa's coattail or mama's skirts and he's up there trying to buy tickets. He'd turn around and leave, maybe without picking up his change, but there was always a detective there to grab him and say, 'Here mister, you pick up your change, we won't touch it.' That's the kind of outfit they ran."[101]

In some respects, it is curious that Barnum, Bailey, Cody, and the Ring-lings marketed themselves as taciturn models of self-discipline and fru-gality when their productions ballyhooed excess. (Indeed, in *Ragged Dick* (1868), Horatio Alger included Barnum's Museum as part of the "tip-top" world of "low" amusements that drained Dick's scant income from boot-blacking; only when Dick stopped going to Barnum's and the Bowery, opened a bank account, saved his money, took regular baths, and dressed neatly did he begin to succeed.)[102] Advertisements for the railroad circus trumpeted its extravagance: its splendid cacophony of three rings, two stages—all "too big to see at once"—and its dizzying array of people and animals. Route books and newspaper articles raved about profligate con-sumers on Circus Day who spent their hard-earned cash on ephemeral stuff—cotton candy fluff, pink lemonade, games of chance, and sideshow displays—even though the circus itself was a well-oiled model of human (and animal) discipline, set to music, that incited its audiences to behave in decidedly undisciplined ways.

LABOR AND HIERARCHY

The social structure of the railroad circus was built upon an occupational hierarchy akin to a caste system, in which musicians ate and slept with mu-sicians, and candy "butchers" with candy butchers. An outfit's size magnified this caste system and ultimately made the labor process more efficient. At big railroad circuses, often workers barely saw one another because they traveled and performed their duties at different times of the day. As we have already seen, advance men traveled several weeks before the actual show; tent polers, transportation men, railroad gangs, and other working-men arrived at the grounds several hours before the performers and pre-pared for departure while the evening big-top show still played. Al Mann, a Wild West concert rider with the Ringling Bros. in 1923, and his wife, Irene Mann, a trick roper, rider, and aerialist, observed this social system in action. Both knew few sideshow acts, because the big-top and sideshow players worked at opposite ends of the show grounds. Consequently, ac-cording to Al, they "didn't go up there and hang around."[103] Because work-ers spent a great deal of time with people performing the same labor, they forged close bonds within their given occupation, which, in turn, further reinforced the division of labor at the circus.

The geography of the grounds enhanced the hierarchy. In the dining tent, canvas side walls ran down the center, dividing the tent into two sections with two separate entrances. Owners, managers, and performers in the principal big-top acts sat on the right side of the partition, while laborers sat on the left. In both areas, tables were arranged to reflect the hierarchy. The owner's table was at the front of the tent, followed by show managers and star acts. Secondary players sat in the back of the tent.[104] Each occupation had its own table: ushers and ticket sellers, different groups of band musicians, Wild West actors, sideshow players, and so on. The bands were divided into a descending hierarchy comprising the big-top band, the after-show concert band, and the sideshow band, also called the "nig show" because African American musicians worked in it.[105] Each worker was assigned a specific seat at his or her occupational table. One waiter served approximately twenty-four people.[106]

Designated by job, race, and sex, sleeping arrangements on the circus train further augmented the occupational hierarchy. The advance cars, which traveled independently weeks before the departure of the actual circus train, housed only male workers. The floor plan of the advance cars reinforced the hierarchy within the advance team itself: billposters slept together in separate berths in a communal area of the car and shared a bathroom. The head advance staff and additional professional management staff shared a stateroom containing a desk, a safe, a sofa that could be extended into a double berth, an upper berth, and a private bathroom.[107]

In the circus train itself, single male and female employees occupied separate Pullman cars, while married couples slept in the same car, either together or in different berths, depending on the size of the berths. Star big-top acts and managers were privileged enough to sleep in spacious staterooms. Fred Bradna, an acrobat, rider, and later the equestrian director for the Ringling Bros.' circus in 1915, and his acrobatic equestrienne wife, Ella, fondly remembered the day they moved into a stateroom:

No one can realize, unless he and his wife have spent every night of their married life in an upper and a lower berth, the feeling of release, of exhilarated spaciousness, resulting from removal to a stateroom with twin beds instead of bunks, windows to curtain, a bath and, most wonderful of all, privacy. I carried Ella over the doorsill of this, our first home. She wept when her trunk was hauled in and the delivery boy asked, "Where shall I put it?" Imagine such capaciousness that there was more than one place to put a comb and brush, let alone a trunk! She rushed downtown the next day, and returned with curtain materials, needles and thread, paint and brushes. For half a morning she walked about, actually able to take five steps in any direction without stumbling over something or someone, babbling to herself in her happiness.[108]

Sleeping assignments within the cars were an index of occupational status. Fred Bradna noted that after eight years of trouping, he and Ella no longer had to sleep over the wheels in the married couple's car. By 1911 they had "moved up" to the center of the car, where the ride was smoother.[109] If the circus possessed more than twenty cars, then the show might have an "owner's car," usually the train's last car, which was elegantly appointed with staterooms, private bathrooms, and a spacious dining area. In contrast, workingmen slept in densely packed bunks or hammocks, often converted from day coaches. Laborers also slept in stock cars or on flats, close to the areas in which they worked. The elephant men, for instance, slept in a narrow space above their animals, which meant that they were with them virtually nonstop. In 1895 some 300 Barnum & Bailey laborers occupied three sleeping cars that were each designed to hold fifty to sixty people, or half the number of people actually sleeping there. Not surprisingly, these crowded quarters were potentially dangerous. In 1884 sleeping quarters for the male laborers were so cramped on the Orton Brothers' circus train that bunks were erected in front of the doors once they were closed for the night. When a trash can started burning one night, sixty men were trapped inside the burning car—the only exit point was a tiny window at the front of the car, and at least eight men died.[110]

Ballet girls, like male laborers, lived in congested conditions on the train. The Single Ladies Car (commonly called the "virgins' car" by circus workers) often held four berths per section, although each section was designed to contain just two berths. Tiny Kline, a Barnum & Bailey ballet girl in 1916, recalled that women occupying the upper berths all wore pajamas to bed because none could execute the bodily contortions needed to wiggle in and out of the cramped quarters in a nightgown without exposing herself.[111] The lavatories in this car had no walls; show managers had them removed to prevent workers from monopolizing the bathrooms. The porteress kept a vigilant eye on the lavatory and made certain that passengers did not wash their bodies below the neck and arms. (Performers would have the opportunity for a full body bath once they reached the dressing tent before the first show each day: they braced themselves clean using buckets of icy water.)[112] To maintain order, a monitor was stationed at each car—male or female depending upon the passengers. Four washbasins were positioned at each end of the car, and each person had one hook in a giant closet for hanging his or her clothes. Everyone was required to tip the car porter twenty-five cents a week; the porter polished shoes every day, took care of the laundry once a week, and cleaned the berths.[113]

Circus workers maintained and policed the caste hierarchy among themselves. Employees often treated top players with deference. Josephine De-Mott Robinson, a star bareback rider from the 1880s to the 1920s, wrote that a female performer's spot in the women's dressing room reflected her importance within the show. In the days before electricity, Robinson's dressing area was directly under a chandelier, the brightest area in the dim tent.[114] Photographs of the star aerialist Lillian Leitzel from the 1920s reveal a scene of even greater privilege: a maid primps Leitzel's hair and makeup in a private dressing tent, nicely appointed with chintz-covered chairs, a table with linen, and fresh flowers.[115]

In the big-top dressing tents, trunks sat in the same place throughout the show season, arranged end to end in four straight rows. The location of each trunk revealed a performer's position in the hierarchy. In the women's big-top dressing tent, the first row was the most spacious and private position in the tent, indicating where the star players—usually bareback riders and aerialists—dressed. Trunks positioned further down this "queen's row" and into the second and third rows belonged to women with less status. Finally, the lowest members of the female hierarchy—ballet girls, statue girls, and performers in other secondary acts—had their trunks placed in the fourth row, set against the outer side wall of the tent, where they were exposed to constant foot traffic and to the wind and rain that often seeped through the tent. Male players' trunks were also positioned to reflect their status. In the men's dressing tent, clowns (with the exception of those who were famous) generally received the least desirable place to dress.[116]

A worker's wages reflected his or her position within the circus caste system. Not surprisingly, those who received the most deference were among the best paid, while the large team of roustabouts lived in crowded train cars and were paid the lowest wages at the show (fig. 9). All circus employees received free room and board in addition to their actual wages; consequently, their total compensation package was much higher than their monetary earnings—a fact that one must bear in mind when comparing circus jobs with other jobs.

Circus work was usually seasonal, and artists' wages varied greatly, depending above all upon the public appeal of the individual or group act. (Employees working as part of a group were paid as a group, not individually, and records generally do not reveal how group earnings were collected

Figure 9. "Payday," Barnum & Bailey, Brockton, Mass., 1903. Although the circus's social structure was characterized by a caste hierarchy, all waited in the same line each week to receive their pay check. (Frederick Glasier Collection, neg. no. 1403; black-and-white photograph, copy from glass plate negative, 10 × 8 in., museum purchase, Collection of the John and Mable Museum of Art Archives, Sarasota, Fla.)

or divided up.) The pay range for circus workers ranged dramatically and was generally determined by three factors: skill, gender, and race. Lastly, rates of pay differed from circus to circus; the largest railroad outfits usually paid higher wages than smaller circuses.

Advance agents were generally Euroamerican males who were fairly well compensated by the standards of the day. Because some advance agents had legal training, one might compare their earnings to the earnings of contemporary professional workers. In 1892 a bank accountant earned $16.63 a week[117] on average.[118] Although data for advance agents come from the years 1902 to 1912, wages in real dollars differed very little from 1892 to 1912, because the value of the dollar fluctuated by a maximum of only nine cents throughout this twenty-year period.[119] Detailed financial records exist from 1902 to 1912 for the Gollmar brothers circus, another outfit of five brothers based in Baraboo, Wisconsin (cousins to the Ringling brothers, in fact), that took to the rails in 1903 and eventually consisted of over twenty cars. At the Gollmar brothers, an advance agent's wage ranged from $10 to $50 a week,[120] depending on the year and the agent's rank. Billposters, who accompanied the advance agents, received considerably less money, earning about $5 to $6 a week[121] in 1902.[122] At Barnum & Bailey's circus, the advance agent paid the billposters their weekly wages. To make sure that the billposter stayed for the entire season (and that he refrain from swearing, drinking, gambling, or anything else "immoral"), the advance agent retained part of his weekly wage, called a "gratuity," or "holdback," which the billposter did not receive until the end of the season. This practice extended to other jobs in the circus business.[123]

Big-top acts were usually paid larger salaries than sideshow performers of similar fame. In 1917 Lillian Leitzel received $200 per week[124] during the Ringling Bros.' long Chicago engagement, and $165 per week[125] during her season on the road. Her contract stipulated that she receive a stateroom during the traveling season and a small dressing room tent, "if desired." In contrast to less popular performers, whose contracts stated that they must also "make themselves generally useful," Leitzel was only required to execute a "first class aerial ring act," and she did not have to appear in the parade.[126] At the sideshow that same year, the long-standing Sumatran player Krao, "the Missing Link" Farini, was paid $50 a week,[127] or a fourth of Leitzel's salary. Farini's contract stated that she "Exhibit self in Side Show and as required, same as 1916. First class costumes etc. . . . Not required for entrees, specs, parades. Can have ladies artist size trunk 20″ × 22″ × 28″."[128] Because Farini was well known, she was better paid than most sideshow acts, and she received a larger trunk, a notable special privilege at the nomadic circus. Unlike many big-top actors, sideshow players (Farini included) usually did not appear in opening parades and grand entries because their marketability depended upon their curious bodies—if audiences were able to preview the freaks, there would be little reason to pay ten to twenty-five cents extra to gaze at them at the subsequent "kid show" (sideshow in circus lingo). The concept of "skill" was integral to explaining the wage disparity between big-top and sideshow stars: managers deemed that athletic prowess constituted a higher skill than sheer bodily exhibition at the sideshow.

Popular female big-top acts occasionally received bigger wages than their male coworkers. The bareback rider Lizzie Rooney, of the famous Riding Rooney family, made $50 a week[129] in 1906, whereas her brother Charles, also a bareback rider but less known, was paid $15 a week[130] that same year.[131] As part of husband-and-wife or brother-and-sister teams, many women were not individually compensated for their labor, and therefore it is often difficult to determine the actual distribution of wages to female members of a family troupe. Harry Brandon, the principal clown with the Gollmar Brothers, received $35 a week [132] along with his wife in 1904, but in the following year he was paid $25 a week [133] when performing without his wife.[134]

But secondary women under the big top (i.e. ballet or chorus members) made less money than their male counterparts. Work contracts from the Ringling Bros. circus from 1900 to 1910 reveal that minor female performers earned approximately $7 a week [135] on average, compared with earnings of $10 to $15 a week [136] for male players in secondary roles.[137] These figures are generally commensurate with the wages of working women in other fields. In New York City, for instance, 56 percent of female factory workers earned less than $8 a week,[138] and most women earned less than $7.50 a week [139] in retail trade.[140] Yet, as noted earlier, free room and board were included in the total compensation package at the circus, and therefore the effective wages paid to female circus workers were actually much higher than those paid to women in other jobs.

The circus also maintained a large, inexpensive, and expendable unskilled labor force of workingmen. Because of their transience, only the boss of their departments and the cashier in the office knew their real names. Typically they were known by their town or state of origin (e.g. as "Boston" or "Kansas"), by their ethnic group ("Frenchy," for instance), or by a defining physical characteristic (such as "freckles" or "blackie"). If a roustabout bore a strong resemblance to George Washington, Daniel Boone, or any other notable, then he assumed that name.[141] The workers were virtually interchangeable. Fred Gollmar, the stage manager for the Gollmar Brothers circus, noted that a few days of rain could wipe out three-quarters of the workingmen, and that he spent much time in Chicago and other large cities securing new work gangs.[142] James Bailey also stated that these faceless laborers were "dispensable," adding that it was cheaper simply to add new workers at every stop rather than entice workingmen to stay the season by paying higher wages and providing better working conditions.[143] In 1892 the Walter L. Main circus, a medium-sized outfit with twenty to twenty-five cars, paid its "inexperienced and cheaper class of labor" $3 a week [144] for jobs not specified, with no mention of additional hold-back pay at the end of the season.[145]

Experienced workingmen commanded higher pay than other roustabouts. The Walter L. Main show recorded a detailed list of its salaries for "first-class experienced men." In 1902 this list included the following big-top assembly jobs: canvasmen, ring makers, stage men, seat men, and chandelier men (who set up and tore down the open-flame gas or oil lighting containers attached to the center poles); train polers, or polemen; and others. These positions paid $15 a month,[146] with an additional $5 or $10 a month[147] held back until the end of the season.[148] The rates of pay for experienced workingmen were similar to those of laborers in other industries, if hold-back pay is taken into account. In 1892 an American laborer made approximately $23.67 a month,[149] excluding room and board, which were included for all circus workers.[150]

Circus owners described the workingman's labor as good for one's health. Managers for Barnum & Bailey in 1908 told the journalist Harriet Quimby that hundreds of men with lung problems applied to become workingmen in order to regain their vitality: the applicants wanted to work in the open air and escape the confinement of indoor factory labor.[151] Impresarios even extended to laboring circus animals their association with labor, health, and solid character. In a defense of "working" animals that countered the logic of contemporary animal welfare activists, Dr. William T. Hornaday, a naturalist and Director of the New York Zoological Society (who also worked closely with several circuses), asserted that hard work was wholesome for both animals and humans in his "Wild Animals' Bill of Rights": "A wild animal has no more inherent right to live a life of lazy and luxurious ease, and freedom from all care, than a man or women has to live without work or family cares. . . . Human beings who sanely work are much happier per capita than those who do nothing but grouch. . . . [I]t is no more wrong or wicked for a horse to work for his living—of course, on a human basis— either on the stage or on the street, than it is for a coal carrier, a foundryman, a farmer, a bookkeeper, a schoolteacher or a housewife to do the day's work."[152]

But roustabout labor—for humans and animals—was grueling. Working hours were long and intensive, filled with the physical stress of constant travel and little sleep. Ceaselessly moving and lifting heavy materials, machines, and animals was also dangerous. Route books document in gruesome detail frequent accidents involving workers who were crushed, maimed, or knocked unconscious on the job. This entry from the Ringling Bros.' 1892 route book is typical: "Poleman Phillips breaks his leg under ticket wagon as it comes down the run from the train."[153] Additional entries describe roustabouts dying from their injuries. Indeed, as I will explore later, dangerous labor was also a form of masculine spectacle. Unquestionably, the dangers of roustabout labor help explain the high turnover rate among workingmen.

Senior workingmen were generally Euroamerican males. But the Sells Brothers circus was an exception. Based in Ohio (which historically had a relatively large free black population), the Sells Brothers employed scores of African Americans in the 1880s. Still, other big railroad outfits did not hire blacks until the turn of the century. The boss canvasman Bill "Cap" Curtis recalled that when James Bailey combined the Sells and Forepaugh circuses in 1896, Bailey reversed the Sells Brothers' practice by no longer hiring black canvasmen.[154] George Bowles, a press agent for Barnum & Bailey, explained that the circus in 1903 hired African American canvasmen (to work together in a segregated workplace) only in response to a shortage of white workers:

> Times were so prosperous that any man with a good pair of biceps could not only get a job, but would have people bidding for his services, and many employers who wanted husky boys overbid the circus, whereupon the canvasman, figuratively speaking, folded his individual tent and silently stole away. These disertions were so frequent that the circus for about six weeks was constantly in more or less trouble. We sent everywhere for men. . . . The problem was solved only when, for the first time in the history of the circus, Mr. Bailey imported a large force of Virginia negroes, who were greatly pleased with all the excitement and novelty of circus life. He tried to avoid this move, but there was too much doing for white men, to leave any other recourse.[155]

For decades, African Americans had been employed as circus acts — sideshow musicians, "savage" freaks, or occasional big-top players, often at good pay. In a closely knit working community, these jobs were nonthreatening to white laborers because African American entertainers were paid to play roles reinforcing racial stereotypes that confirmed white supremacist ideologies. African Americans were also hired to work in the dining tent as waiters, a position that complemented pervasive stereotypes about black servitude. The presence of black and white laborers working the same jobs, nevertheless, threatened to undermine prevailing racist norms. One can surmise that James Bailey was reluctant to hire African Americans as workingmen because white workers interpreted the presence of blacks as a threat to their own social and economic standing. The perceived threat of the black (or Chinese, or Chicano) worker "taking" white jobs for lower pay was an integral part of contemporary Euroamerican unionist discourse and helps to explain the failure of the Knights of Labor and other labor groups to create a racially united labor movement in the Gilded Age. Those shows that hired African American workingmen paid them substantially less than white workers. The salary list for the Walter L. Main circus duly chronicled, "if should go south and use Darkies," black manual laborers were to be paid $2 a week,[156] with no additional hold-back pay. Euroamerican common laborers, in contrast, made $3 a week.[157] On the job, African American laborers generally remained segregated from white coworkers in separate work gangs.

Although the circus comprised an international array of people, the big-top program was generally divided along racial lines. Nonwhite big-top acts typically performed with members of their own race and were often described as family members (even when unrelated). At the Gollmar Brothers circus in 1903, the three Japanese acrobats who made up the M. Ando tumblers were collectively paid $40 a week,[158] a fairly typical wage for minor big-top acts.[159] Nonetheless, few African Americans performed under the big top. Eph Thompson, a black animal trainer with the Adam Forepaugh circus in 1888, found scant employment opportunities at the American circus during the late nineteenth century; as a result he moved to Europe, where he became a successful elephant trainer in London and with Carl Hagenbeck's menagerie. As late as 1966, John Ringling North, owner of the Ringling Bros. and Barnum & Bailey circus (and nephew of the five original brothers), sparked an uproar in the circus community when he hired Priscilla Williams, an African American aerialist apprentice.[160] By and large, only certain positions were available: as workingmen, as waiters, and exoticized performance roles. Given the circus's racial division of labor, opportunities to move into more lucrative positions as big-top acts or managers were effectively closed to African American circus workers.

The black circus owner Eph Williams was the rare exception to this rule—but only because he set out on his own, away from an established show hierarchy. As a bartender and barber at the Plankinton Hotel in Milwaukee, Williams began training horses and dogs in his spare time during the 1880s. From 1888 to 1902 he owned a show that played under a variety of names, including Prof. Williams's Consolidated Railroad Shows, and traveled first by wagon and then on two to five railroad cars. His intrepid circus roamed the rough, booming timber and iron range frontier of northern Wisconsin and the upper peninsula of Michigan, and into Minnesota and the Dakotas. He quit the circus business in 1902 and bought a tented minstrel show, Silas Green from New Orleans, which made him wealthy.[161]

"LIKE A BIG FAMILY"

Despite the constant racial, gender, and class divisions within the circus caste system, show people still saw themselves as part of a closely knit traveling community. Circus people often perceived themselves to be removed from the rest of society because home, for most circus folk, was on the road.[162] Al Mann remembered that "circus was a family," that even John Ringling, or "Mr. John" as he was known among Ringling workers, visited freely with performers as they waited to enter the ring: "I talked to him a lot. He used to visit and talk to me and the same time he'd be watching everything they's doing with loading and tearing down at night during the concert . . . [H]e was right there with his black cigar, standing on that platform and he'd talk to me while I was waiting to get on the bucking horse."[163]

Circus workers maintained a sense of solidarity within their particular occupational group.[164] Carrie Holt, a fat lady, felt a fierce sense of camaraderie with her fellow sideshow actors: "As for the sideshow, the public don't know anything about us from just seeing us on our platforms—on exhibition. Why, we're just like a big family." Holt and her racially and physically diverse coworkers formed a social club, complete with dues, rules, and a big party at the end of the season: "I had a grand time. I was in everything even if I was fat."[165] Other circus workers expressed this same kind of solidarity. George W. Stevenson, a billposter, referred to himself as a "Brother Paste." His ditty, "Only a Bill Poster," written in 1896, playfully illustrated his consciousness of his social position: "There's a class in this world of the Miss Nancy Kind / Who turn up their noses (the largest part of their mind) / For he's 'only a Bill poster' without any brain / Crowding through life for positive gain."[166]

Circus workers referred to noncircus folk as "outsiders," "gillies," or "rubes," and frequently mentioned in their autobiographies that they were only comfortable with other show folk. They spoke a language peppered with jargon unfamiliar to the "gillies" that widened the gulf between insiders and outsiders. For example, an elephant was commonly known as a "bull," "punk," or "rubber mule," big-top acrobats were "kinkers," zebras were "convicts," the sideshow was the "kid show," and a circus's off-season grounds were its "winter quarters." If spectators became violent on the show grounds (typically outside the tents), a worker would shout "Hey

Rube!" and all employees would drop what they were doing, grab a tent stake or any other handy object, and descend upon the offending party. The cry "Hey Rube!" was a show of solidarity, and an effective means of maintaining order for circus folk working among thousands of potentially violent audience members, especially when local law enforcement was ineffective or lacking.

"Hey Rube!" also demonstrated that this collective spirit could (temporarily, at least) transcend racial divisions. A newspaper noted this potential at the show grounds of Buffalo Bill's Wild West in Brooklyn, New York, after a cowboy named "Wild Bill" accidentally bumped into an Italian peanut vendor and the enraged vendor stabbed the cowboy with a stiletto knife. The *Brooklyn Citizen* observed: "The other cowboys, the two Indians who were in their war paint ready for the afternoon performance, the two Cossacks, and the Mexican vaqueros *forgot their race differences for the moment* and rushed to Wild Bill's assistance. One of the Mexicans let go a lasso he carried. He missed the Italian, but he caught the peanut stand, which, with a crash, was overturned and deposited its load on the Halsey street car tracks" (emphasis mine).[167]

A "Hey Rube" could nonetheless leave a circus vulnerable. Lawsuits inevitably followed a melee, and circus folk were invariably blamed. W. E. "Doc" Van Alstine, a canvasman (among other jobs) with several big railroad shows at the turn of the century, remembered a couple of particularly violent "Hey Rubes" in an interview for the WPA in 1938: "I was in a Hey Rube in Lincoln, Illinois, once. It was one of the toughest battles I ever seen. The town boys was coalminers and some of the toughest customers I ever seen. We strung out in a circle around our stuff and stood 'em off with 'laying out pins' [used to set up the tents] and whacked 'em with 'side poles,' finally giving 'em the run, but they sure could take it. Another Hey Rube in Ann Arbor, Michigan, was started by a gang of students from the University of Michigan, for no good reason at all except perhaps they thought it was funny. It cost the circus I was with more than $35,000 in lawsuits and damage to equipment. In a Hey Rube, most of the lawsuits that follow is usually by some innocent bystander who gets hurt in the scramble."[168]

During the show season, circus workers were essentially homeless, living out of a trunk. Many did not keep their money in banks. Instead they bought loose diamonds and kept them around their necks in chamois pouches called "grouch" bags. Diamonds could be quickly transformed into cash when necessary.[169] Workers also wore diamond jewelry for the same purpose. While the Ringling Bros. circus stayed in Adrian, Michigan, in 1892, a hotel clerk commented, "There's enough diamonds worn by that gang to set up a jewelry store."[170] Even during the off-season, many players were nomads because they worked on the vaudeville circuit, in the rodeo, or in carnivals, or traveled south to work in circuses in the southern United States or Latin America. During his early years with the circus in the 1920s, Emmett Kelly bounced from show to show as a clown and in a trapeze act with his wife, Eva; they often spent their off-seasons apart, each working at a variety of jobs: at police circuses and state fairs, performing manual labor in a glove factory, and working as an industrial painter. Meanwhile, the "Aerial Kellys" continued to polish additional acts, the "iron jaw" act among others, to make themselves more employable as a team.[171]

The news media and juvenile fiction romanticized the act of running away and joining a circus, but the hiring process was generally more systematic for performers. Such circus folk were born into the business, solicited employment from proprietors by mail, or were discovered by circus agents while they were employed elsewhere. Some acts were hired by big railroad outfits based on their reputation from their work at carnivals, fairs, dog-and-pony shows, and smaller railroad circuses.

Female big-top players, in particular, were often born into the business, which enhanced their caste status. Josie DeMott Robinson, a top bareback rider in the late nineteenth century and the early twentieth, was born into an illustrious circus family. She knew no other life than the circus, and spent her childhood learning equestrienne acrobatics and living on trains. Some women joined the circus because their husbands were members of a show. Irene Mann's experiences provide a later example of this practice. She became a trick roper after she married Al Mann, in 1928. During the first year of their marriage, Al traveled with the Ringling outfit, while Irene remained at their home in Wisconsin, practicing roping and riding stunts. When Al joined the John Robinson circus in 1930, a manager noticed Irene's picture on Al's trunk and, as Al puts it, said, "We need girls as pretty as her with the show, send for her."[172] Within the month, Irene Mann joined her husband at the circus.

After seeing employment advertisements in entertainment papers like the *New York Clipper* and *Billboard*, many circus players sent solicitation letters to potential employers that described their talents and salary requirements. If interested, proprietors enclosed a contract with their reply, and the hiring process was completed by mail. Some acts were discovered in auxiliary fields. While working as a cattle foreman in Montana, Al Mann became a Wild West rider after a circus performer saw him win a rodeo contest. Then Cy Compton, who managed the Ringling Bros.' after-show concerts and Wild West shows, wired Mann an offer of employment and free travel to New York City from Lander, Montana, for the opening of the Ringling Bros. and Barnum & Bailey circus at Madison Square Garden in March 1923.

The hiring process was also systematic for American Indian acts.[173] Wild West agents generally hired Native Americans at Indian reservations, after getting approval from the Bureau of Indian Affairs. At the Pine Ridge reservation in South Dakota, Black Elk, an Oglala Sioux shaman, was hired for a season in 1886, when he was twenty-three, by *Waisichus* (white men) working for *Pahuska* (William F. Cody).[174] Standing Bear, an Oglala Sioux chief, was hired at Pine Ridge for Buffalo Bill's Wild West during its European tour from 1902 to 1906 as a translator and entertainer.[175]

Still, "running away" was a common way to join the circus, particularly among workingmen. A circus worker until 1917, W. E. "Doc" Van Alstine recalled that his family in Kinderhook, New York, wanted him to become a doctor like his father, a surgeon. Van Alstine had other plans, however: "At an early age, I had a yearning for the show business. School didn't interest me a bit. I hated books. I wasn't a danged bit interested in reading about what somebody else did, or where they went, or what they saw. I wanted to go, do, and see things for myself, and I couldn't think of any better way to satisfy my ambition than to join up with a circus. . . . Come a day, once, when I was a young gaffer in my early teens, I had a chance to run away with the Mighty Yankee Robinson Circus. The lure of sawdust and spangles was much stronger than family ties or the red schoolhouse, so off I goes."[176] Van Alstine worked as a block boy (someone who helped set up and tear down the general admission bleacher, or "blues," seats) for four days before his family dragged him home. He returned to school and studied medicine— hence his later nickname, "Doc." Finally, he persuaded his parents to let him leave, and he ran off to the circus for good. He stayed there for the next sixty years.[177]

CONDUCT IN A TRAVELING COMPANY TOWN

The workplace at the transient circus was all-encompassing, cohesive yet stratified. Employees led lives, both on and off the job, that revolved around their work, much like workers living in established industrial company towns such as Pullman, Illinois: they slept, ate, and worked with circus folk. Circus managers structured workers' leisure time by arranging circus baseball teams and fancy holiday celebrations. The large railroad circus was an early example of welfare capitalism: to quell potential unrest and union activism, circus owners provided food, lodging, and leisure activities for their workers.[178] Managers attempted to keep groups of workingmen racially segregated off the job as a way to build a cohesive (yet paradoxically divided) company town culture. This was not true of the performers' more racially integrated work and leisure culture but was generally true among the workingmen. In 1907 a newspaper in Decatur, Illinois, mentioned a baseball game between a team of black canvasmen, the "Lucky Sevens" and another black team, the "Black Diamonds," who maintained the animal menagerie.[179]

To heighten productivity, circus owners imposed strict rules of conduct on their employees that rivaled those of any industrial assembly line. Pinkerton agents monitored workers' behavior—with varying degrees of effectiveness—in addition to unruly spectators.[180] Circus contracts dictated how players dressed off stage, and how (and with whom) they spent their leisure time. A press release from Buffalo Bill's Wild West in 1899 informed the public that "absolute neatness is an imperative" among its employees.[181] As a way to maximize efficiency, some circuses prohibited workers from marrying each other during the show season. Fred Bradna became the equestrian director for the Ringling Bros. after the previous director, William E. "Bud" Gorman, was fired for eloping with a ballet girl named Gladys during the 1915 season.[182] During the first decade of the century, Ringling Bros.' work contracts listed fifty-one conduct rules, including the following:

> Be cleanly and neat in dress and avoid loud display. . . . Gambling, especially in the cars or near the cars, on or near the show ground, is strictly prohibited. . . . No pet animals, revolvers, intoxicants or imflammables allowed in the sleeping cars. . . . Loud talking, singing, playing upon musical instruments, or disturbing noises in or near the [railroad] cars must stop at 11 P.M. . . . Do not clean teeth at wash-bowls. Cooking is prohibited in the cars. . . . Do not sit "cross-legged" on floats or tableaux wagons [in the parade]. . . . Button up coats, etc. . . . Absolutely, do not chew (gum or tobacco) or smoke in parade. . . . Do not make remarks to anyone while in parade or talk to employees who are ahead of you or follow you in parade. . . . Do not nod to friends or acquaintances who may be in the audience. . . . Avoid arguments with other employees. Be agreeable and promote harmony.[183]

Given their focus on self-restraint and efficiency, it is little surprise that the biggest railroad circuses prohibited the consumption of alcohol among employees during the show season. P. T. Barnum, a temperance advocate, was the first railroad proprietor to market his circus successfully as a respectable entertainment, based largely on his enforcement of a dry workplace. Circus owners used the holdback system of pay in part to curtail potential drunkenness — even though instances of alcohol consumption occurred regularly. Contract rules dictated that workers found drinking on or off the job would not receive their holdback pay at the end of the season, and they could be fired, depending upon the circumstances.[184] The Gollmar Brothers organized intramural baseball teams to keep employees physically fit and alert during their leisure time and deter them from drinking and gambling.[185] Elsewhere, welfare capitalists organized employee baseball and bowling teams, picnics, and sing-a-longs, to promote company loyalty across ethnic lines and occupational ranks.[186] Circus managers generally focused their enforcement of temperance rules on manual laborers, even though some performers (not to mention William F. Cody) were alcoholics.[187] Like turn-of-the-century temperance advocates, proprietors looked to alcoholism as a working-class problem.

By contrast, alcohol had flowed freely at the mid-nineteenth-century circus. As a young business manager for William Lake's circus in the 1860s, James A. Bailey kept careful records of the show's weekly saloon expenses.[188] Mid-nineteenth-century circuses did not have to meet tight railroad schedules; in addition, these outfits generally employed a score or two of workers whose jobs often overlapped — in contrast to the multitude employed at the highly specialized turn-of-the-century railroad circus. In the mid-nineteenth century, before the biggest circuses adopted full-time railroad travel, the boundary between work and leisure blurred considerably. The movement of the overland show was determined by natural factors: length of daylight, weather, and seasons. With the advent of the railroad, impresarios placed greater emphasis on individual sobriety and discipline on the job and effectively "Taylorized" the workplace, thus prompting the journalist Charles Theodore Murray to proclaim that "the railroad has civilized the circus man."[189]

Proprietors and newspapers both described the railroad circus as a popular-culture counterpart to a modern army. Given its ability to travel quickly and feed and house some 1,200 employees in makeshift quarters, the huge American railroad circus resembled a sprawling military encampment. Buffalo Bill's Wild West programs described the outfit's operations in terms of preparing for war. Both Colonel William F. Cody and the show's co-owner, Nate Salsbury, were war veterans, qualified to oversee the "manoeuvers of so many troops, horses and guns."[190]

In fact, beginning in the 1890s, the U.S. War Department periodically sent army officers to travel with the circus in order to observe how show managers coordinated massive numbers of people and animals. Early in the decade several army officers from Fort Leavenworth, Kansas, spent a week with Barnum & Bailey studying the circus's transportation methods as a way to improve the logistics of the army's own artillery service.[191] On May 15, 1906, the quartermaster general of the U.S. Army, G. F. Humphrey, wrote to George Starr, general manager of Barnum & Bailey's circus, to inform him that Major I. W. Littell, quartermaster, U.S.A., would accompany the circus from Baltimore to Canal Dover, Ohio, to learn "up to date methods of moving men, animals and baggage."[192] That same day, Henry G. Sharpe, commissary general of the War Department, also wrote to Starr, summarizing the instructions that Captain James Addison Logan Jr. had received from the secretary of war: "[Logan is] to proceed to the points at which you give exhibitions as far as Wheeling, West Virginia, for the purpose of investigating the methods of obtaining supplies and serving food to the members of your company. Any assistance that you can render him in this particular will be deeply appreciated."[193] Paradoxically, then, the three-ring circus, which was fast becoming a synonym for wholesale disorder, was also a valuable model for day-to-day U.S. military operations.

The specialized division of labor at the railroad circus created the illusion of a huge, living, military machine. The smooth operation of the circus was dependent upon a cooperative workforce, aided by the circus's totalizing culture: its workers ate, slept, worked, and spent their leisure time together in an efficient, tightly knit, yet socially stratified traveling company town. General Leonard A. Wood, who engineered the McKinley administration's policies in Cuba during and after the Spanish-American War, asserted that watching the circus—with its cooperative and efficient division of labor—was an instructive, patriotic act: "No real American can resist the temptation to watch a circus unload, and seeing the erection of the tents."[194] After World War I, Wood noted that the Walter L. Main circus had more than seventy-seven service stars in its flag; four circus workers had been killed in the war and eight veterans had rejoined the show, "[now] better men and more advanced in education and discipline."[195] Echoing turn-of-the-century military officials, Woods observed that the "great American army machine" would benefit by copying the transportation and labor systems of the "great American circus machine." Woods judged the American circus to be an instructive institution, composed of exemplary patriots and industrious workers.

Despite its machinelike quality, the disciplined circus "army" occasionally rebelled against its highly regimented work environment. Still, the forms of resistance that these workers took cannot be measured by the yardstick of traditional labor activism. Roustabouts were so scattered and marginal to the world of organized unionism that they did not participate in institutionalized forms of resistance. They did not join unions, because their jobs were essentially invisible to the world of organized labor. The American Federation of Labor, the nation's dominant union at the turn of the century, was founded by a cigar-maker, Samuel Gompers, in 1886; its brand of "unions, pure and simple" was for skilled workers only. The AFL excluded unskilled workers, women, and people of color from its ranks, arguing that skilled white male workers were the vanguard of the labor movement and that to include others would only weaken their already precarious position. Although Bill Haywood and Eugene Debs tried to build "one big union" for all industrial workers in the Industrial Workers of the World (created in 1905), the AFL's exclusionary ideology dominated the day.[196] Circus laborers were unskilled and commonly illiterate (many of them signed their paychecks with an "X"), and they moved quickly and anonymously from job to job. The practice of retaining "holdback" pay until the end of the season speaks to the speed with which roustabouts left their jobs. Indeed, the most potent—and commonest—form of resistance for these workers was simply to quit.

Circus workers resisted in other, subtler ways as well. Their tactics occurred in the realm of what the historian D. G. Robin Kelley and the anthropologist James Scott have called "infrapolitics," as part of everyday life at the workplace rather than at the union hall.[197] As a way to relax, circus workingmen often sneakily imbibed alcohol on the job, despite official regulations. When Buffalo Bill's Wild West toured Europe at the turn of the century, Native American artists protested long working hours by taking whiskey breaks.[198] Sideshow workers also created their own strategies of resistance. When people crowded too close or became obnoxious, the Ringling Bros. fat lady Carrie Holt would pretend to sneeze: "That makes 'em move on. I suppose they think my germs must be as much bigger than ordinary germs as I am bigger than ordinary people." Holt and her sideshow comrades broke the monotony of their work by secretly poking fun at the "freaks" in the audience: "When I see a real funny one, I say, kind of careless-like, to Miss Gilmore, the snake charmer who sits next to me, 'I hope there's an extra platform!' That's a tip for her to look for a freak down in the crowd. Or maybe Miss Gilmore says to me, 'Well, Carrie, put on your things and go home. You're going to lose your job.' Then I know there's an awful fat woman in the crowd, and I begin to look for her." [199]

Pronounced forms of employee resistance were risky because workers had little recourse for airing their grievances. Employees remained without a union until 1937. At the turn of the century, circus contracts absolved proprietors and railroad companies of any liability for employees who were killed or injured during the show season. If workers were injured on the job or became ill during the season, they received no wages or sick-time benefits while they were unable to work. [200] Occasionally, a circus short on cash would "red-light," or strand, its workers without pay or transportation back home. [201]

Although circus workers had little protection, some occasionally walked off the job in protest against low wages and rough working conditions. In May 1903, for example, 150 Barnum & Bailey canvasmen temporarily stopped working because managers would not grant them a raise of $5 a month. [202] The workers protested that their duties had become increasingly difficult; that year, James Bailey replaced all the thousands of big-top seats with heavy iron orchestra seats—each containing a footrest—which made the big top's set up and disassembly, and the train-loading process, cumbersome. Managers responded to the canvasmen by replacing them. [203] Showmen generally ignored workers' demands because they were confident that these nonunionized workers would be disorganized and unable to create a collective protest. [204] Indeed, circus workers were unable to sustain a widespread strike until 1938, when unionized workers at Ringling Bros. & Barnum & Bailey walked off the job in Scranton, Pennsylvania, in protest against the company's decision to cut wages that season.

Circus press agents pointedly deemphasized instances of employee resistance. During the canvasmen's walkouts in 1903, press agents issued a flurry of releases to newspapers in towns along the show's route, flatly denying labor troubles and bearing headlines like "No Labor Troubles with Barnum & Bailey's Circus," "Circus Employees Did Not Strike: No Truth in Story that Men Went Out Yesterday Morning," and "Why Canvasmen Do Not Strike." [205] Proprietors justified delays by stating that the circus had simply become "too big," too "stupendous" to move quickly, or that constant bad weather had caused problems. Moreover, these same articles steadfastly claimed that the workingmen were "quite satisfied" with their jobs.

Showmen mocked workers' dissatisfaction by fabricating pointedly comical instances of resistance in press releases. Press agents for Barnum & Bailey crafted a nationwide "rebellion" among sideshow acts just as the 1903 circus season opened—the same season that canvasmen walked off the job. A series of articles reported that these sideshow workers demanded to be called "prodigies" instead of "freaks." Ostensibly, they had formed a "union" called the Sunday Order of the Protective Order of Prodigies and threatened to strike and destroy circus billboards if their wishes were ignored. Reports of the "freak revolt" were suspiciously riddled with oxymoronic images and puns. For instance, at a committee meeting the armless man wrote the minutes and the fat girl Emma "gave weight to the argument," while the Living Pin Cushion accidentally stabbed himself with a penknife. In 1907 sideshow players attempted to unionize again: the "Glass-Eater is Chewing on the Plan, and [the] Armless Wonder Writes of It."[206] Press agents also used clowns to poke fun at workers' activism. In 1909 a newly unionized group of clowns supposedly traveled to Washington to meet with members of Congress about lowering a tariff on clown white. One article observed that the clowns were "unusually interesting and solemn."[207]

At the beginning of the twenty-first century, the term "three-ring circus" is synonymous with chaos in the American lexicon. ("Media circuses" like the white Bronco flight and trial of O. J. Simpson in 1994 or the death of Princess Diana of Wales in 1997 immediately spring to mind.) But as this chapter has shown, the physical operation of the three-ring railroad circus that came to life during the Gilded Age was quite the opposite—even if the audience was so disorderly as to resemble a true "three-ring circus" in today's terminology. The fin-de-siècle railroad circus was a labor show of dazzling proportions, a logistical spectacle of sheer numbers: people, zebras, elephants, yards of canvas, eggs eaten, and so forth. Astonished customers arrived before dawn and left after dark just to take it all in. Showmen publicly emphasized the tremendous cost of their productions as part of their advertising campaigns—an essential feature of their identity as Gilded Age corporations. Profiting from the popular mythology of equal opportunity, circus proprietors marketed themselves as ordinary men who had "made good" through economy, diligence, and abstinence. The elaborate performance of the "army" of labor at the railroad circus visually reinforced these ideals because of its totalizing aura of perpetual industry. Yet, as will be seen in chapter 4, the presence of nearly nude female circus performers made the showmen's claims of decorous entertainment perplexing and problematic.

4
RESPECTABLE FEMALE NUDITY

In 1896 the New Woman came to the circus. That year, Barnum & Bailey's program contained "Three Graceful, Original and Interesting Equestrian Novelties," which playfully suggested that the "New Woman" might create a new social order dominated by women: "The New Woman supreme in the arena for the first time anywhere. Novel and picturesque exhibition of the assertion of the rights of the twentieth-century girl . . . in the circus. A positive usurpation of the ring in which man has no part. Progressive maidens fascinatingly conducting an entire equestrian act in up to date costumes."[1] During this act and other "New Woman numbers," women, clad in "becoming" bloomers, "of the most trim fitting, advanced new woman dress reform pattern," played all roles in the arena: ringmaster, groom, and object holder. Press releases announced that "no man is allowed to occupy that sacred ground of territory."[2]

In the collapsible canvas world of the circus, these "New Women" seemingly erased corporeal boundaries between the sexes. Women performers proudly displayed rippling bodies while demonstrating impressive feats of strength and handling dangerous animals. In 1911 Barnum & Bailey provided detailed muscle measurements of a German weightlifter, Katie Sandwina, and declared: "She Tosses Husband about like Biscuit. Frau Sandwina is Giantess in Strength."[3] At the circus, some women wore full, flowing beards. The lady giantess towered over the curious crowds at the sideshow.

African, Asian, Latin American, and Australian women at the "ethnologi-cal congress" easily defeated men in athletic contests. Moreover, the cir-cus was a comfortable space for women who felt alienated by social norms. Mabel Stark, a tough big-cat trainer whose body was, in the words of one colleague, "a network of scars" from frequent cat bites, knew as a teen-ager that she did not share her female classmates' predilections for dating and socializing.[4] After seeing a circus as a child in Princeton, Kentucky, Stark knew that she would eventually join a traveling show. Thereafter, she spent her spare time at the zoo watching the animals; while working as a nurse as a young adult, Stark "ran away" to California, met the showman Al Sands, and in 1913 joined the Al G. Barnes circus.[5] In an era when a ma-jority of women's roles were still circumscribed by Victorian ideals of do-mesticity and feminine propriety, circus women's performances celebrated female power, thereby representing a startling alternative to contemporary social norms.

These "New Women" were also nearly nude (fig. 10). Thousands of litho-graphs saturated the site of each future show, portraying barely dressed women in a range of bodily attitudes: on the trapeze, with snakes, lions, horses, or clowns, or en masse as members of a giant chorus. In perfor-mance, lithe, scantily clad acrobats and bareback riders freely twisted and contorted their bodies. Wearing a short skirt and nearly sleeveless top, the "Lady Hercules" lifted prodigious weights (plate 2). Spangled female ani-mal trainers wrestled "man-eating" tigers, while spray-painted, virtually naked women posed topless, nearly bottomless, and motionless as nymphs, Venuses, or maidens in the "Act Beautiful" or statue act, a turn-of-the-century variant on the antebellum tableaux vivants, or living pictures genre.

But showmen were keenly aware of circus women's transgressive poten-tial. As a result, they repositioned these strong, athletic, traveling women into traditional gender categories: as models of domestic womanliness, and as objects of titillation. In their elaborate advertising campaigns, propri-etors used gender, race, class, and representations of empire to create an ir-resistible sexual striptease under the guise of "clean" family entertainment. Arguing that there is no such thing as generic nudity, this chapter examines how nudity in some contexts was "respectable" and in other contexts sala-cious—the distinction was often created by racial stereotypes. Nudity itself is a historical construction and will be considered here in the context of dress standards at the turn of the century. Because American women gen-erally wore full skirts and long-sleeved shirtwaists at this time, virtually

Figure 10. "Statue Girls," Barnum & Bailey, Brockton, Mass., 1903. Wearing messy white or bronze greasepaint, these virtually nude women posed motionless as they imitated high art. (Frederick Glasier Collection, neg. no. 639; black-and-white photograph, copy from glass plate negative, 10 × 8 in., museum purchase, Collection of the John and Mable Museum of Art Archives, Sarasota, Fla.)

anything short of that coverage could be construed as "nude," including the wearing of leotards, tights, or short-sleeved dresses above the knee.

Proprietors presented white women as quintessential models of civilized, athletic womanliness, while they exhibited women of color (or Euroamerican women in racial disguise) as live, educational artifacts, whose nudity was an integral part of their racial "authenticity." Although this chapter focuses on the representational strategies of circuses owned by Euroamericans, one should not assume that only white circus folk were concerned about respectability during the Progressive Era. People of color in the amusement industry were equally attuned to the tensions between propriety and sexual display. For example, a poster for the African American blues singer Ma Rainey and the "Smart Set" promised "The Greatest Colored Show on Earth . . . The Biggest Bevy of Singing and Dancing Girls You Have Ever Seen . . . Everything Clean, Moral and Refined."[6] In all cases, however, these claims of propriety were made with a wink. Showmen's constant emphasis on female performers' lives, loves, and body-hugging tights became yet another way to talk about sex.[7]

THE HISTORICAL CONTEXT

Impresarios' focus on circus women's propriety at the turn of the century was particularly striking because they had downplayed the presence of female players just fifty years earlier. In antebellum America, reformers' responses to female circus players — indeed, to the circus as a whole — was unfavorable. In Chillicothe, Ohio, and Rochester, New York, as well as other communities, the Presbyterian and Methodist clergy condemned the circus's celebration of the body, its connection to the theater, and its omnipresent shell games. Dating back to the colonial period, antitheater laws were often rooted in Enlightenment thought. William Penn, among other colonial figures, argued that the cosmos was fixed and that external appearances revealed eternal truths. According to the media scholar Robert Allen, the theater's emphasis on mimicry, spectacle, and inversions of gender and class sharply confounded the rationalistic idea that the world was what it seemed.[8]

Critics also charged that neither circus workers nor the class of people whom the circus attracted engaged in productive labor. The editor of a religious periodical in Lexington, Kentucky, lambasted Joshuah Purdy Brown's circus in 1831 because it encouraged "idleness, intemperate drinking, profanity, a taste for low company [and] boisterous vulgarity."[9] Connecticut, home to P. T. Barnum, outlawed the circus altogether until 1840. The state legislature forbade any unusual feats of the body for monetary gain. Circuses continued to travel to Connecticut, but they could not advertise in newspapers for fear of arrest.[10] During the Second Great Awakening in Rochester, New York, town clergymen in this part of the "Burned-Over District" led the movement to close a permanent circus building and turned it into a soap factory.[11] From 1824 until the early twentieth century, the Vermont legislature virtually taxed the circus out of the state instead of banning it outright.[12] P. T. Barnum himself explained why the antebellum circus deserved to be censured:

> In those days the circus was very justly the object of the Church's animadversions. [I]n afterpiece, "The Tailor of Tamworth" or "Pete Jenkins," . . . drunken characters were represented and broad jokes, suited to the groundlings, were given. Its fun consisted of the clown's vulgar jests, emphasized with still more vulgar and suggestive gestures, lest providentially the point might be lost. Educational features the circus of that day had none. Its employees were mostly of the rowdy element, and it had a following of card-sharpers, pickpockets and swindler generally, who were countenanced by some of the circus proprietors, with whom they shared their ill-gotten gains. Its advent was dreaded by all law abiding people, who knew that with it would inevitably cause disorder, drunkenness and riot.[13]

Antebellum circus audience members targeted circus women as disreputable. Most objections centered on costuming. Although tights and leotards were not worn until after the Civil War, antebellum players wore stockings under knee-length skirts—a far cry from the proper, heavily corseted, long-sleeved, floor-length dress of the period.[14] One woman recalled that as a child in 1857, her grandmother forbade her to go to a circus: "[Grandmother] said it was all right to look at the creatures God had made, but she did not think He ever intended that women should go only half dressed and stand up and ride on horses bare back, or jump through hoops in the air."[15] Purity reformers and audiences both frequently thought that all female entertainers were prostitutes because they exposed their bodies for pay. Even women audience members were often treated with suspicion. During the first half of the nineteenth century, American theater managers commonly reserved the third tier of seats, or upper gallery, for prostitutes and their customers as a site of sexual exchange.[16]

Because they were often banned by law for their bodily spectacles, circuses deemphasized their women players; some even excluded women altogether. In 1840 an advertisement for the Raymond, Waring and Company circus guaranteed that its production in Philadelphia would contain no women: "[T]he introduction of Females into an Equestrian Establishment is not calculated to advance [the Chestnut Street Amphitheatre's] interests, while they not unfrequently mar the harmony of the entertainments, and bring the whole exhibition into disrepute. It never was ordained by Nature that woman should degrade the representatives of her sex which are not calculated for any other than the stalwart male."[17]

The circus was hardly the only nineteenth-century entertainment to censure the display of seminude women. Robert Allen has written that burlesque was a battleground where nineteenth-century Americans fought over shifting attitudes about gender and class. When the English actress Lydia Thompson and her burlesque troupe of "British Blondes" arrived in New York City in 1868, thousands of New Yorkers thronged to Wood's Theater in lower Manhattan, where the thinly clad actresses performed male roles and verbally poked fun at social norms. Local authorities tolerated the troupe only at this working-class venue; when the popular "British Blondes" moved to a reputable theater, Niblo's Garden, the press, local government officials, and reformers condemned the troupe. Unlike contemporary ballet, in which seminude women played silent, otherworldly nymphs and fairies, or melodrama, in which actresses typically played pious, sexless roles, in burlesque the performer talked and leered openly at her audience. Consequently, theaters catering to decent male and female audiences banned burlesque, which rapidly moved to the male world of the concert hall and increasingly was characterized by acts of mute bodily exhibition.[18]

In distinct contrast to burlesque's downward historical trajectory, the circus became more reputable over time. Circus proprietors increasingly used women to sell their productions as decent. Posters portrayed well-attired white women and families as part of the turn-of-the-century audience.[19] Impresarios adopted this sales strategy, in part because they knew that their patrons had changed. In antebellum America, men constituted the great majority of the circus audience. The virtual absence of women at the circus mirrored larger limitations on women's presence in public life. The prevailing ideology of separate spheres — whereby women's sanctioned role was in the home while men engaged in paid labor in the public world — helps explain why few women frequented antebellum circuses and why circus women were commonly censured. Although poor women participated in the paid labor market throughout their lives, bourgeois women became increasingly privatized, owing largely to changes in the American economy. Before the nineteenth century, the American labor market was characterized by a patriarchal system of household and shop production in which male family members worked as artisans or farmers; the home and the workplace were the same. By the mid-nineteenth century, paid work occurred away from the home because the market economy and its attendant systems of factory and wage labor had generally proletarianized the artisan and small farmer. As a result, the distinctions between public and private life became more sharply drawn.[20] Many women were publicly active in church-based voluntary associations connected to abolitionism, temperance, marital property reform, and women's suffrage, but their activism was predicated upon their maternal authority. Novels and new women's magazines like *Godey's Lady's Book* (1836–98) stressed the primacy of women's domestic role as the moral guardian of the home.

By the beginning of the twentieth century, Victorian distinctions between public and private spheres were crumbling. As part of this sea change, a growing number of families attended the circus, and female acts became increasingly visible and socially acceptable.[21] Circus women's heightened presence occurred just as more women were participating in paid labor and public activism. At mid-century, the majority of female paid laborers had worked in the private, familial setting of domestic service; its decline in the second half of the nineteenth century paralleled the ascendancy of factory production.[22] Concurrently, with the growth of corporations and service industries during the Gilded Age, young and single native-born women worked as office clerks (especially after the advent of the typewriter in 1873) and in retail sales at sprawling new department stores. Although many young women gave their parents a share of their earnings, they also had some disposable income and congregated at a variety of public amusements. Kathy Peiss argues that the rise of urban factory labor with distinct hours helped create a heterosocial leisure culture in which young, unmarried working-class men and women shared their nonworking hours together.[23] But the specter of young women "loose" in the streets was unsettling to some. Jane Addams, a founder of Hull House, noted, "Never before in civilization have such numbers of young girls been suddenly released from the protection of the home and permitted to walk unattended upon city streets and to work under alien roofs."[24] From 1889 onward, new settlement houses and clubs served as spaces where young working women could engage in "uplifting" cultural activities that would distract them from the "tantalizing" world of popular amusements out in the streets.

Women also became increasingly visible in public life by participating in a raft of Progressive reform movements. Women activists across the country worked to ameliorate poverty, social unrest, and racism through diverse organizations such as the Woman's Christian Temperance Union (1874) and the National Association of Colored Women (1896). Like members of antebellum voluntary associations, women reformers during the late nineteenth century argued that their domestic and maternal "nature" gave them special authority to push for social reform outside electoral channels, because women did not yet have the vote. But unlike the antebellum reformers, who effected social change through private institutions and "moral suasion," Progressive women embraced the public sphere by turning to the state. Before the Nineteenth Amendment for women's suffrage was finally ratified in 1920, women suffragists held street parades, open-air meetings, and pickets outside the White House, using tactics they had learned in part from the flamboyant English suffragists Emmaline and Christabel Pankhurst.[25]

The suffragists' colorful tactics mirrored the spectacular display of the female body at the circus. Josie DeMott Robinson, a bareback rider, played an active role in the suffrage movement; at rallies, she posed atop her rearing horse for publicity photographs.[26] Circus day, with huge crowds, was a highly visible occasion on which to promote the vote. On the Fourth of July in 1912, members of the Wisconsin Woman's Suffrage Association drove an automobile to the Ringling Bros. circus grounds in Racine, where they were well received by circus employees. The suffragists spent their day distributing literature to the circus crowd.[27]

Women's public activism occurred alongside growing popular references to sexuality. In 1913 the magazine *Current Opinion* proclaimed that it was "Sex O'Clock in America!"[28] At dance halls across the nation, young, unmarried women and men danced closely, doing the turkey trot and the slow shimmy with abandon. Built in 1897, Coney Island's Steeplechase amusement park offered young couples plenty of opportunities to kiss and hug in the dark, meandering "Tunnel of Love."[29] The vaudeville actresses Eva Tanguay and Gertrude Hoffmann portrayed themselves as worldly "personalities."[30] In his first visit to the United States in 1909, Sigmund Freud postulated that sexuality was the defining aspect of the human experience, from infancy to old age. After 1910 a group of young intellectuals known as the Greenwich Village sex radicals denounced state-sanctioned monogamous marriages in favor of multiple sexual partnerships.

Progressive purity reformers were called to action partly because of new attitudes about female desire. During the nineteenth century, Victorian so-

cial theorists believed bourgeois women to be asexual; consequently, reformers saw prostitution as a "necessary evil," a way to quell the potentially
dangerous sexual appetite of the white male. But by 1900 social theorists
viewed all women — regardless of race or class — as sexual beings, capable of
amative feelings and able to satisfy their husbands' passions, which rendered
the prostitute's services obsolete. The English sexologist Havelock Ellis
(*Studies in the Psychology of Sex*, 6 vols., 1897–1910) and the Swedish theorist
Ellen Key (*Love and Marriage*, 1911), argued that women's sexual passions
equaled men's. In this intellectual context, the "necessary evil" had become
the intolerable "social evil." Local red-light abatement acts (beginning in
1909) and the federal Mann Act (1910) attempted to legislate prostitution
out of existence.[31]

The rise of the physical-culture movement also challenged older ideals
about female sexuality. Just as women's public activism contradicted prevailing notions about separate spheres, women's athleticism at the turn of
the century confounded the standard of the neurasthenic, asexual woman.
After she suffered a series of emotional breakdowns, Charlotte Perkins Gilman's husband subjected her to the monotonous "rest cure." But Gilman
regained her strength and sanity only when she went back to her writing
and her exercise routine at the gymnasium.[32] Susan Cahn has noted that
the women's physical-culture movement experienced some of its growth at
new women's and coeducational colleges. The number of women attending college increased from 11,000 in 1870 to 85,000 in 1900 to 283,000 in
1920.[33] In a collegial setting, young women played basketball, baseball, ten-
nis, and golf. Wage-earning women also participated in athletic activities.
Factory managers extolled women's physical-education programs at the
workplace to foster increased productivity and company loyalty. Settlement
house workers, YWCA chapters, and local members of the Playground Association of America (1906) organized athletic events for urban children
to promote better physical and moral health. Furthermore, the Progressive advocates of the playground movement may have designed urban play
areas with the circus in mind: children could build their bodies and spirits
by twirling on the Roman rings, or swinging on a tiny trapeze, just like
well-toned circus acrobats.

Physical-education reformers posited that female athletic activity was
crucial to moral, physical, and even "racial" well-being. During the "bicycle craze" of the late 1880s and 1890s, thousands of women took up the
novel pastime of bicycling. Frances Willard, leader of the Woman's Christian Temperance Union, learned how to ride at the age of fifty-three. New

women's magazines frequently mentioned women's athleticism with exciting stories about exploring the world, training animals, or climbing mountains: "How I Climbed a 14,000-Foot Mountain," by Dora Keen, triumphantly recounted Keen's dangerous adventures on the Weisshorn in Switzerland.[34] In the context of women's participation in physical fitness, and the concurrent growth of a leisure culture in which a greater number of women became tourists, the lady thrill act—in which a woman rode a bicycle on a high wire or climbed a mountain—demonstrated women's public physicality in American culture.

But female athleticism had its share of critics. Despite the medical profession's general praise for the bicycle, several authorities claimed that cycling caused tremendous reproductive damage, specifically uterine displacement. Moreover, bicycle saddles reportedly allowed women to masturbate while riding, thereby reducing procreative desire and hastening the decline of native-born fecundity.[35] Still others charged that repeated cycling could create an ugly "bicycle face," characterized by a hard, clenched jaw and bulging eyes. Similarly, one commentator asserted that "circus face" was a malady brought on by excessive female athleticism under the big top.[36] Physicians and intellectuals called the female athlete and the "New Woman" a danger to traditional notions of domestic propriety.[37] The scientific community often represented them as "mannish," a liminal "third sex" neither female or male.[38] Theodore Roosevelt and the president of Clark University, G. Stanley Hall, both claimed that the progress of American civilization depended in part upon preserving sexual differentiation. They flatly stated that women should spend their reproductive lives as wives and mothers while men should dominate public life. Roosevelt, for one, growled, "When men fear work or fear righteous war, when women fear motherhood, they tremble on the brink of doom; and well it is that they should vanish from the earth."[39] In line with this logic, the New Woman was seen as largely responsible for "race suicide," because she delayed or refused motherhood in favor of higher education, paid labor, and public activism.[40] But the magazine *Ladies Home Calisthenics* (1890) argued that athletic activity in fact diminished the threat of race suicide, because female athletes were strong, healthy, and able to bear larger families.[41]

In the milieu of the women's physical-culture movement, audiences could read circus women's meager dress as a function of wholesome athleticism. Physical-fitness advocates argued that women's bodies should be free of tight, cumbersome clothing. An anticorset movement began in the 1830s and gained wide acceptance among physical-culture proponents by the end of the century.[42] Nineteenth-century anticorset advocates included Catharine Beecher, Dioclesian Lewis, and Thomas Wentworth Higgins, who charged that this fashionable device was responsible for respiratory difficulties, bodily malformations, and reproductive failure. The traveling phrenologist and nineteenth-century impresario Orson Fowler claimed that "tight lacing" caused shortness of breath, sexual excitement, and delirium: "Who does not know that, *therefore*, tight-lacing around the waist keeps the blood *from* returning freely to the heart, and *retains* it in the bowels and neighboring organs, and *thereby inflames all the organs of the abdomen*, which thereby *excites amative desires?*" (emphasis in original).[43]

Around 1900 visual images of uncorseted female athletes were increasingly common, both in publications of the physical-culture movement and on the pages of the *Police Gazette.* Bernarr Macfadden, an athlete, publishing mogul, and huckster, placed photographs of unbound female athletes throughout the pages of his popular magazine *Physical Culture* to boost circulation rates. When Macfadden assumed management of the magazine in 1899, its anemic circulation was approximately 3,000. By 1901 Macfadden's vigorous marketing tactics and pictures of nearly nude women had caused circulation to skyrocket over 100,000![44]

The genesis of the American empire provided an important sociopolitical context in which circus proprietors could promote female nudity as instructive. Women's increased participation in the public sphere and the rise of physical culture enhanced circus women's public viability. But they alone do not sufficiently explain why showmen focused on the female performer as a symbol of the circus's propriety, because women in other areas of popular culture were still regulated. Anthony Comstock, for one, arrested Macfadden shortly before the Physical Culture Show of 1905 in New York City for presenting "lewd" pictures of reclining athletic women dressed in union suits and a man dressed in a leopard-skin breech cloth.[45] But popular images of exotic nudity in toys, games, storybooks, and ethnological exhibits became increasingly commonplace once the United States gained control of noncontiguous territories after the Civil War and the Spanish-American War.

World's fair organizers, the publishers of *National Geographic* (1888), and circus impresarios alike used nonwhite women's bodies to make educational claims. Racial "color" defined the degree of nudity that was deemed appropriate for display. *National Geographic*, for one, in 1896 first published photographs of bare-breasted black women.[46] Euroamericans easily accepted such photographs of women of color as edifying, while topless white women were found only at seedy carnival cooch shows and nascent strip joints—not on the pages of a decent magazine or big "Sunday School" circus. *National Geographic* first photographed topless white women in the 1980s—and then only from behind![47] Turn-of-the-century impresarios, however, still drew public attention to seminude white female bodies, but they used different strategies to do so.

"THE LADY DAINTY" UNDER THE BIG TOP

How did turn-of-the-century showmen transform the female circus athlete into a highly publicized "queen of the arena"? Looking at posters and press releases, one might conclude that women dominated the circus, because female aerialists, rope walkers, bareback riders, animal trainers, and acrobats were omnipresent in circus advertising. But these marketing efforts were disproportionate to women's actual numbers. For instance, in 1891 Adam Forepaugh employed thirty male and twelve female big-top performers.[48] Barnum & Bailey's circus in 1896 had sixty-two male and thirty-four female principal big-top players, in addition to approximately 1,000 chorus members and many sideshow acts.[49]

Proprietors employed sentimental discourses of domesticity to neutralize the sexualized presence of strong, seemingly placeless circus women, who publicly exhibited themselves for pay in front of huge crowds. The circus woman supposedly abhorred modern life and shunned crowded cities during the off-season.[50] Circus press agents paid special attention to the origins of big-top women as a way to mitigate the possible public impression that circus women were anonymous, roving exhibitionists. Yet circus promoters paid little attention to the origins and social standing of the male acts. Circus media emphasized that the female performer never traveled alone; during the show season, her parents, brother, or husband invariably accompanied her, often appearing with her under the big top. Hardly a woman on the loose, she remained under the protective gaze of her family.

Showmen billed the typical female big-top player as a member of an old, distinguished family troupe, preferably from Europe (fig. 11). In contrast to the itinerant, tented American circus, the European circus was consistently respectable; intimate, one-ring productions took place in elaborate, permanent circus buildings in front of royalty and upper-class audiences. As a child, the American female big-top player reportedly learned her craft from her parents, who had performed for the crowned heads of Europe. One press release noted that over two-thirds of Barnum & Bailey's acts were

Figure 11. "Grand Equestrian Tournament," Barnum & Bailey, 1894. Crisp, fully clad poster scenes of "expert high school" equestrienne skills—cross-country jumping, fox hunting, and dressage—underscored showmen's claims that circus women were refined. (Lithograph courtesy of Circus World Museum, Baraboo, Wis., with permission from Ringling Bros. and Barnum & Bailey,® The Greatest Show on Earth,® B+B-NL38-94-1F-1)

European, the majority of them English, and that the women of the circus were "ladies . . . of good breeding."[51]

Three circus stars of the early twentieth century, Josie DeMott Robinson, Lillian Leitzel, and May Wirth, came from old circus families. Robinson, an American bareback rider from the 1880s to the turn of the century, was the product of a long line of illustrious French riders and horse trainers who fraternized with Napoleon Bonaparte; Robinson grew up around the circus because her parents owned the DeMott and Ward Circus.[52] Born in Breslau, Germany, in 1892, Leitzel became an aerialist as a child and performed with her mother, Elinor Pelikan, a Czech aerialist. As part of her mother's troupe, the Leamy Ladies (named after their American manager, Edward T. Leamy), Leitzel came to the United States in 1908 and debuted with the Ringling Bros. circus in 1915 as a solo artist.[53] May Wirth, an Australian bareback rider born in 1894, was the daughter of impoverished, itinerant circus players. After May's parents separated when she was seven years old, she was adopted by the Wirths, a well-established Australian circus family. When May opened as a center-ring star for Ringling Bros. in 1912, press releases did not dwell on her indigent early childhood (which was well known in Australia, where some speculated that she was an aboriginal), but instead depicted her as a member of an old foreign circus family.[54]

Circus media stressed female players' class status. Unlike many male big-top stars, or circus owners who promoted themselves as industrious "rags-to-riches" characters, female circus stars were supposed to be born into respectability. Upward mobility was a trope of the male capitalist, not the female performer. Barnum & Bailey's aerialist Nettie Carroll was reported to be a member of an aristocratic Ohio family. Beautiful "in face and form," Carroll was from the "smartest set," yet she chose the travel and excitement of circus life, despite her mother's desire that she settle down and get married.[55] Isabella Butler, an American who performed the "Dip of Death" with Barnum & Bailey from 1906 to 1908, was a "refined" student studying medicine at Vassar College.[56] Programs described Miss Lotta Jewel, a rider with the Carl Hagenbeck circus in 1906, as a paragon of elite American womanhood: "Miss Lotta Jewel is a splendid type of Gibsonesque American beauty and an ardent devotee of the invigorating and health-giving-out-of-door-life. A personal fortune has made possible liberal indulgence in her favorite pastimes—riding and driving—and she is the proud possessor of the finest stable of privately owned roadsters and saddle horses in New York City."[57]

The unmarried woman performer was reportedly on the threshold of marriage. Days before a circus arrived, press agents like Barnum & Bailey's Tody Hamilton flooded local papers with titillating "inside" stories about circus women, such as "Quits Ballet for Fortune: Romance of an English Girl Who Married against the Wishes of Her Parents," "The Women of the Circus: How They Live and Love," and "Big Circus Tents Cover a Very Pretty Circus Romance: Fair Italian Acrobat Wears Her Lover's Picture on Her Collar: He Holds a Trusted Position."[58] In addition to heightening the audience's curiosity, these stories presented the single female big-top player as a decent lady, who desired domesticity over the transient and potentially liberating life of sawdust and spangles.

Marriage was good publicity fodder, not limited to circus women; press agents frequently advertised impending nuptials for animal stars as well. The tantalizing prospect of sexual activity in captivity motivated animal "weddings" and drew big crowds to the menagerie. Primates, in particular, were popular subjects. A new scientific category in 1758, the order Primates had long been a metaphor, as the historian Donna Haraway has shown, for the politics of gender, race, class, and empire.[59] The union of two chimpanzees, Chiko "the $10,000.00 Chimpanzee"[60] and Johanna, dominated Barnum & Bailey's press releases in December 1893 while the menagerie was stationed at Central Park. Although the chimp couple never got closer than eight feet from each other,[61] press agents wrote detailed accounts about their physical interaction: Chiko was sexually aggressive while Johanna was chaste; yet she had a "grip like steel," and the couple fought furiously at their first meeting. One headline fairly shouted: "Chiko Wanted to Shake Hands but When Johanna Resented His Familiarity, He Nearly Tore His Lady Chimpanzee's Ear Off."[62] Johanna humorously "aped" human conventions: "[S]he does whatever she sees people doing about her, as if anxious to get into the 'swim' of human society. She wears skirts and housewrappers, smokes cigarettes, stirs her toddy with a spoon, and drinks it off like a seasoned old justice."[63] Circus writers waxed melodramatic when Chiko abruptly died at the end of the 1894 season: "His widow now bemoans her fate / He was so cute and slick, O! / And wears her mourning 'up to date,' / For was he not her Chiko?"[64] Johanna's keeper, Matt McKay, observed that the grieving chimpanzee covered her eyes "with almost perfectly shaped hands" for nearly one month after Chiko's death.[65]

At the circus, animals were templates for human desire, for social norms and transgressions. Johanna, the strong, earthy chimp, caricatured the strict gender constructions that constricted her human coworkers: she was stronger than her husband, smoked cigars, and drank alcohol. But race shaped gender stereotypes, even in the "raceless" animal world, because press agents used the same stereotypes to describe Johanna that they used to advertise nonwhite women: Johanna was immodest and was brawnier and bolder than men (fig. 12). For that matter, Johanna and Chiko were caged next to people of color at the ethnological congress.

In another section of the same menagerie tent, the "Happy Family" presented a startling scene of domestic bliss (fig. 13). First presented at traveling menageries, this group of mortal animal enemies positioned together in

[*Respectable Female Nudity*]

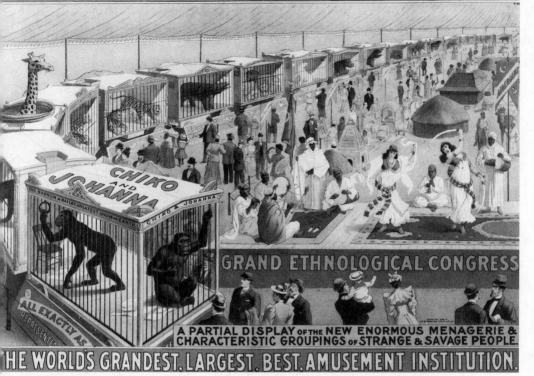

THE Barnum & Bailey Greatest Show on Earth

CHIKO AND JOHANNA

THE BARNUM & BAILEY GREATEST SHOW ON EARTH CHIKO & JOHANNA

ALL EXACTLY AS REPRESENTED

GRAND ETHNOLOGICAL CONGRESS

A PARTIAL DISPLAY OF THE NEW ENORMOUS MENAGERIE & CHARACTERISTIC GROUPINGS OF STRANGE & SAVAGE PEOPLE.

THE WORLD'S GRANDEST. LARGEST. BEST. AMUSEMENT INSTITUTION.

Figure 12. "Chiko and Johanna, Ethnological Congress," Barnum & Bailey, 1894.
Caged next to people of color from around the world, the two primates were racially
scripted to represent a living evolutionary continuum. (Lithograph courtesy of
Circus World Museum, Baraboo, Wis., with permission from Ringling Bros. and
Barnum & Bailey,® The Greatest Show on Earth,® B+B-NL38-94-1F-5)

the same cage became popular at P. T. Barnum's American Museum (1841–
68). There, on the fifth floor, Barnum exhibited his own "Happy Family": a
mélange of drowsy, well-fed monkeys, dogs, rats, cats, pigeons, owls, por-
cupines, guinea pigs, cocks, and hounds, all in the same large cage in front
of thousands of self-styled "happy families" each day.[66]

In contrast to the animals, the placeless Euroamerican female circus stars
purportedly preferred life at home. These women reportedly had tea and
sewing clubs, and according to one writer, "The thoughts of many of them
as they go flying through hoops, or whirling through the air on a trapeze,
are in some faraway home with their children."[67] In a press release, Mrs.

Figure 13. "Happy Family," or "Monkey Play," P. T. Barnum's Circus, Museum, and Menagerie, 1887. A staple of traveling menageries since the eighteenth century, this scene of domestic bliss inverted ordinary relationships of predator and prey: a bear cradles a pig and cats sit next to mice. (Interior program illustration courtesy of Buffalo Bill Historical Center, Cody, Wyo., MS6.Davidson.10)

George O. Starr, a former bareback rider, and "Zazel," a cannonball stunt artist who was married to a Barnum & Bailey manager, advertised the upright circus woman:[68] "The domestic instinct is very strong among circus women, for the reason that they are deprived of home life a great part of every year. She finds an outlet in many little ways, one of which is an appeal to the chef in charge of the dining car to be allowed to bake a cake. . . . [In some instances] [i]t isn't all unusual for them to go to one of the houses along near the track and ask the woman who lives there to let them use her kitchen. Almost always they get permission and afterwards pay her for it. They sew too, and many do pretty exceedingly fancy work."[69]

In an interview with troupe members of the Barnum & Bailey circus in 1908, Harriet Quimby characterized the backstage environment as "wholesome," filled with modest females, many of them mothers whose children were also part of the production. Barnum & Bailey's press agent Dexter Fellows claimed that backstage one would "see the domestic side of a circus."[70] Dixie Willson, a secondary big-top player who rode in the opening spectacle, or tournament, with the Ringling Bros. and Barnum & Bailey in 1921, spent the $22 that she earned in her first paycheck[71] on curtains, pictures, a potted plant, silk pillows, and a rose-colored lamp for her four-by-six-foot home on the circus train.[72] May Wirth, wearing her trademark pink bow in her bobbed brown hair as she performed a full forward somersault mid-gallop atop her horse, was portrayed as a sweet, shy young woman (fig. 14). After her marriage to Frank White (who took her surname) in 1919, not only was May the greatest woman rider in the world, she was reported to be a devoted wife, a sentiment captured nicely in this radio interview from 1942, five years after Wirth retired from the circus: "It was quite another kind of accident which took me out of the circus. . . . I fell in love. . . . Now I have a home in Forest Hills. We still have the horses and I still ride once in a while. But my main interest is discovering appetizing salads, planting successful flower gardens and tending to spring housecleaning."[73]

Despite well-publicized claims of happy, nomadic family life, several circuses prohibited young (i.e. nonperforming) children from traveling with the show. In 1916 the bear trainers Emil Pallenberg and his wife were forced to leave their two-week-old son Emil Jr. on a farm in Connecticut for Barnum & Bailey's show season.[74] Jules Turnour, a clown, sadly recalled the death of his beloved son faraway in New York while Turnour was on the road. Just before going on to a packed house, he received a telegram from his wife saying that his son was gravely ill. "There I stood in fool's garb, with the hot tears streaming down my make-up. I heard a voice say merrily: 'Come, Jules, we're waiting for you.' So I had to go out into that crowded arena with a breaking heart, and disport myself that the mob might laugh — playing with a dummy [rag] child while my own lay dying."[75]

Publicity pieces invariably mentioned the husbands of female stars to prove that women's primary loyalties lay with their husbands. Yet the reality contradicted these images. In 1906 Barnum & Bailey managers hailed Josie DeMott Robinson as a courageous heroine: Robinson had been retired from the circus from 1890 to 1905 after marrying Charles Robinson, son of the showman John Robinson, and reportedly returned as a noble way of easing her husband, a former politician, out of debt. In the words of one headline: "From Home of Riches to the Bareback Ring: Left Circus Ring as Rich Man's Bride: Returns to Aid Husband: Josie DeMott, Somersaulting Equestrian, Aiding Husband, the Son of Showman Robinson to Retrieve Losses."[76] But in reality, Robinson came back to the circus because she found married life as a "gillie" (circus outsider) suffocating. Through marriage and retirement, she had become a "mummy," "choked and imprisoned by corsets and fashion."[77] While in retirement, she took up bicycling to regain her strength

Figure 14. May Wirth, ca. 1924–27. A circus star in America since her center-ring debut with Ringling Bros. in 1912, Wirth engaged in strenuous bareback riding, contradicting showmen's attempts to market her as dainty and demure. (Photograph courtesy of Circus World Museum, Baraboo, Wis., BBK-N45-WTHM-28)

and to avoid the corset, but ultimately Robinson was set free only when she left her husband and returned to the circus ring. Robinson felt that the circus was a haven: "I knew that world. I loved it, and I felt safe there."[78]

Images of Annie Oakley were equally paradoxical—domestic and normative on the one hand, subversive on the other. Playing "Little Sure Shot" at Buffalo Bill's Wild West (1885–1902), Oakley was reported to be "the sweet heart of the entire male population of the country."[79] Proprietors advertised her as a homebody and true patriot who performed pro bono for American troops during World War I.[80] Her brother recalled that Oakley was "quick as a cat" in shooting, but nonetheless: "She was always a lady, always reserved and always modest. Wouldn't tie her shoes in front of a stranger. She was always a Christian, and she said her prayers every night on her knees, wherever she was, just like a trusting little child."[81] Press agents depicted Oakley as a happy, devoted wife to her husband, Frank Butler, who frequently performed with her. Yet as the theater historian Tracy Davis points out, endangering her husband was a central part of Oakley's act. During her seventeen years with Buffalo Bill, Oakley frequently used Butler as her target—she shot apples off his head, or razed the ashes off his cigarette while he smoked. Furthermore, Oakley chose to perform in full, stereotypical western wear— cowboy hat, long skirt, vest, long-sleeved blouse, and boots—rejecting the brief garb of the circus woman. Press agents rarely noted Butler's part in Oakley's act, instead focusing on Oakley's good character and legendary self-taught shooting skills.[82] Although showmen likely intended such promotions to diffuse the unsettling implications of a wife symbolically shooting her husband in public, these containment strategies were sometimes unsuccessful. For one, Hearst newspapers reported in 1903 that Oakley was a cross-dressing thief. But Oakley triumphed in the end after she sued Hearst for libel and won.[83]

Impresarios extended women's love of domesticity to female animal trainers. Articles delighted in revealing that many lady lion tamers were afraid of mice and spiders.[84] Contemporary magazines acknowledged that the image of a "gentle" woman handling wild beasts was arousing: "When we go to see a woman run these risks, we give secret play to barbaric emotions which in spite of years of civilization are yet latent in us."[85] Female animal trainers and handlers were generally more physically interactive with their beasts than male trainers. Lucia Zora, whose husband Fred Alispaw taught her how to train animals with the Sells-Floto circus, performed several daring acts with her elephants and big cats. Her most spectacular act involved riding atop the tusks of Snyder, the killer elephant, while he stood on

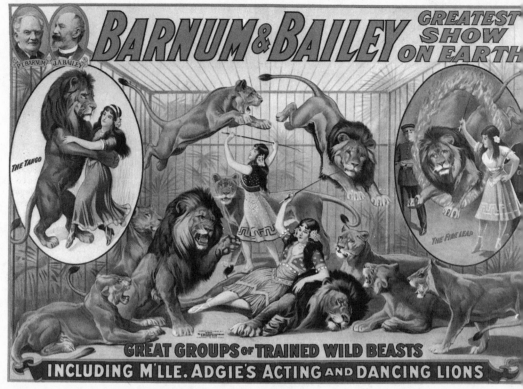

Figure 15. *"Great Groups of Trained Wild Beasts," Barnum & Bailey, 1915. Sexually suggestive circus posters commonly depicted women trainers like Mademoiselle Adgie as more physically interactive with their nonhuman charges than male trainers. Here she performs the "tango" with one of her lions. (Lithograph courtesy of Circus World Museum, Baraboo, Wis., with permission from Ringling Bros. and Barnum & Bailey,® The Greatest Show on Earth,® B+B-NL39-15-1F-5)*

his hind legs.[86] Mabel Stark recalled that she got her first circus job with the Al Barnes circus in 1913 because the manager, Al Sands, thought that Stark, a petite blonde, would look alluring to audiences while she handled nasty tigers.[87] Stark created the first circus wrestling act with a tiger, in which she and Rajah (whom she raised from a cub) rolled around the sawdust arena to music.[88] Although women trainers wore protective paramilitary clothing during their acts, lithographs often depicted them in bare, glittering garb (fig. 15). Like their male cohorts, women animal trainers proudly recounted their stoicism. Stark recalled many instances of finishing her act with blood pouring into her boots from her lacerated thighs and calves while smiling calmly at the audience. Stark asserted, "I am not afraid. I like the challenge of [the cats'] roaring defiance. . . . I know that my will is stronger than their rippling muscles."[89] Stark's bravado notwithstanding, programs referred to her as "the Lady Dainty in a den of ferocious tigers."[90] And despite the "masculine" strength and cool calm required for performing animal acts with elephants, bears, and big cats, showmen used these stunts to heighten prescribed gender differences. Women usually worked with smaller, non-threatening animals such as birds and dogs. The equestrienne and trainer Ella Bradna performed the "Act Beautiful," in which a bevy of trained birds, horses, and dogs assisted Bradna, the "Lady Dainty of the arena," in a "most beautiful and altogether delightful display of color and charm."[91]

Circus women also performed with automotive "brutes" and, in the 1920s, airplanes. The brothers Charles and J. Frank Duryea were bicycle engineers who built the first workable American gasoline-powered automobile in 1893; in the early twentieth century Henry Ford created affordable American cars based on European models like the 1901 Mercedes. Ford was so deluged with orders for his 1906 Model N that he was motivated to introduce his Model T in 1908, a more economical and reliable car.[92] In this technological environment, the circus coupled this new mode of transportation with another nascent technology: flight. From 1905 to 1908 Barnum and Bailey featured the "L'Auto-Bolide, 'The Dip of Death,'" a chilling, highly dangerous automobile act in which the driver raced down a steep track, ending in a free-falling somersault before the car landed upright on a different track (fig. 16).[93] The stunt required complicated machinery: a four-story dromedary-shaped steel structure which held a forty-foot chasm—to be crossed by the somersaulting auto. The car was raised to the top of the platform with a cable in front of the audience, so that the crowd could "see the machine and study its construction." The driver, a French woman named Mademoiselle Mauricia de Tiers, entered the car at the top of the platform, sped down the steep track, performed the loop-de-loop, and landed upright on the other side of the track.[94] At San Francisco in 1905, "most of the crowd were satisfied that the quiet little French girl was flirting with the ferryman of the Styx when she trusted herself entirely to the laws of inertia. There was a universal sigh of relief when the dip of death was accomplished safely."[95]

These acrobatic female stunt drivers were ubiquitous in circus advertisements. Tiers and Isabella Butler, the American woman who replaced her in 1906, were the subject of hundreds of Barnum & Bailey stories published throughout the United States and Canada. In all, impresarios described these women as small, beautiful, and well-bred "heroines," who were vulner-

Figure 16. "L'Auto Bolide," Barnum & Bailey, 1905. Transforming the already novel automobile into a flying object, this thrill act, as the poster suggests, was incredibly dangerous and expensive. (Lithograph courtesy of Circus World Museum, Baraboo, Wis., with permission from Ringling Bros. and Barnum & Bailey,® The Greatest Show on Earth,® B+B-NL39-05-1F-4)

able yet fearless New Women. To prevent audiences from becoming bored with the lone flipping automobile, Barnum & Bailey's proprietors expanded the perilous act for the 1908 season: now two women, the sisters Caroline and Nettie Rague, drove separate cars down a 60 percent grade on separate tracks, one underneath the other. One car somersaulted across a twenty-foot chasm while the other simultaneously shot out straight beneath it. Both landed on the other side of the track at the same time.[96]

Women's costuming was another vehicle for domestication and eroticism. Circus media constantly justified bare apparel with stories about healthy, wholesome female circus athletes. Brief clothing was also critical to the safety of circus women, who somersaulted atop galloping horses, twirled through space from the flying trapeze, gyrated madly on the Roman rings, and pranced around on the high wire forty feet aloft (fig. 17). In addition, performers and proprietors justified scant dress as evidence of good character. Julia Lowanda, a popular Cuban-born bareback rider at the turn of the century, reasoned that the circus woman's athletic activity made her more seemly than the society woman: "We never keep the shockingly late hours of the society maiden or matron, and the fact that we live almost entirely in the open air makes us strong and healthy. In fact, instead of being dyspeptic and irritable, we find life worth living and we are cheerful."[97] A Hagenbeck-Wallace circus program also pointed to female circus athletes' vigor: "Circus women . . . keep in good health by taking two or three cold baths daily in the dressing tents and by plenty of physical exercise out of doors."[98]

In the ring, the nearly nude female body was both sexually attractive and strong. Yet outside the ring, showmen classified these muscular bodies into normative categories. Josie DeMott Robinson recalled that upon leaving the ring in her scanty costume, she was required to don a long skirt immediately: "Of course skirts would endanger our lives when we were performing so scanty attire was the thing, but the minute the act was over out came the long skirt or cloak immediately, and I was told not to be so immodest as to stand around half naked. There was no sense in it."[99]

Impresarios used well-publicized employee conduct rules to bolster their claims of female propriety. Many work rules applied exclusively to women and dictated their off-duty dress and bedtime. Robinson remembered that even allowing a man to help her put on her coat was forbidden.[100] However, men were generally exempt from such rules. As a bachelor, the trick rider Al Mann freely fraternized with local women while the circus was in town. He dated several women and was often invited to family dinners.[101] Among circus women, ballet girls received special scrutiny. The work contract for Kathy Edwards, a ballet girl for the Ringling Bros. in 1913, detailed a series of supplementary regulations for ballet girls that did not apply to male employees. If Edwards violated these rules, she could have been fired and easily replaced: "3) Ladies should be in sleeping cars at a reasonable hour after the night performance, as per instructions from Ballet Master. 4) Board and sleeping accommodations being furnished by Ringling Bros., all ladies are expected regularly and orderly to avail themselves of same. 5) Male companions during hours when not on duty, strictly prohibited. 6) Flirting and boisterous conduct at all times and places, prohibited."[102] Other contracts specified that ballet girls be "neat and modest in appearance. . . . Do not

Figure 17. May Wirth (left) and Lillian Leitzel, 1924. Tights and short dress were sensible clothing for circus performers, allowing them complete freedom of movement. (Photograph courtesy of Circus World Museum, Baraboo, Wis., BBK-N45-WTHM-14)

dress in a flashy, loud style; [g]irls must not stop at Hotels at any time.
. . . The excuse of 'accidental' meetings [with male members of the Show
Company] will not be accepted."[103] These regulations were marked with
an endnote where the Ringling Bros. management justified its policies: "If
some of the rules seem harsh and exacting, please remember—experience
has taught the management that they are necessary. It is intended to protect
the girls in every possible way. Good order and good behavior are necessary,
if you are to be comfortable and happy."[104]

Big-top women also enforced unwritten codes of conduct among themselves. While working as a statue girl for Barnum & Bailey in 1916, Tiny
Kline observed that nude bathing was "taboo" in the ladies' dressing room.
"[A]ccording to the rules of those ladies—the aristocrats—[star big-top
acts] who dictated the 'laws'; no self-respecting female would disrobe completely, without the shield of a kimono or bath-robe which was kept over
her constantly during her bath. She would remove her outer garments and
immediately get under the robe, which, hanging from her shoulders, served
as a private bath-house while removing her under garments, then fasten-
ing it under the arms she would proceed with her bath."[105] Kline and the
statue girls, however, bathed completely nude in a cramped area partitioned
off from the rest of the ladies' dressing room: the successful removal of the
sticky greasepaint required total nudity so that everything nearby would be
kept free of paint. Kline observed that the self-imposed mandate on modesty
was vigorously enforced, suggesting that codes of virtuous conduct were
not simply imposed by domineering proprietors but were internalized by
the players themselves.[106]

To demonstrate to the rest of the world that circus women were well
policed, proprietors freely publicized their stringent female conduct rules.
An article in the *Evening Telegram* in New York City referred to the rules in
a long defense of the circus woman's high scruples: "Circus women belong
to that vast majority that will have 'Misunderstood' put on as headstones.
It is a curious thing that the general impression of these women of the sawdust is that they live as high as they swing or jump and that figuratively
speaking their existence is one prolonged vortex of spangles and tights. As
a matter of fact, nothing could be more erroneous. . . . Don't you see that no
women could lead more protected lives? . . . And when you remember that
a circus woman is almost invariably married, and that her husband is with
her, you will see that the moral standard of the profession is high."[107]

Posters occasionally depicted women acts in long, flowing gowns and
puffy, high-collared shirtwaists—clothing which was extremely dangerous

if worn in performance. Months before a show date, rural and urban residents saw these curious poster scenes depicting fully skirted women whirling off the trapeze, or poised atop a horse. A few acts, such as the Meers sisters (Ouika and Mari), did perform equestrian acts in street dress: a Barnum & Bailey program in 1896 promised a performance "with grace and skill while costumed in long skirts, all the difficult feats executed by the famed male riders."[108] That same year, a lithograph advertising the Arrigosi Sisters, an iron jaw and trapeze act with the Forepaugh & Sells Brothers circus, depicted the sisters twisting, bending, and flying through midair in long dresses: "First American Appearance of the Flying Wonders. The Arrigosi Sisters. In Their Astonishing High Trapeze. Long Skirt Evolutions, Leaps and Dives" (fig. 18).

Although such dangerous costuming was uncommon, press agents for Barnum & Bailey waged a media campaign about a fictional "gowning revolution" from 1904 to 1906. Press releases stated that several principal female performers, led by the star equestrienne Rose Wentworth, preferred the latest Paris fashions—complete with hair worn long in bows and combs—to traditional circus garb. From all accounts, this trend was a smash with audiences.[109] Newspapers reported that Wentworth mastered wearing the cumbersome long skirt in the arena by eliminating the hoops and learning how to jump a horse in her street clothes.[110] These articles, however, variously identified the founder of the "movement" as Wentworth, Josie DeMott Robinson, or someone else, thus making the entire "gowning revolution" suspect. All accounts stressed the sudden absence of tights at the newly clothed circus: "There are tights, plenty of them, but they again, are almost the exclusive trademark of the merry . . . burlesquer."[111] Circus press agents sought to distance their productions from "tawdry" burlesque shows featuring "bold" talking women who jeered at their audiences. In contrast, the circus lady was almost always silent during her act.

But these fashionable, fully clad circus women were provocative in many respects: presented as high-society women, they had athletic skills that were heightened by their ability to perform daring feats in long skirts—a humorous inversion of normative behavior for women. Placed in an entertainment space where bareness was the norm, formal street dress was unusual and provocative—the circus woman's constant jostling on the trapeze, high wire, or horse provided audiences with constant opportunities to "peek" underneath her skirts. The gowning campaign kept audiences guessing about what the circus woman really wore, creating anticipation for a potential striptease: Would she be fully dressed or not? Or might she actu-

Figure 18. "The Arrigosi Sisters," Adam Forepaugh and Sells Brothers, 1896.
In an attempt to market their shows as "moral," showmen occasionally depicted women
performers in full gowns. Note the iron-jaw lady performing the "slide for life" down
a wire, and the vigorous trapeze work in the foreground. (Lithograph courtesy of
Circus World Museum, Baraboo, Wis., with permission from Ringling Bros. and
Barnum & Bailey,® The Greatest Show on Earth,® A4PS-NL33-96-1F-2)

ally undress in front of the circus audience? Some contemporary vaudeville
companies and smaller circuses presented "disrobing" acts on the trapeze
and high wire; female aerialists wore layers of clothing which they speedily
abandoned until they were dressed in fleshlings and a leotard. On January 19, 1912, Lillian Leitzel's mother, Eleanor Pelikan, graced the cover of
Vanity Fair performing an iron jaw disrobing act. The magazine claimed that
the "Aerial Venus" would "carefully remove every part of her costume," yet
she finished the act in basic circus garb.[112]

The circus press abruptly stopped its "revolution in gowning" campaign
in December 1906. Barnum & Bailey's agent W. D. Coxey announced that
management had issued a moratorium on long skirts for the 1907 season.

In line with other advocates of the cult of the strenuous life, Coxey now argued, once again, that the well-muscled female body was healthful and inspirational:

> The human form, when it reaches a high form of symmetry and muscular development, is something to be proud of, whether in man or woman. This old puritanical idea, that because a woman displayed her figure in tights, she was outside the pale of society, has given way to a sane and sensible admiration for the physically perfect woman. There is no reason why tights should not be preferable to gowns in the circus arena. No matter how clever the performer may be, long skirts interfere with the free movement of the body. It is also a question of whether they are not a serious detriment from the standpoint of health. Tights for women in the circus are proper, modest, attractive, and hygienic, and a return to them will be welcomed by the public.[113]

Men also wore tights, but circus media generally ignored them. Yet articles used tights to describe ballet girls in prurient terms, as a "prolonged vortex of spangles and tights."[114] One headline screamed, "A Circus in Undress."[115] Tights figured heavily in reportage about Miss Evetta Mathews, an acrobatic English lady clown who performed with Barnum & Bailey in 1895. A proponent of physical culture, Mathews participated in "all of the new woman's fads" (including wearing bloomers) and "does everything a man does to keep herself in proper trim" (plate 3).[116] But her most riveting attribute was her "shocking" pink fleshlings. Allegedly, she strolled out of the center ring at Madison Square Garden during her act and sat down in a group of men. The ringmaster offered her $5 to return to the ring, but she replied that a young man had offered her $10 to stay,[117] so why should she leave? The press release then teased audiences into coming to see what Mathews might do next: "There is nothing in her appearance to indicate that she is a clown or a lady, but the appearance of pink tights at such short range created almost a panic on the south side of the house."[118] Mathews did not see herself in such lurid terms, however. She thoroughly enjoyed her life with the circus and viewed her success as evidence of women's ability to excel in traditionally "male" fields:

> I believe that a woman can do anything for a living that a man can do, and I do it just as well as a man. All of my people laughed at me when I told them that I was going into the ring as a clown; but they do not laugh now when they see that I can keep an engagement all the time, and earn

as much money and more than they can in their branches of the business. I like the work and try to put ideas into it. Every day I think out something new and the management usually gives me pretty wide latitude. My chief difficulty is making myself heard, but then nobody ever listens to what a clown says; everything depends on the antics.[119]

Although circus women like Mathews gained a real sense of empowerment from their work, impresarios diminished their subversive potential by positioning them back into more standardized representations of the erotic female body. One could find this sort of normative female spectacle across the show. In the free morning street parade, proprietors primarily picked young, comely women to ride in the mounted sections. Older women often appeared in "oriental" costumes but were told to wear veils over their faces so as to hide their age from the spectators.[120] Some showmen doctored posters of women to make them more salable. In the 1920s the Ringlings' press agent Roland Butler used photographs of Ziegfeld Follies dancers in place of the well-chiseled female circus stars in programs. Butler altered the dancers' faces so that they would resemble actual circus stars, while retaining the curvaceous Ziegfeld bodies. Butler even substituted a full-length picture of a "Follies" performer for that of a big-top performer, the famous bareback rider Flora Bedini, who consequently did not speak to Butler for nearly twenty years. "[Flora] was real circus," Butler stated, "but she was about as photogenic as a bagful of gravel. We couldn't run anything like that in the program."[121]

The normative exhibition of the female body reached its fullest form with the ballet girl. Despite the pretensions of high culture that the title "ballet girl" implied, the job required little skill other than the ability to look attractive in scant costumes. Harry Conlon, an electrician seeking employment with the Ringling Bros. in 1903, requested work as a ballet girl for his wife, whom he described as "of very good appearance and very quick to learn anything." Managers responded by asking Conlon to send a picture of his wife, along with details about her height, weight, and complexion; they did not inquire about her skills as a performer.[122] One former circus owner, Fred Pfening Jr., recalled that during the days of the American Circus Corporation outfits (1921–29), circus managers used to refer to the day when ballet girls were hired as "choosing day," because owners and managers would each "choose" a ballet girl to be their companion during the season.[123]

Also known as "chorus girls," and "oriental dancers," sometimes over

1,000 ballet girls were hired each year to participate in the morning parade, dance in the opening spectacle or tournament, and perform other unspecified duties "as needed." Supposedly, thousands of "aggressive and attractive" women stormed Bolossy and Imre Kiralfy's auditions in order to work at the circus.[124] Press agents lured potential audiences by crafting teasing stories about vulnerable chorus girls who gazed directly at male audience members: "A . . . French [ballet] girl can find more accidental ways of looking a man in the eye than are in the imagining power of most men, nor does the young woman of the Kiralfy chorus glance quickly away when her glance is met. She has nothing to be ashamed of. She was merely looking your way. There's no reason why she should stop, and then . . . around the corner of her mouth . . . and from somewhere near her paint-encircled eyes, a friendly smile, and a sigh betokening helpless loneliness and suffering. And the stranger finds himself right up in the blue [cheap seating area] line of arena hustlers working their shoulders one way and another to get a place for a better view."[125]

Whether portraying a Moroccan concubine in "Columbus and the Discovery of America" (1892), a South Asian *nautch* dancer (temple dancer) in "The Durbar of Delhi" (1904), or the "Wizard Prince of Arabia" (1914), the ballet girl wore heavy eyeliner, lipstick, a dark wig, and filmy, pseudo-oriental costumes, all to render her a more "authentic" Other (fig. 19). Owners consciously chose to costume the ballet girl in skimpy dress, though in fact Islamic women were often garbed from head to toe, in accord with the tenets of female modesty in the *Shari'a* (Islamic law). When planning the Ringling Bros.' spectacle "King Solomon and the Queen of Sheba" (1914), Al and Charles Ringling recognized that the spec would be more culturally authentic if the ballet dresses were floor-length, but they nevertheless agreed to shorten the dresses to the knee, because, according to Al, doing so would "make the ballet look better."[126]

At first glance, it might seem that the circus's emphasis on the ballet girl's sensuality would undermine its claims to highlight respectable white womanhood. But as a disguised character, the ballet girl complemented the racist writings of contemporary European and American theorists and novelists. The best-selling author Thomas Dixon wrote that nonwhite races were both promiscuous and violent.[127] At the same time, African American activists, including Addie Hunton, fought against these virulent stereotypes in articles like "Negro Woman Defended."[128] In *The Souls of Black Folk* (1903), W. E. B. Du Bois undermined racist explanations of black sexual "immorality" by analyzing the structural reasons — tenancy, debt, and other

Figure 19. "The Wizard Prince of Arabia," Barnum & Bailey, 1914. As part of a
literal cast of thousands, Euroamerican ballet girls played an exotic, indeterminate racial
Other in these extravagant orientalist productions. (Lithograph courtesy of Circus World
Museum, Baraboo, Wis., with permission from Ringling Bros. and Barnum & Bailey,®
The Greatest Show on Earth,® B+B-NL39-14-1F-3)

forms of institutionalized racism—that forced African American women to
delay marriage and bear children out of wedlock.[129]

The racially masked ballet girl also resonated with broader imperial
imaginings. Edward Said observes that the sexually charged "oriental"
woman was a common character in nineteenth-century European fiction
who personified the alluring, geographically imprecise "Orient" as a literary sexual playground for repressed bourgeois European protagonists.[130] In
a similar spirit, Imre Kiralfy's circus spec "Columbus and the Discovery of
America" (1892) depicted the ersatz Moroccan women as lascivious: "Our
tableau opens with King Boabdil El Chico, surrounded by his wives, favorites and slaves. . . . Presently music greets the ear, the female slaves begin
the slow, sensuous movements of oriental dances, while songs by female

Figure 20. Albert Hodgini as "The Original Miss Daisy," 1907. Hodgini was such a good drag bareback rider that he reportedly received marriage proposals from European noblemen. (Photograph courtesy of Circus World Museum, Baraboo, Wis., BBK-N45-HDGA-7)

slaves are heard accompanied by the wild, weird, mysterious music of quaint instruments, and the scene gradually becomes one of splendour." [131]

The ballet girl's persona was a model for other forms of popular culture. The early movie producer Richard Fox paid white actresses to become explicitly provocative with makeup and scant dress—the tools of racial masking. In the years surrounding World War I the actress Theodosia Goodman, a Jewish woman from Cincinnati, became "Theda Bara," an anagram for "Arab Death." Bara's birth was fictionalized and exoticized: she was born "on the desert sands in the shadow of the Sphinx." Her steamy performance in *A Fool There Was* in 1915, based on Rudyard Kipling's poem "The Vampire," created the word "vamp." [132]

Gender masking was another element of disguise at the circus. Big-top "women" who were really men were costumed and marketed in virtually the same manner as "dainty darlings." Reported to be extremely desirable, they were also from good families. Albert Hodgini, sporting a cascade of thick, long, brown curls, smooth skin, and a small, corseted waist, played the "Original Miss Daisy" with the Ringling Bros. from 1908 to 1914 (fig. 20). English by birth, Hodgini's circus family tried to make itself more intriguing and salable by changing its name from Hodges to Hodgini, to capitalize on Anglo and American stereotypes of Italians as a racial Other. To establish his credibility with the audience as a woman, Albert Hodgini began his act by riding sidesaddle. Eventually, he performed a series of handstands atop a horse, somersaulted madly, and then juggled a bunch of bottles and plates while still riding the same horse. Reportedly, Hodgini attracted wealthy male suitors who were later embarrassed once they found that Miss Daisy was a man. [133] In some respects, Hodgini's act was radical because it demonstrated that gender was a fluid, performative category: wearing the correct clothing, makeup, and wig, Hodgini could play female.

But Hodgini's potential ability to cross gender boundaries was limited, paradoxically, by the very gender norms that his act denaturalized. Hodgini's act deftly brought potential gender transgressions back into the rubric of respectability, because his gender play suggested that there was a single standard of appropriate female appearance and comportment. Although his identity as Miss Daisy was secret, press releases constantly teased audiences about Hodgini's "real" identity by stating that Miss Daisy performed stunts that a woman had yet to accomplish. When his maleness was eventually "discovered" by the circus press—ever hoping to increase a performer's salability—Hodgini's publicity photographs confirmed his identity as a decent family man with a wife and two children (fig. 21). As Judith Butler suggests, drag is not intrinsically subversive; instead, showmen could use it to bolster social norms.[134]

By 1920 the female big-top performer's paradoxical image of domesticated eroticism made her a perfect candidate to sell products in the new modern age of mass consumption. Several female circus performers, particularly the big stars of the decade, May Wirth, Lillian Leitzel, and Bird Millman, endorsed various products from sheet music to face cream. In programs, advertisements for the Harry von Tilzer Music Publishing Co. splashed a photograph of the Australian American bareback rider May Wirth and her "great big hit" "When My Baby Smiles at Me" across the page. Advertisements for the Leo Feist sheet music company described how the singing wire walker Bird Millman performed her act to the popular tune "Peggy." The aerialist Lillian Leitzel pitched a cornucopia of products: Lysol, the Walworth Stillson wrench, and Grandma Brown's ginger tea tablets.[135]

A highly visible presence in circus advertising during the 1920s, Lillian Leitzel led an exciting romantic life that enhanced her popular appeal for American consumers. She was married three times and divorced her first two husbands—an anonymous property man and Clyde Ingalls, the Ringling Bros.' flamboyant sideshow manager. In 1927 she and her third husband, Alfredo Codona, the handsome Mexican trapeze artist, were hailed as the Douglas Fairbanks and Mary Pickford of the circus. At the turn of the century, when remnants of Victorian prudery remained in force, Leitzel's turbulent marital life might have been kept secret. But in the 1920s, the details of Leitzel's romances were well known, enhancing her magnetic "personality" and her ability to sustain public interest.[136] Discussions of female sexuality and desire were moving increasingly into the mainstream: educated Americans widely read the writings of Sigmund Freud and Havelock Ellis, greater numbers practiced birth control, and fertility rates continued falling. Contemporary advertising and movies brimmed with images of independent, flirtatious "flappers"—pictured occasionally in swimsuits, especially after Gertrude Ederle swam across the English Channel in 1926 in record time.[137] In this social environment, press agents easily marketed Leitzel as both a dynamic "personality" and a paragon of domesticity, devoted to her home on the road, pampering her bulldogs, nursing injured animals, playing the piano, and telling stories to small children.[138]

Figure 21. "Mr. and Mrs. Hodgini," undated. From drag performer to family man, Hodgini easily played both at the circus. (Photograph courtesy of Circus World Museum, Baraboo, Wis., BBK-N45-HDGA-8)

Unlike the big top, where thousands of audience members sat far from the constant action in the ring, the sideshow tent was small enough that patrons could ask performers questions.[139] At the sideshow, the player's body—not her athletic skill—was the source of her marketability. As a rule, showmen categorized sideshow acts as either "born" (for the congenitally deformed or racially exotic), or "made" (when the acts involved conscious bodily disfigurement, specifically tattooing or glass eating, among other stunts). The sociologist Robert Bogdan notes that in an era before the professionalization of medical science, the sideshow's appeal lay in the ability of audiences to judge for themselves what the freaks "had."[140] Likewise, the literary scholar Rosemarie Garland Thomson suggests that the freak show itself was a product of the "tensions of modernity," because it threw into sharp focus two competing worldviews: an older, religious one that viewed the freak as a product of the supernatural, and a newer, rational, professionalized scientific one that pathologized physical abnormality.[141]

Located inside the menagerie tent, the ethnological congress of "strange and savage tribes" physically collapsed human and animal boundaries in spectacular acts of Otherness like the sideshow (fig. 22). Zoological proprietors pioneered these systematic, full-blown exhibits of exotic people with animals.[142] In 1874 the German animal dealer and zoo owner Carl Hagenbeck first incorporated "natives" into his foreign animal displays, after a friend, the animal painter Heinrich Leutemann, suggested that a family of Lapps would render a display of reindeer "most picturesque."[143] In his memoirs Hagenbeck recalled, "This seemed to me a brilliant idea. . . . My optimistic expectations were fully realized; this first of my ethnographic exhibitions was from every point of view a huge success. I attributed this mainly to the simplicity with which the whole thing was organized, and to the complete absence of all vulgar accessories. There was nothing in the way of a performance."[144] Banking on the proven success of this sort of premeditated "naturalness" and absence of "performance," Barnum and London presented their first ethnological congress in 1886, at which "even the best informed and most intellectual had something to learn when visiting the show."[145] James Bailey, who planned the exhibit in 1885, sent two American agents around the world to collect human specimens for future productions. Much like Hagenbeck, Bailey claimed that he originally conceived of the ethnological congress as scenery for his exhibit of sacred cattle from Thailand.

Figure 22. "Ethnological Congress" (companion lithograph to fig. 12), Barnum & Bailey, 1894. Note the presence of the well-appointed Euroamerican families viewing the display of the "strange and savage tribes" as an instructive exercise. (Lithograph courtesy of Circus World Museum, Baraboo, Wis., with permission from Ringling Bros. and Barnum & Bailey,® The Greatest Show on Earth,® B+B-NL38-94-1F-6)

By surrounding the cows with authentic Siamese people, Bailey hoped to render the cattle more "authentic."[146] Inspired by the great financial success of similar ventures at international exhibitions—from Paris to Chicago— Barnum & Bailey proprietors in 1894 expanded their ethnological displays of people of color.[147]

From fat ladies, bearded ladies, and lady giantesses to armless and leg-less ladies, Euroamerican women freaks were seemly ladies in press re-leases, in programs, and on postcards (which virtually all freaks sold to supplement their income). "Fat Marie" Lil, a fat lady for Barnum & Bailey, presented herself as an innocent English school girl, wearing prim outfits laden with ruffles. Articles detailed her "delightful" complexion, her vast daily diet, her sensitivity and modesty: Lil always took her dinner in private after people laughed at her inability to fit her knees under the table at a restaurant.[148] Newspaper articles described Lil's love of food and her desire to gain weight: at 389 pounds, Lil was still lighter than her mother, who weighed 420 pounds. Her father, in contrast, was "a little, slim man."[149] Press releases might exaggerate a fat lady's size or appetite, but she was always presented as innocent, never sexually potent.

But the marriage of two freaks—whether real or staged as a publicity stunt—kept audiences wondering about the bodily logistics of sexual activity, much in the same manner that people speculated about sex between circus animal "couples." Bodily deformity, like animality and racial non-whiteness, was a license, an acceptable avenue through which to discuss sexuality.[150] Press agents hinted at the possibility of sexual activity through the conventional trope of marriage. Circuses commonly presented opposites together: the Skeleton Man and the Fat Lady, or midgets with giants, like Ella Ewing, the eight-foot, four-inch "lady giantess," with the twenty-three-inch "Great Peter the Small" (fig. 23).

In circuses throughout the country, the fat lady commonly "married" her sideshow coworker, the Skeleton Man. Hannah Battersby, who reportedly weighed 600 pounds, was married to her fellow Barnum & Bailey sideshow performer Jonathan Battersby, a "Living Skeleton" who weighed seventy pounds.[151] Like Barnum's highly publicized union of the midget couple Lavinia Warren and Tom Thumb in 1863, marriages of this kind in all likelihood enabled audiences to imagine the Fat Lady in sexual situations, particularly with the willowy Skeleton Man, whom she might crush to death during sexual intercourse.

Outfitted in Victorian dress, motherhood and refinement defined the Euroamerican bearded lady. Impresarios noted that Madame Josephine Fortune Clofillia, a Swiss bearded lady (and one of P. T. Barnum's first hirsute acts in the 1850s), possessed official medical certificates confirming her two children's births.[152] Programs noted that Grace Gilbert was desirable and had numerous marriage proposals, despite her full, flowing beard. On stage since the age of twelve months as the bearded "Infant Esau"[153] (later the "Lady Esau"), Annie Jones was an accomplished musician of good character.[154] Divorced and then widowed, Jones spent her entire life as a bearded performer. After her death at thirty-seven in 1903, Barnum & Bailey publicized Jones's supposed dying wish that her husband, the press

[*Respectable Female Nudity*]

Figure 23. "Two Living Human Prodigies" (Ella Ewing and Great Peter the Small),
Barnum & Bailey, 1897. Sideshow acts were animated by the juxtaposition of opposites.
Despite Ewing's financial independence and very public career, she praised marriage and
motherhood as the "proper" roles for women. (Lithograph courtesy of Circus World
Museum, Baraboo, Wis., with permission from Ringling Bros. and Barnum & Bailey,®
The Greatest Show on Earth,® B+B-NL-44-97-1U-2)

agent, Theodore Bower, marry her bearded successor—despite James A. Bailey's objections.[155] Even after her death, Jones was reportedly generous and selfless—concerned only with her husband's future happiness as she lay dying.

But burlesque lurked near the surface of the bearded lady's womanliness. While proprietors advertised hirsute women as the "genuine article," they also tickled potential audiences into coming to the sideshow to judge the authenticity of the bearded woman for themselves, by recounting stories about hirsute women whose beards had been yanked off by the doubting public. In one report, a horseman strode into a restaurant and tore off a woman's fake beard while she dined with her fellow freaks.[156]

Newspaper commentary jokingly observed that the beard was evidence of the hirsute woman's efforts to appropriate male secondary sex characteristics—not just political equality. A Boston newspaper exclaimed, "Not content with claiming the right to vote, and laying siege to our nether garments (a la bloomer), our beards are actually in jeopardy. Heaven forefend!"[157] Similarly, the Italian anthropologist Cesare Lombroso argued that women were "undeveloped men" because of their small size and lack of facial hair.[158] In line with this sort of thought, bearded women might actually be men. Showmen were quick to tantalize the audience with the theoretical possibility that bearded women might be male. Robert Bogdan observes that in publicity photographs, bearded women were often staged with their husbands and children as a way to subvert prim conformity with gender ambiguity.[159]

Ella Ewing, a Euroamerican "Lady Giantess," also represented a standardized "womanly womanliness," despite her towering size. Born in Gorin, Missouri, in 1872, Ewing was six feet nine inches tall when she was ten years old, and grew—reportedly—to be eight feet four inches. When Ewing was eighteen, a proprietor of a museum in Chicago wanted to hire her. Because she felt that she was the "burden of her poor father's care," Ewing accepted the showman's offer. She spent the rest of her life as a sideshow attraction, touring county fairs, the vaudeville circuit, and famous circuses under the management of her parents. She was exhibited at the World's Columbian Exposition in Chicago in 1893 and traveled with Barnum & Bailey, Buffalo Bill's Wild West, and the Ringling Bros. until her death in 1913 from tuberculosis.[160] During the off-season, Ewing returned to Gorin, where she lived in a large, specially designed house with high ceilings. Extremely sensitive about her size, she wore multiple rings to hide her long fingers and long skirts to cover her big feet.[161] As a testament to

Ewing's moral character, press releases emphasized her frugality, reporting that she saved $30,000 during her career.[162] Although Ewing received several offers of marriage, she dismissed them as "business propositions," designed to cash in on her celebrity. She announced flatly, "I would not live in a loveless married life. When I am in business I am in business for Ella Ewing. As for marriage, I believe my views in regard to it are the same as those of any other truly womanly woman. Wife, mother and housekeeper are the three things woman's being requires to make her life complete. God created her for that sphere. But my size will prevent me from marrying."[163]

Like other white female stars, Ewing was reputed to be pious. One account spoke glowingly of her ability to "master even the most abstruse Bible questions, and to surpass even her teachers."[164] These images of saintly femininity were supposed to quell the freely publicized rumors that Ella was actually a monster who feasted on women and children: "But [Ewing is] a giantess fully equipped with all the modern improvements in the direction of sweetness and light and so startlingly different from the giantesses you have read about that you will have to reconstruct entirely your conception of the word. Miss Ewing never ate a baby in her life, nor transformed a princess into a calf, nor chained a lovely maiden in a dungeon. . . . On the contrary, she is a singularly lovable woman and the most popular doe in Gorin, and her home life . . . is altogether charming . . . . [She] is remarkably proficient in domestic matters, supervises the care of the house and helps with much of it."[165] Despite her life of public performance, constant travel, financial independence, and desire to remain unmarried, Ewing touted a life of privacy and domesticity as most appropriate for a woman.

Press agents placed physically diverse women freaks into the rubric of traditional womanhood by using the visual trappings of normality: gowns, husbands, parlors, and love of home. Proprietors' portrayals of fat women, bearded women, and lady giantesses performed two contradictory functions. Impresarios normalized physical abnormality by staging these women in normative settings. These representations also helped reify a single standard of ideal womanhood, because showmen marketed each of the women as quintessentially a "real woman" at heart. Still, proprietors used standard representations of marriage, motherhood, elaborate dress, and the parlor to poke fun at contemporary gender norms through these visibly abnormal bodies.

The snake charmer engaged in racial rather than gender disguise. Although her skin remained pale, the snake charmer manipulated her racial identity by wearing thick eye liner, lipstick, filmy, diaphanous clothing, and of course snakes. She was staged in humorous publicity shots draped with snakes and clothed in corseted dress, but in performance she wore brief "oriental" garb with snakes slithering against her bare skin. Because animals gave an act a certain degree of sexual license, press agents described the charmer's stunts with openly suggestive language: "To see her lithesome figure, her strong muscular arms and shapely limbs bravely caressing the huge squirming boa constrictors, never fails to produce a great impression."[166]

In drawing audiences to the sideshow tent, circus media freely admitted that this seemingly mysterious foreigner was a "home-grown" Euroamerican woman. The racial disguise became a racial tease, the woman's "real" identity being openly masked as she slipped into the meager garb of the fictitious Other. Ida Jeffreys, a snake charmer for Barnum & Bailey's circus in 1888, was advertised as a "Hindoo" with supernatural powers, able to stun snakes with a glance. Yet a newspaper press release revealed her true identity: "Her eyes are as blue and soft as a baby's, neither does she charm [snakes] with low, soft, soothing tones on a piccolo like the Hindoo magicians you hear about, or yank them around in her herculean grasp. She is a cool-headed New York girl, Ida Jeffreys, off the stage, and she handles snakes for pay as calmly as an artist handles his brush."[167]

At the turn of the century, virtually all snake charmers were women. As an activity that required less training than acrobatics and bareback riding, snake charming was commonly performed by managers' wives (notably Lou Ringling, known on stage as Inez Morris), who were entitled to free room and board only when they actually worked for the circus (fig. 24).[168] The snake charmer draped herself with a limp collection of boa constrictors and indigenous snakes which, if poisonous, had been defanged. Yet the work was dangerous at times: Lulu La Tasca, a Dutch woman who worked for Barnum & Bailey in 1891, told how she kept a sponge soaked with ether in a little oily silk-lined pocket stuffed into her corset: "When the snakes get too frisky I thrust the sponge into their eyes, and they hush up quick, I can tell you. . . . I have brads in my slippers, sharp, stout steel ones, which I stick clear into them when they don't behave. It is very, very funny to me, when I think of my smiling to the audiences as if it were real fun to charm the snakes, and all the time, I am bradding them and etherizing them and shaking in my skin for fear they will tighten their coils and be too much for me."[169]

Despite the secret battles that the charmer waged with her snakes, she feigned great pleasure in her work. Her costuming and writhing move-

Figure 24. Lou Ringling, also known as "Inez Morris," ca. 1887 (not 1919 as written on the card). An unflappable snake charmer and equestrienne, Ringling was married to Al Ringling, director of exhibitions. (Howard Gusler Collection, P-N45-RGLL-3; cabinet photograph courtesy of Circus World Museum, Baraboo, Wis.)

ments were nevertheless a far cry from those of the fully clad South Asian male snake charmers who performed their rituals on religious occasions.[170] Set against a backdrop of fake or painted foliage to create a fertile, tropical scene, the circus lady snake charmer was advertised as intimate friends with her "slimy pets," sharing her home with them and even letting them sleep loose on her sofa.[171] Damajante, a snake charmer with Barnum & Bailey in the 1880s, explained her snakes' weekly routine in a subversion of domestic ideals: "I give [my snakes] as much attention as a mother does a child. Regularly every Saturday night they are washed in lukewarm water and wrapped up in blankets."[172]

The tattooed lady also performed a racial masquerade. Her body was colored with paisley prints, tropical scenes, flags, battle scenes, pictures of U.S. presidents, queens, and Mother Mary, and more. One tattooed woman, Lady Viola, displayed six presidents on her chest, the Capitol on her back, and, by the 1920s, ten movie actresses on her arms, and Babe Ruth, Charlie Chaplin, and Tom Mix adorning her legs and thighs.[173] A sideshow fixture from the 1880s onward, the tattooed woman performed almost naked to afford the best view of her elaborately ornamented body.[174] In an age of European and American imperialism, the tattooed lady carried the colorful marks of her fictional contact with faraway Pacific lands. Proprietors highlighted the fictional circumstances in which the tattooed lady became "marked," breathlessly chronicling how she had been captured and forcibly tattooed by Native Americans or "savage" South Sea Island men. Beginning in the 1880s, Cesare Lombroso studied the purported link between tattooing, sexuality, and criminal behavior, asserting that criminals and prostitutes were much more likely to be tattooed than law-abiding folk. Proprietors diffused the tattooed woman's potential image of criminality by promoting her as a victim of the "primitive" practices of nonwhite men at home and abroad.[175]

With the advent of the electric inking process around 1900, procuring a tattoo became relatively easy. As a result, the market became glutted, wages fell, and tattooing lost its novelty. Showmen tried to make the act more salable by presenting tattooed families, tattooed dwarves, tattooed motorcycle riders, a congress of tattooed men, and even tattooed cows and dogs (the Tattooed Great Dane, for one), but by the early 1900s the act had become ordinary.[176]

At some sideshows, "gentlemen only" could pay an extra twenty-five cents to stand in a small enclosed area and watch "oriental" dancing girls perform a brief dance.[177] Tiny Kline, who worked as a cooch-show dancer with Arlington & Beckman's Oklahoma Ranch Wild West in 1913, remembered vividly how the talker would attract customers with the following spiel: "'Gather 'round me a little closer, men, 'don't want the ladies to hear this, but you are about to get a little treat inside this curtain . . . only one quarter—twenty-five cents.'"[178] Wearing red or green tights under a brief beaded costume which hung in heavy fringes from the chest and hips—"to accentuate the movements of those parts"—Kline preceded another woman who wore a short, ruffled "oriental" dress. In her old age, Kline recalled the entire act:

> The place filled up in no time; we could hear the wise-cracks and otherwise 'smart' remarks, from behind another curtain—our dressing-room —as they gathered in anticipation of seeing [in the talker's words] "those muscles shake and shiver like a bowl of jelly in a gale of wind; the dance that John the Baptist lost his head over!" On a short, shrill note of the flageolet—the signal,— I came out first, climbed up to the platform which was roped off all around for protection against the impudence of the standing audience who might make a grab at our limbs (which they sometimes tried anyway) I went into my dance, a short routine of about two minutes duration, doing high kicks and the 'split' which was then, considered "naughty." There wasn't anything in that music to inspire dance spirit within me, I could never "feel" the mood, nor figure out the timing; always against tempo, but I finished with a fast "fouetter" a twist-kick spin, and climbed down. Then came Helen, the other girl. . . . She did—what in Algiers might be considered a sedate parlor dance, but here in America they called "Hootchy-kootchy." The most outstanding feature of it was the way she could make her head slide from side to side while looking straight at you, just like a serpent.[179]

In 1913 one newspaper commented that the Gollmar circus cooch show was "an immoral performance, and many did not hesitate in saying that it should not have been tolerated by the authorities or those in charge of the fair grounds."[180] Yet the evidence suggests that no turn-of-the-century community banned the cooch show, perhaps because its racially disguised performances gave it a certain degree of immunity against censure. Sometimes the masquerade was sexual as well: the male audience received a big surprise when the cooch dancers turned out to be a group of taunting, raucous male clowns in drag, wearing exaggerated foam breasts and buttocks along with their usual collection of oversized shoes, noses, tiny hats, and loud, ragged clothes. The male audience, in its quest for anonymous titillation in the crowded little cooch show, was suckered.

Sideshow women of color were represented as preindustrial "primitives" and as animals. For instance, an African American girl who suffered from vitiligo, a skin disorder which causes spotting, was named "Louise the Leopard Girl." The characters played by women of color were linked to the process of imperialism: foreign women of color were supposed to represent "newly discovered" races from newly colonized countries. The literary critic Anne McClintock suggests that "commodity spectacles" like the Crystal Palace Exhibition (1851) gave their audiences the impression that culture could be consumed at a glance, and that only western imperial powers were capable of gathering the world's cultures under one roof neatly for systematic inspection. Like these national expositions and technologies of panoptic surveillance such as photography (which offered the promise of being the "monarch of all I survey"), circus exhibits of nonwhite women told audiences that the world was knowable through frozen images, photographic or live.[181]

Krao Farini, a Laotian woman, performed in several roles at the circus (fig. 25). She arrived in the United States as a child working as a "gorilla girl" with the John B. Doris circus in 1885; later, with the Ringling Bros. circus, she also played a "missing link" and a bearded lady. Newspaper articles and talkers recounted her anachronistic origins as a "specimen" of "ape-humanity": Krao was allegedly "caught" in a Laotian forest at the age of seven by a Norwegian explorer, Carl Bock, who captured Krao's father first and then Krao herself, after which her mother surrendered. When separated from their daughter, Krao's parents reportedly cried plaintively, "Kra-o," "Kra-o," which became the girl's moniker at the sideshow. After her father died of cholera and her mother was forbidden by the king to leave Laos, Krao and Bock traveled to Bangkok and then to London, where she became an exhibit for the showman G. A. Farini at the Royal Aquarium. Krao took Farini's last name, which she kept for the rest of her life.[182] Press releases reported copious testimony from scientific authorities willing to verify Krao's authenticity as a "missing link" between apes and humankind. One such "expert," the ethnologist A. H. Keane, described Krao in racially animated primatological language that sharply contradicted her appearance:

> [Krao's] whole body is . . . overgrown with a . . . dense coating of soft, black hair about a quarter of an inch long, but nowhere close enough

Figure 25. Krao Farini, posed against a fake jungle scene, 1885. Portrayed as a "gorilla girl" and a "missing link," Farini, a Sumatra native who spoke seven languages fluently, made her home in Bridgeport, Connecticut, where she tutored children at the local library. (Photograph courtesy of Circus World Museum, Baraboo, Wis., SID-N45-KRAO-2)

to conceal the color of the skin, which may be described as of a dark olive-brown shade. The nose is extremely short and low, with excessively broad nostrils, merging in the full pouched cheeks, into which she appears to have the habit of stuffing her food, monkey-fashion. Like those of the anthropoids her feet are also prehensile, and the hands so flexible that they bend quite back over the wrists. The thumb also doubles completely back, and of the four fingers, all the top joints bend at pleasure independently inwards . . . the beautiful round black eyes are very large and perfectly horizontal. Hence the expression is on the whole far from unpleasing, and not nearly so ape-like as that of many Negritos.[183]

At Barnum & Bailey's circus in 1903, Krao was featured next to "Johanna, the Live Gorilla." Unlike many sideshow players, Krao was not physically deformed. By juxtaposing her with the chimpanzee, proprietors invented a tradition of evolutionary continuity between the "gorilla" and the "Gorilla Girl." The copy on the back of one of Krao's postcard from 1922 described her as a "Laotian monkey girl," using language that is nearly identical to the passage quoted above: "Krao . . . has some abnormal peculiarities and some points of resemblance to certain species of the monkey tribe; the distribution of hair is one, as it grows like that of a monkey, in similar waves, that on the forearm pointing upwards from the wrist to the elbow. The fingers are very supple, being capable of being bent completely back. The cheeks are pouch-like and like monkeys." [184]

Krao wore skimpy, ruffled costumes, and was presented against a backdrop of painted fronds as a "mysterious vestige of prehistoric humanity." Yet over time, Krao's persona changed: although still playing a "missing link," she also became known as a "civilized primitive," whose exposure to European and American civilization had "uplifted" her. She spoke seven languages fluently, and had "faultless" manners. When Barnum & Bailey wintered at Bridgeport, Connecticut, Krao volunteered as a tutor at the local library.[185] Throughout her career, she performed in minimal dress as an affirmation of racial "authenticity" and, not by accident, as a way to draw audiences. Popular throughout her long sideshow career, Krao earned a comfortable living, although her public persona was that of a "savage" "gorilla girl," whose "arrested" evolutionary development would forever keep her a juvenile in the public's eye. Krao, like her nonwhite colleagues, held a contradictory position at the circus: on the one hand, she was able to make a good income in a racist society where there were few lucrative employment options for a person of color. Moreover, she maintained close friendships with sideshow players. The fat lady Carrie Holt characterized Krao as "the sweet-est and loveliest lady I ever met . . . a good deal more refined than most of the crowd that stares at her." [186] Yet Krao's job required her to perform ide-ologies about nonwhite savagery that circumscribed people of color in all areas of American life.

In 1894 Barnum & Bailey's new ethnological congress included entire families, echoing the Midway Plaisance at the World's Columbian Exhibi-tion in Chicago in 1893. Barnum & Bailey's 1894 route book describes the "family style" ethnological congress as a sort of intellectual fast food allow-ing whole cultures to be "eaten" at a glance: "What gave the Congress an added interest was the fact that nearly all the natives were accompanied by their women, wives and families, who brought with them all the domestic utensils, used when in their native countries . . . so that a complete and com-prehensive idea could easily be had at a glance of just how these people lived in their own countries." [187]

Foreign women and children played a primary role in creating an authen-tic domestic landscape. In its formative years, a circus or museum might have hired an individual South Asian female *nautch* dancer (temple dancer). But she would have been presented next to something irrelevant, say a tiger skin, or an ossified walrus penis. By 1906 the *nautch* dancer was part of Carl Hagenbeck's Grand Triple Circus East India Exposition, which contained one hundred Hindoos, including women and children, from the "great black empire." The program stated: "They are a strange and wonderful people from any view-point; — strange in contour and character; in their dress (or lack of it). . . . Housewives will be shown at their duties, baking, cooking, washing and sewing in their own Oriental and primitive way; children will romp and indulge in their native games and play and the beautiful *nautch* girls will pose for the time being in all their bewitching and be-jeweled splendor. Competent interpreters will be in attendance at all times and there will be frequent lectures of an instructive nature." [188]

Two common themes characterized the presentation of women from "savage" societies: sexual promiscuity and participation in physical labor. Although the photographic evidence suggests that members of the ethno-logical congress danced or simply sat during their act, colorful lithographs depicted women working hard. The circus's spectacular live presentations of race and female gender were already familiar to Americans steeped in nineteenth-century travel narratives. Herman Melville, for one, wrote lurid, fictionalized versions of his contact with different cultures while working on a whaling ship in the 1840s. In one such tale, *Typee: A Peep at Polynesian*

Life, Melville crafted a wild account of imprisonment and eventual escape from a cannibalistic society on Typee, an island in the Marquesan chain in the South Pacific. In this tropical, mountainous setting, flirtatious, topless women, "fancifully decorated with flowers," beckoned Melville "with faces in which childish delight and curiosity were vividly portrayed. . . . But in-spite of all their blandishments, my feelings of propriety were exceedingly shocked, for I could but consider them as having overstepped the due limits of female decorum." [189] Melville also depicted the Typees as noncultivators, living lazily in a land in which breadfruit and coconuts simply fell into one's lap. [190]

Just as tights were a code for eroticism under the big top, lack of cloth-ing characterized the "lady savage." Programs lured audiences, promising a scene of "wild, weird and strange picturesqueness, the bright colors of the dresses of some contrasting with the brown naked skin of their neigh-bors." [191] Tropical zone players were reportedly proud of their lack of cloth-ing. Memene, a Fiji princess "Cannibal Girl" with Barnum & Bailey's ethno-logical congress in 1895, pronounced American women's fashions prudish: "And the [American] dresses must cost so much money . . . why do the women wear such very big sleeves . . . it must take very much material to make them. Why do the skirts spread out so? Do American women put on so much clothes to hide their figures? Are their figures so very bad? In my country, if you have a good form, you are proud of it. You do not seek to cover it up. Yes, the dresses are very pretty, but I do not like them." [192] Memene's seemingly extemporaneous commentary was simply a publicity device: by allowing a performer to speak, seemingly in her or his own voice, showmen made an act more tantalizing and attractive to potential audi-ences, particularly when the performer spoke candidly about public nudity and the consumption of human flesh.

In circus programs, press releases, and posters, women labored while men lolled. Such representations had a long history in American culture. In colonial Virginia, English settlers wrote that Indian women did the daily chores—gardening, food preparation, housekeeping, and childcare—while Indian men were "lazy" because they fished and hunted only occasionally. [193] Over time, this stereotype became universalized in colonial discourse, serv-ing as a justification for imperial expansion under the guise of the "white man's burden." Cesare Lombroso argued that the "Law of Non-Labor" was the essential condition of female existence. Cynthia Russett posits that the scope of women's work was central to contemporary constructions of race and female gender: "With the possible exception of sexual laxity, female labor represented the most striking difference in gender relations between savage societies and [the Victorians'] own. Savage women toiled; civilized women did not. It was self-evident, therefore, that the path of progress for the feminine half of humankind involved an increasing emancipation from productive labor."[194]

In their quest to highlight the contrasts between "civilization" and "savagery," showmen showered their media with images of hard-working women of color. One headline proclaimed, "Women Are the Workers in Siam," while another argued that the equal division of labor in Papuan society proved that supporters of female equality such as Susan B. Anthony and Charlotte Perkins Gilman were wrong to postulate that labor equality among men and women was evidence of an advanced civilization. The article discussed the culture's practice of polygamy and the women's nudity, stating in a subheading that European imperialism had made native women "modest," through British enforcement of prudery: "British Influence Is Responsible for Women's Skirts Made of Palm Leaves."[195]

At the sideshow and ethnological congress, the inversion of contemporary gender norms continued, as impresarios presented women of color as stronger, faster, and fiercer than men. To draw audiences to its revamped congress "of strange and savage people" in 1894, Barnum & Bailey held an exposition of Australian Aboriginal boomerang throwers on Manhattan Field. One woman, Tagara, was the standout, and "could throw better than any tax-paying resident on Manhattan Island."[196] In 1895 Barnum & Bailey exhibited female Gilbert and Fiji Islanders who "take to the water like ducks"; they were reminiscent of the sensuous women swimmers of Melville's *Typee*, swimming tirelessly to meet the whaling boats, "their jet-black tresses streaming over their shoulders, and half enveloping their otherwise naked forms."[197] The most extreme inversion of Euroamerican gender roles came with Barnum & Bailey's group of Dahomey women, who played "blood-thirsty Amazons": "[They] are conspicuous with their almost naked black and shiny skins and scarred breasts and faces. These are probably the only true Dahomey Amazons ever known to leave their native fastness, and are fine specimens of those Fierce and Savage Black Female Warriors that have defied the armies of civilized nations. Reared from infancy in bloody scenes of war, with every female instinct annihilated, skilled in the use of weapons, they are as Ferocious in War as Wild Beasts."[198]

The Euroamerican female audiences who gazed at the Dahomeys perhaps felt united in their shared whiteness, despite their own ethnic differences. Vibrant circus posters depicted well-appointed Euroamerican fami-

Figure 26. "Ubangi Savages," Ringling Bros. and Barnum & Bailey, 1930. Ritualized disfigurement became the basis for racial novelty and spectacle at the circus in an age of radio, automobility, and the expansion of the motion picture industry. (Lithograph courtesy of Circus World Museum, Baraboo, Wis., with permission from Ringling Bros. and Barnum & Bailey,® The Greatest Show on Earth,® RBB-NL37-30-1F-4)

lies gazing and pointing at nonwhite acts (see figs. 12 and 22). Jim Crow seating arrangements and concessions solidified the white spectators' shared privilege of witnessing together the display of what they collectively were not.[199] Black audiences likely shared this same sense of ethnic distance from the performers, even though they were often characterized as being of the same "race" as the players in the ring.

By the early 1930s, in an age of movies, radio, and increased magazine readership, the spectacle of seminude women of color from around the world engaged in "typical" activities had lost its novelty at the circus. To meet their audience's demand for newness, circus proprietors hired foreign women of color whose bodies had been ritualistically disfigured. Arriving in the United States on March 31, 1930, eight Congolese women, known as the "Ubangi Duck-Billed Savages," became an instant sensation at the Ringling Bros. and Barnum & Bailey circus (fig. 26). Briefly clad in short, colorful

[*Respectable Female Nudity*]

cotton skirts, the Ubangi women's principal draw was their practice (starting at the age of six) of wearing large wooden plates inside their lips, which stretched the lower lip to a diameter of over nine inches in adulthood.[200] By 1930 the term "French Congolese" no longer sounded as mysterious as it had in an era before increasingly sophisticated mass media; consequently, the Ringling Bros. press agent, Roland Butler, further exoticized the women by coming up with the name "Ubangi" after studying maps of Africa and finding a remote district so named, hundreds of miles away from the tribe's real locale.[201] After stints at the Paris Zoo and in Rio de Janeiro, the women joined Ringling Bros.' Congress of Freaks. In addition to performing at the sideshow, the women, accompanied by their husbands, walked once around the big-top arena, each smoking a pipe and playing the drums while the bandmaster, Merle Evans, conducted modern jazz melodies which had, in his words, the "strong underlying beat of jungle rhythm."[202]

The presence of a sham professor who "explained" the Ubangis to American circus audiences was a critical part of the "savage" persona. The Ubangis' manager, "Professor" Eugene Bergonier (a cheat who stole their salary of $1,500 a week[203] and allowed them to keep only the proceeds from their postcard sales), spieled about the origins of the lip-stretching practice: supposedly it began years ago to make Ubangi women unattractive to pirates, and over time the result became a mark of beauty.[204]

The huge lips of Ubangi women, who were presented as "monster-mouthed . . . savages . . . strangest people in all the world,"[205] were central to the showmen's construction of their sexuality: bodily disfigurement was a means for Ubangi men to keep their women "safe" from "marauders." The women's lips also served as a metaphor for engorged labia, a visual image surely not lost on Euroamerican audiences steeped in stereotypes about black women's supposed sexual availability. At the beginning of the nineteenth century, British imperialists, in the name of scientific "objectivity," named the fictively large African labia and buttocks the "Hottentot Apron," after a Khoisan woman, Sara Bartman, who was dubbed the "Hottentot Venus." Bartman was abducted by an Englishman in South Africa and exhibited in England and France from 1810 until her death in 1815 at twenty-eight. While on stage, she was dressed in faux "native" garb composed of ostrich-shell beads and a short, tight cotton skirt to enlarge her buttocks. Curious spectators jostled and probed Bartman as she struggled to keep herself covered. After her death, a group of French scientists dissected Bartman and pickled her genitalia, which they put on exhibit at a museum in Paris. Throughout the nineteenth century and the early twentieth, other women of color from Africa and South America were also exhibited as Hottentot Venuses.[206]

Anne McClintock suggests that Africa, the Americas, and Asia represented the "porno-tropics" for the western imagination, which reduced human beings to oversized genitalia.[207] Ringling Bros. and Barnum & Bailey workers remembered the Ubangi women in a similarly sexualized manner. Tom Barron, the "World's Tallest Clown," recalls that the Ubangis, "didn't want to wear any clothes. . . . [T]hey had this Frenchman [Bergonier] who was sort of their 'chief.' He had a hell of a time trying to control them . . . every once in a while they'd take all their clothes off."[208] Frequently juxtaposed with "dainty" circus women and occasionally topless, the Ubangi women were marketed as the antithesis of "womanly" beauty, their nudity a signifier of racial inferiority.

During the 1930s the "Giraffe-Neck Women of Burma" also exhibited their ritually disfigured bodies at the circus (fig. 27). Performing for Ringling Bros. and Barnum & Bailey in 1933 and for Hagenbeck-Wallace in 1934, three women demonstrated the stunning results of their cultural practice of gradually elongating their necks with heavy, solid brass coils. A $3 program booklet,[209] "Interesting Facts and Illustrations of the Royal Padaung Giraffe Neck-Women from Burma," marketed the coils to American audiences as a sign of beauty and sexual attractiveness. It explained that the women's mothers had slowly stretched their necks beginning in early childhood, adding new coils every year until each girl's neck was approximately sixteen inches long. The women also wore coils around their legs which, combined with those around their necks, weighed between fifty and sixty pounds. The booklet noted that the women's desirability for marriage was measured by the weight of the brass that each carried and described how American doctors got in the act of explaining what the Burmese women "had" by x-raying them to explore the physical consequences of ritualized neck stretching.[210]

"The Last of the Unknown People of the Earth" were supposedly isolated relics, untouched until now by modern industrial society. Showmen reported that Mu Kaun, Mu Proa, and Mu Ba came from villages in remote, mountainous terrain several hundred miles north of Mandalay, still traveled by elephant, and spurned paper currency. The women were persuaded to travel to the United States only after Ringling agents presented their relatives with axes, knives, tins of fish, bright cloth, and silver rupees. In 1934 the Hagenbeck-Wallace program and route book contained a photograph of an elephant pulling a plow, captioned: "In the jungles in the opposite side

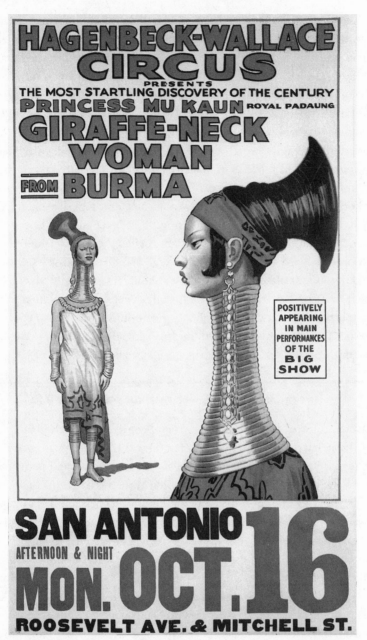

*Figure 27. "Giraffe-Neck Women from Burma," Hagenbeck-Wallace, 1934.
Showmen stressed that the Padaung women performed physical labor in Burma, a
characteristic that defined them as "savages." (Lithograph courtesy of Circus World
Museum, Baraboo, Wis., with permission from Ringling Bros. and Barnum & Bailey,®
The Greatest Show on Earth,® HW-NL41-33-1U-1)*

of the world, the 'two elephant' roadster is always an up-to-date model."[211]
Yet another photograph suggested that modern American consumer culture had already changed these women: here, all three sat in the front seat of a brand-new 1934 knee-action Pontiac that they hoped to take back to Burma.

Evoking turn-of-the-century circus constructions of gender and race, pamphlets noted that the Padaung women performed heavy labor in Burma, in spite of their fragile, weighty necks. They chopped wood, cultivated rice paddies and carried buckets of water over rough terrain. American audiences got to peek at the Padaungs' private lives by purchasing their publicity booklet, full of pictures of the women sleeping together on an American bed, playing cards, drinking tea, singing.[212] Although the Giraffe-Neck Women, or "long necks" as their coworkers called them,[213] considered American women's bare necks indecent, they were provocatively presented, clothed in filmy, sarilike wraps with their shoulders exposed.[214] Irene Mann, a rope twirler who worked with the Giraffe-Neck women in 1933, remembered them sadly in 1994: "You didn't get to know them . . . [they didn't talk]. . . . That's a pretty awful thing, to bring them all the way over and put them through that. After all, they are human beings."[215]

WALKING THE TIGHTROPE OF PROPRIETY

At the beginning of the twentieth century, proprietors used multiple strategies to restrict circus women's power: they marketed female seminudity and physical variability as simultaneously erotic, proper, and instructional. In many respects, they succeeded in selling the contradictions between titillation and respectability, because the circus escaped state regulation in an era when virtually all public amusements faced some form of censure from purity reformers. Progressive reformers targeted the saloon, dance hall, movie theater, skating rink, and ice-cream parlor as sites of salacious activity, where men and women mingled and prostitution might flourish. Andrea Friedman observes that in early-twentieth-century New York City, activists and government regulators attacked specific popular amusements only after they became "mass" entertainment with a heterogeneous class base. In contrast, reformers generally ignored "hard core" forms with small markets, like pornographic "french postcards."[216]

At the turn of the century, antiobscenity activists rallied across the country. Mocking this wave of purity reform, *Billboard* reported—in jest—that in Pawtucket, Rhode Island, in 1898 the Young Men's Christian Association covered "indecent" piano and chair legs with fabric.[217] From Minneapolis to Boston, diverse social reform groups, notably the Woman's Christian Temperance Union and its "Department for the Suppression of Impure Literature," the Federation of Women's Clubs, and local government officials (all characterized by *Billboard* as "prowling prudes") tried to ban the use of women wearing tights in advertising. Minnesota and Pennsylvania lawmakers introduced bills in 1890–91 to prohibit tights from being worn at public exhibitions.[218] In 1898 the Board of Aldermen in Boston ordered theatrical managers to "tone down their representations of women in tights and skirts."[219] Moreover, that same year, the Mayor of Somerville, Massachusetts, proclaimed: "I am not disposed to make any attack upon the portrayal of the nude, or partially nude human body in proper places and under suitable conditions. . . . But the highly-colored caricatures of the female body, displayed on some of the billboards of this city, they stand for nothing but obscenity and appeal only to prurient tastes."[220] However, these official efforts to corral tights generally failed.

Furthermore, no state law or statute specifically addressed the conduct or dress of circus women at the turn of the century.[221] State agencies occasionally fined the circus for violating state child-labor laws. But the only repercussion a circus might face for its explicit bodily exhibitions was the occasional press afterblast (a critical newspaper editorial after the show's departure).[222] Individual social purity reformers also ignored the circus, even as they simultaneously regulated sexual materials and practices throughout American society. The federal Comstock Act (1873)—often personally enforced by the prudery zealot Anthony Comstock himself—outlawed the circulation through the U.S. mail of pornography, as well as birth control information and devices. Yet Comstock did not protest the circus, even as he harassed dime-museum proprietors and vaudeville owners for their erotic content. In 1887, using his authority as a special agent of the post office, Comstock arrested the well-known New York City art dealer Alfred Knoedler for selling photographic copies of paintings of French nudes, because, according to Comstock, the photographic process had rendered the paintings more prurient.[223] Even P. T. Barnum, whose "Greatest Show on Earth" profited from the exposed female body, indirectly participated in the purity movement. As Bridgeport's representative to the Connecticut legislature, Barnum in 1879 chaired the Joint Committee on Temperance, which approved a bill forbidding all trafficking in "obscene" literature and materials dealing with sex or reproduction, and the "use" of any drug or instrument for the purpose of preventing conception. Although

Barnum personally did not support the bill, he did not block its eventual passage, thus underscoring the circus's paradoxical connection to the world of contemporary purity reform.[224]

Antivice and purity reform movements focused on the relationship between "obscene" amusements and the "declining" morals of vulnerable youth. Nicola Beisel suggests that this "focus on the family" gave the purity reform movement its power among the Euroamerican middle class.[225] But contemporary magazines, children's books, and newspapers depicted the circus's seminude bodies as "wholesome" fun for "children of all ages." These media presented the circus as a fanciful, adventurous part of the childhood imagination. An unidentified missionary, for example, recollected that as a child around 1900, she had three career ambitions: to become a circus performer, a missionary, or Santa Claus.[226] When Ringling Bros. played an extended run at Chicago in 1901, the pastor at the nearby Grace Episcopal Church, the Reverend Ernest M. Stiles, had only one objection to the circus: its planned Sunday night performance, a license for which he opposed: "Of course I could do nothing but refuse. It is one thing to be continually fighting Sunday amusements in general, which are bad enough under any circumstances; it is quite another to have the responsibility thrust upon you of having a circus at your church door. As we had no service at the church on Sunday afternoon I did not object to the circus performance at that time."[227]

Not only did state officials ignore the circus's spectacle of seminudity, they actually condoned it. After an inspection of the wages and working conditions of circus "girls" with the Ringling Bros. circus in 1914, the Factory Department of the State Department of Illinois concluded, "The girls with the circus receive higher wages, perform easier duties and enjoy more wholesome physical and moral surroundings than girls working in Chicago department stores and factories."[228] Consequently, it would seem that the impresarios' elaborate sexual containment strategies were successful.

Still, audience members often rejected such claims of propriety. Some spectators interpreted circus women's scant dress as a sign of sexual availability. Circus workers frequently recorded instances of voyeurism on the show grounds. Harry Webb, a rider with Buffalo Bill's Wild West in 1910, remembered an instance in which William F. Cody himself brawled with a "husky" thirty-year-old whom Cody had caught peeking through a rip in the women's dressing tent. That same season, Webb also witnessed fellow Wild West workers pummeling two war veterans who had been caught sexually assaulting a couple of female audience members.[229] Al Mann recalls that townsmen—usually middle-aged men—felt that they had a right to

Plate 1. *"Wild and Weird Indian Scenes," Adam Forepaugh Shows, 1889.
This "original Wild West exhibition" at the circus demonstrates the
interconnectedness of these two amusements. (Lithograph courtesy of Circus
World Museum, Baraboo, Wis., A4P-NL4-89-1F-4)*

Plate 2. "Madam Yucca," Barnum & Bailey, 1892. Capable of lifting men, horses, and multiple cannon balls, the "champion American female Hercules" undermined notions of neurasthenic Victorian womanhood. (Lithograph courtesy of Circus World Museum, Baraboo, Wis., with permission from Ringling Bros. and Barnum & Bailey,® The Greatest Show on Earth,® B+B-NL4-92-1F-1)

Plate 3. *"Evetta the Only Lady Clown," Barnum & Bailey, 1895. Here posing in clownish, patriotic bloomers in front of an ardent admirer, the English performer Evetta Mathews was also an accomplished acrobat and contortionist. But she was best known during her year with the American circus in 1895 for her clown work in pink fleshlings. (Lithograph courtesy of Circus World Museum, Baraboo, Wis., with permission from Ringling Bros. and Barnum & Bailey,® The Greatest Show on Earth,® B+B-NL44-95-1U-2)*

Plate 4. "Jumbo," Barnum & London, 1882. Although reportedly reluctant to leave England in 1882, Jumbo soon became a profitable feature at Barnum's circus. Killed by a train in 1885, Jumbo still toured—in skeletal form—after his death. (Lithograph courtesy of Circus World Museum, Baraboo, Wis., with permission from Ringling Bros. and Barnum & Bailey,®

Plate 5. "Startling and Sublime Exhibition of Savage Wild Beasts and Domestic Animals," Barnum & Bailey, 1894. Animal trainers showed their "mastery" over gun-toting bears, unfathomable cliques of dogs, tigers, and the biblically evocative teams of lions and lambs lying next to each other. (Lithograph courtesy of Circus World Museum, Baraboo, Wis., with permission from Ringling Bros. and Barnum & Bailey,® The Greatest Show on Earth,® B+B-NL4-94-1F-7)

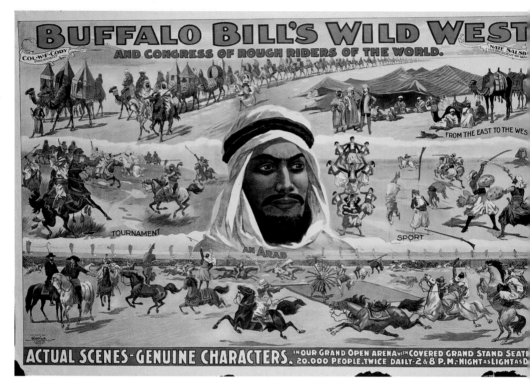

Plate 6. "An Arab," Buffalo Bill's Wild West and Congress of Rough Riders of the World, 1896. From trick riding to circuslike contortionism and acrobatic gun play in the desert, Arab "cowboys" performed as "typical" representatives of their race. Note also that in 1896 the show promised "night light as day" with its brand-new electric spotlighting system. (Lithograph courtesy of Circus World Museum, Baraboo, Wis., BBWW-NL48-96-1F-3)

(opposite)

Plate 7. "1776, Historic Scenes and Battles of the American Revolution," Adam Forepaugh Shows, 1893. This spec highlighted the long roots of American exceptionalism, celebrating the peaceful democratic transfer of power and civic virtue as a model for all nations. (Lithograph courtesy of Circus World Museum, Baraboo, Wis., A4P-NL450-93-2SU-2)

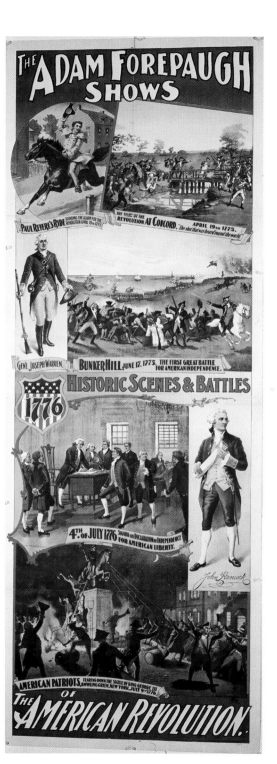

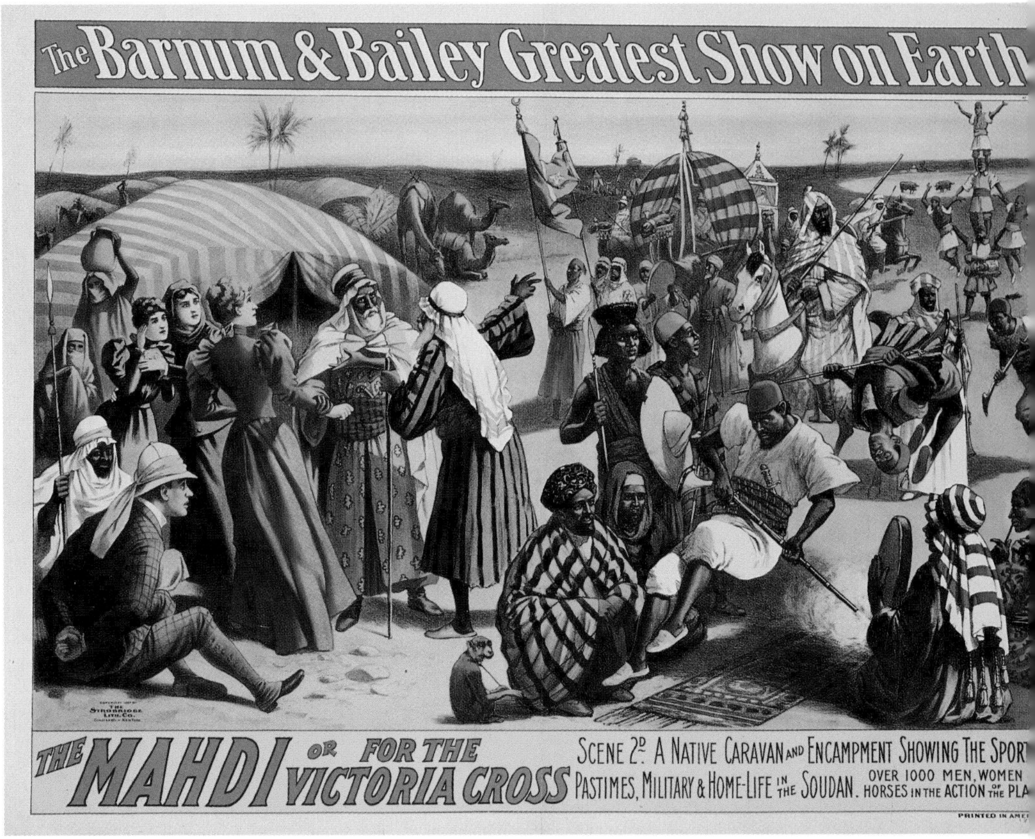

Plate 8. "The Mahdi," Barnum & Bailey, 1898. While abroad, Barnum & Bailey remained au courant with English audiences. At the circus, England's ongoing war in the Sudan became an orientalist spectacle, complete with acrobatic natives, reversed pyramid building, pageantry, veiled women, and terrified white captives. (Lithograph courtesy of Circus World Museum, Baraboo, Wis., with permission from Ringling Bros. and Barnum & Bailey,® The Greatest Show on Earth,® B+B-NL38-98-1F-5)

cut a hole in the women's dressing tent and peek inside. Female performers would alert Mann and the rest of the male workers that a peeping Tom was on the premises. The men would sneak up on the offending voyeur, kick him in the pants while he was crouched over looking through the hole, and then punch him in the face with leather dress gloves as he whirled around in surprise. Mann noted with disgust: "And [the voyeurs] always held out when asked that was their privilege. And if you looked in their window, of course, they'd have you arrested."[230]

In the transient, crowded environment of Circus Day, male spectators found plenty of opportunities for bad behavior. In some respects, the ephemeral circus offered its audiences a world without consequences. Whereas permanent amusements maintained ties to the communities in which they were located, the circus folded up its tents and moved on. The itinerant character of the American circus provides yet another reason why female performances remained unregulated during the Progressive Era. But, as this chapter has demonstrated, the transience of the circus tells only part of the story. The mobile circus was a staging ground in which multiple shifting American attitudes about gender, race, and the female body were negotiated and contested at the turn of the century. Ultimately, the circus's contradictory impulses toward female nudity demonstrate that societal attitudes toward women's growing participation in public life were ambivalent. In contrast, male performers—Euroamerican big-top players in particular—were relatively absent in circus publicity campaigns. Press releases seldom focused on circus men's background, class status, romantic life, or costuming. In contrast to their extensive marketing strategies for restraining scantily clad circus women, impresarios had little need to justify male players' presence in public life. Instead, impresarios emphasized the startling acts themselves. The next chapter will explore these variegated representations of the male circus body.

5
FROM THE KING OF BEASTS TO CLOWNS IN DRAG

For Andy Wildwood, the circus offered a salvation of sorts. Andy, a fictional, turn-of-the-century orphan boy in a popular novel, was unable to conform to the social mores of small-town American life. In school, he was suspended after he performed a double somersault behind the teacher's back, landed on a desk, and crashed to the floor. He accidentally set a farmer's field on fire after his friend dropped a flaming ring through which Andy and a horse had just jumped. Furthermore, his cruel aunt kept him hostage in his room. When a circus passed through town, Farmer Dale (whose field Andy had burned) told him to escape by joining the show: "I tell you I believe circus is born in you, and you can't help it. You don't have much of a life at home. You're not built for hum-drum village life. Get out; grow into something you fancy. No need being a scamp because you're a rover."[1]

Once Andy joined the show, he worked hard to build his body and mind. He quickly rose through the ranks and became a star acrobat, capturing loose animals and thwarting criminals who robbed the show. In a Dickensian twist at the novel's conclusion, Andy discovers that his deceased father has named him sole heir to the proceeds from a patent worth a fortune. The story of Andy Wildwood contained themes that were common to the ubiquitous genre of circus-boy fiction at the turn of the century: a poor, mistreated, and misunderstood orphan boy realizes his true potential in the exciting, nonjudgmental world of the circus. He rises from a common laborer to a star acrobat or manager, saves the show from violent weather, and easily captures loose animals and con artists.[2]

Proprietors and press agents also marketed the circus and Wild West to young boys as a site of educational adventure. Press releases constantly reported how small boys sneaked into the show, or how they received free tickets in exchange for hard work: "Small Boy Schemes: Seeks Circus Freaks"; "Heaven for Small Boys: In Madison Square Garden Gallery, They Find Perfect Joy"; "Elixer of Youth Provided by Circus Coming: Everybody Going to See the Big Show While Youth of City Rush to Water Elephants for Free Entry."[3] Edwin Norwood's *The Circus Menagerie: True Stories of Interesting Animals Told to a Boy* chronicled a boy's travels with the Ringling Bros. and Barnum & Bailey circus.[4]

Although Andy Wildwood learned normative manly values at the circus—thrift, honesty, discipline, and physical fortitude—he joined the show community because he was a social outsider. At actual (rather than fictional) circuses, male gender norms were exemplified by workers who were often on the edge—nomadic members of American society who drifted, indeed "ran away" to the circus from sedentary community affiliations. In his hometown, Sherwood Anderson recalled that the male members of the local Thompson family took off every summer with the circus. "The Thompsons were a tough lot. . . . When they were at home the Thompsons, father and sons, hung around the saloons and bragged . . . [they] . . . didn't stand so well in town."[5] Male circus workers as a whole were often more liminal than female employees. Women were commonly born into the business as members of established family troupes, while transient men filled the laboring ranks at the canvas city. Nonetheless, as the last two chapters have shown, the world of the circus was far more hierarchical than Andy Wildwood's fictionalized experiences of limitless mobility would suggest.

Proprietors promoted their exhibitions as sites of athletic Euroamerican manliness. Animals and men worked harmoniously together and nonwhite men worked as "missing links," athletic ascetics, or royal "savages." But as this chapter contends, the world of the circus was one of male gender flux, with androgynous acrobats, gender-bending clowns, players in drag, and animals dressed as men. Spontaneous brawls among spectators and workingmen extended these variegated masculine performances to the grounds outside the ring. As a whole, the human menagerie was a place for performative male gender play, even though the circus's deeply entrenched caste system circumscribed occupational and economic advancement within the nomadic community.

MALE GENDER AND MODERNITY

Fin-de-siècle circus performances of male gender reflected the era's constant change. Gail Bederman argues that the rise of an industrial corporate economy driven by wage labor, coupled with cyclical depressions, threatened the position of the "self-made," independent entrepreneur. Women's activism for the suffrage and their growing rejection of the ideology of separate spheres challenged the gender differentiation on which notions of "civilized" Victorian manliness were based.[6] Thus, Bederman's work adds a gendered dimension to Richard Hofstadter's characterization of "status anxiety" which speaks to a causal relationship between late-nineteenth-century socioeconomic upheaval and middle-class participation in reform movements.[7]

These men also sought to reclaim their authority by fortifying the body: they participated in alternative models of male power such as basketball and wrestling, or embraced "primitive" cultural practices, specifically living in the wilderness or hunting wild game.[8] Advocates of outdoor activity, exercise, organized sports, hunting, and adventure (collectively called the "cult of the strenuous life") asserted that modern industrial life had made men of the middle and upper classes "soft." George M. Beard's *American Nervousness* (1881) and other popular classics warned that professional, white-collar "brain workers" were fast becoming effeminate and impotent because they did not engage in physical labor. Social critics extended their gendered critique of industrial modernization to working-class men, too. Recalling his own experiences working at a bicycle factory in the 1890s, Sherwood Anderson bemoaned how the assembly line separated the worker from the fruits of his labor: "[W]hen you take from man the cunning of the hand, the opportunity to constantly create new forms in materials, you make him impotent. His maleness slips imperceptibly from him and he can no longer give himself in love, either to work or to women. 'Standardization! Standardization! Standardization!' was to be the cry of my age and all standardization is necessarily a standardization in impotence."[9]

The school represented one site of manly salvation. New, rapidly expanding secondary and intercollegiate athletic programs focused on the importance of "sportsmanship" in cultivating athletes of manly character. G. Stanley Hall argued that gymnastics should be part of American boys' educational curriculum, to counteract the "degenerative" influence of the city: "It is because the brain is developed, while the muscles are allowed to grow flabby and atrophied, that the deplored chasm between knowing and doing is so often fatal to the practical effectiveness of mental and moral culture. The great increase of city and sedentary life has been far too sudden for the human body—which was developed by hunting, war, agriculture, and manifold industries now given over to steam and machinery—to adapt itself healthfully or naturally to its new environment."[10] In a series of essays and speeches written in 1900, collectively published as *The Strenuous Life*, Theodore Roosevelt claimed that boys needed to balance their school work with physical exercise to become productive citizens. "I believe that those boys who take part in rough, hard play outside of school will not find any need for horse-play in school. While they study they should study just as hard as they play foot-ball in a match game. It is wise to obey the homely old adage, 'Work while you work; play while you play.'"[11]

In some respects, this link between athletic activity and manly character was hardly new. In eighteenth-century Europe, neoclassical intellectuals and artists wedded the well-formed male body to superlative manliness.[12] During the post–Civil War era, evangelical adherents of "muscular Christianity" bonded physical fitness to moral virtue. Yet at the turn of the century, physical fitness advocates now tied male athleticism to critiques of modernization and to scientific racism. The self-styled fitness expert Bernarr Macfadden declared that male potency was the product of vigorous exercise, fresh air, a bland diet, and frequent marital sexual intercourse. A eugenicist, Macfadden asserted that native-born Euroamerican men needed to build their bodies in order to produce large families and "save" themselves from "race suicide."[13] In his aptly titled *The Virile Powers of Superb Manhood*, Macfadden warned: "Lose your sexual power, lose the power to reproduce your species, and, according to the laws of nature, your days of usefulness are past, and decay and death will soon overtake you."[14] In the age of the New Woman, extraordinary immigration rates from eastern and southern Europe, and the stirrings of the Great Migration of African Americans from the rural South to urban centers, this notion of "race suicide" encapsulated contemporary native-born male anxieties.

Outing clubs, summer camps, the Sierra Club, and other "back-to-nature" organizations focused on male bodily fortification. Members of the Appalachian Mountain Club (formed in 1876) marked and maintained over 200 hiking trails, and constructed stone huts and log shelters in the White Mountains. They purchased land to be held in trust for public use throughout New England. One writer waxed, "For all it has meant an opportunity to come in closer contact with the primitive."[15] In the late nineteenth century, the dime novel publishing house of Beadle and Adams sold thousands of copies of biographies of rugged Euroamericans like Daniel Boone, Davy Crockett, and Kit Carson. Alexander Saxton observes that such men were popular subjects because they had been mythologized into "natural aristocrats" who achieved "mastery" over the wilderness by hunting animals and fighting Native Americans.[16] Youthful counterparts to the back-to-nature movement like the Boy Scouts of America, the Sons of Daniel Boone, and the Woodcraft Indians, enabled white boys to assume a temporary nonwhite identity as they dressed up as Native Americans and learned indigenous crafts and camping and survival skills. These organizations paradoxically helped heighten boys' own sense of manly whiteness through the act of what Philip Deloria succinctly calls "playing Indian."[17]

The preservationist John Muir wrote widely about the wilderness as a site of bodily fortification and epiphany. From the late 1860s onward, he often lived an ascetic life, quietly staging his own remarkable death-defying bodily feats: he walked across much of the United States, fasted constantly, and climbed the snow-covered Sierras clad in a woolen shirt, denim pants, and thin shoes. He was jubilant about the natural world and wrote of the surf pounding Cuba's shores as "one great song sounding forever all around the white-blooming shores of the world."[18] Muir argued that the preservation of wild spaces was critical to human health. In addition to writing popular books,[19] he lobbied Congress to create national parks such as Yosemite (1890) and pushed to establish the U.S. Forest Service (1891). He helped found the Sierra Club in 1892 and from 1908 to 1913 unsuccessfully fought a dam project in California's Hetch Hetchy Valley. In his writing and public activism, Muir personified widespread public ambivalence about modernization and urban encroachment.

At the same time, hunters helped lead the conservation movement. Elite eastern men, including Ernest Thompson Seton, a founder of the Boy Scouts in the United States, posited that wild spaces should be set aside for the preservation of bear and bison. Theodore Roosevelt in 1887 became the first president of the Boone & Crockett Club, which advocated the study and preservation of wild animals and their natural habitat. The club's Committee on Parks helped create the National Zoo in Washington. Although

Roosevelt still loved the pleasures of blood sports, he claimed that his primary interest in killing animals was to advance scientific knowledge. His efforts prompted federal legislation to save Yellowstone Park from ecological destruction, to protect sequoia groves in California, and to establish an Alaskan island preserve for the propagation of native species.[20] As president of the United States, Roosevelt added 150 million acres of forest to the national forest system, and established five new national parks and sixteen national monuments.[21]

At the turn of the century, the "primitive" pleasures of hunting large male game were an essential part of masculine renewal. When Roosevelt first entered the New York State legislature in 1882, fellow legislators and the press mocked his privileged background, fancy clothes, and high voice; they called him "weakling," "Jane-Dandy," "Punkin-Lily," and "the exquisite Mr. Roosevelt." But as the historians Edmund Morris and Gail Bederman point out, Roosevelt exploited his experiences of hunting bison and ranching in the Dakotas to transform his public persona into that of a rugged he-man.[22] Fleeing to Dakota Territory in 1884, grief-stricken after his wife and mother had died on the same day, Roosevelt chronicled his transformation from puny asthmatic to vigorous hunter when he shot a huge bison bull: "There below me, not fifty yards off, was a great bison bull. He was walking along, grazing as he walked. His glossy fall coat was in fine trim and shone in the rays of the sun, while his pride of bearing showed him to be in the lusty vigor of his prime. As I rose above the crest of the hill, he held up his head and cocked his tail to the air. Before he could go off, I put the bullet in behind his shoulder. The wound was an almost immediately fatal one."[23] Overjoyed with his prize, Roosevelt danced around the large carcass, shrieking "like an Indian war-chief."[24] Roosevelt, who later became the honorary president of the American Bison Society, construed the hunt as a "primitive" pleasure which revived virile potency in "enlightened" modern men like himself.

Contemporary popular culture also articulated the notion that the wilderness was a metaphor for male renewal in modern society. Jack London's *The Call of the Wild* (1903) conflated animals, the wilderness, and masculine regeneration. Buck, the canine protagonist, was kidnapped from a leisurely, gentlemanly life on a ranch in Santa Clara Valley to the harsh Klondike wilderness, where men needed sled dogs during the gold rush of the 1890s. His muscles became hardened from constant travel over the icy tundra and his disposition was made tough by fighting for scraps of food with other dogs. Forever removed from the "soft" pleasures of civilization, Buck reached the apex of his masculine flowering once he methodically stalked, killed, and ate a bull moose. When his owner was killed by a band of Native Americans, Buck took to the wilderness and joined a wolf pack for the rest of his life.[25] "[I]nstincts long dead became alive again. The domesticated generations fell from him. In vague ways he remembered back to the youth of the breed, to the time the wild dogs ranged in packs through the primeval forest and killed their meat as they ran it down. [His ancestors] came to him without effort or discovery, as though they had been his always. And when, on the still cold nights, he pointed his nose at a star and howled long and wolf-like, it was his ancestors, dead and dust, pointing nose at star and howling through the centuries and through him."[26] In Edgar Rice Burroughs's hugely successful novel *Tarzan of the Apes* (1912) the Englishman Lord Greystoke (Tarzan) was an archetype of "primitive" physicality because he was raised by apes in the "primordial" African jungle.[27]

Popular novels, advocates of the strenuous life, conservationists, and hunters created idealized relationships between male potency, wild animals, and open land at a historical moment when Americans wrote widely about the imminent "loss" of the wilderness frontier. Andrew Isenberg observes that this conservationist vision was uneven, particularly with respect to bison preservation: "The advocates of the preservation of the bison supported both the Euroamerican conquest of the western grasslands and the preservation of the dominant species of the preconquest plains. These contradictory ideals exemplified the dual vision of the North American frontier at the turn of the century: as a progression toward the modern age and as a refuge from modernism."[28] The turn-of-the-century railroad circus articulated this same sort of ambivalence. With trains, flying automobiles, disciplined "industrialized" workers, and confident animal trainers who wielded power over "wild" animals, the railroad circus celebrated modernization. At the same time, showmen publicly mourned urban encroachment, massive immigration, and the imminent loss of the frontier; as such, they marketed the railroad circus as a place where audiences might catch a "last glimpse" at the world's vanishing animals and preindustrial people.

DANGEROUS ANIMALS

In a nascent celebrity culture, impresarios consciously chose adult animal males—Jumbo the elephant, Chiko the "gorilla," and Rajah the Man-eating tiger—to become stars. A circus animal's public profile depended on its connection to the "masculine" wilderness: the more dangerous and distant the wilderness, the more visible the animal. Domestic animals also were an integral part of many circus and Wild West acts, but the proximity of the dog, the duck, the pig, and the goose to the home meant that they never received the same degree of press attention as their more dangerous brethren.

Contemporary scientists wrote widely about the large male animal. In *Tales from Nature's Wonderland*, the naturalist William T. Hornaday idealized the world of an ancient male mammoth who fell into the La Brea tar pits. Hornaday speculated that "Old Ganesa" (named after the elephant-headed Hindu god) and his consort "Constance" became trapped in the "Tar Terror" while trying to rescue their thirsty son "Esau," who thought that the shimmering tarry ooze was water.[29] Despite the matrilineal structure of elephant societies, Hornaday imagined the mammoths living in patriarchal nuclear families: "At [Ganesa's] heels, blindly and obediently following his lead, marched his consort Constance, who during sixty years of good and evil had faithfully followed him all the way from Mount St. Elias to the final Land of No Rain."[30] Old Ganesa, whose giant skeleton stood thirteen feet tall, was a model patriarch.

Although showmen and audiences both imagined ferocious animals as untamed representatives of a vanishing wilderness, the animals had been trained, in most cases, to replicate human movement and behavior. Regarding a twentieth-century elephant pyramid act, the anthropologist Yoram S. Carmeli argues that "these [elephant] bodies are seen as surfaced, as emptied images of real elephants," because the stunt always framed its elephant subjects in human terms. The elephants did not perform as "natural" elephants but rather as "human" elephants because of their posture and bodily configurations.[31] In 1888 the Adam Forepaugh circus introduced "John L. Sullivan, the Boxing Elephant," whose poses mimicked the famous heavyweight champion.[32] Circus animals also mocked nineteenth-century notions of bodily restraint; as they imitated human postures, they behaved unpredictably, sneezing, belching, farting, and defecating without warning, thus mirroring the human body in its most natural, yet least socially regulated form.[33]

Emptied of any connection to the wilderness, tractable animals, including the "porcine prodigy," the rooster, and the rhesus monkey, were supposed to be funny. They played silly, human figures, able to walk, ride horses, play musical instruments, and kick a ball. In an educated animal act at Barnum & Bailey's circus, the Learned Pig thoughtfully picked out the letters P-I-G, stood on his hind legs, and "bowed low and carefully." Later, he and a "pig friend" played cards, "both keeping their tempers perfectly."

Meanwhile, "[d]ogs played dominos with equal politeness." Ponies walked on their hind legs, carrying schoolbooks, and later executed a precise military drill. A trained rat climbed a pole and raised the American flag.[34]

But disorder beckoned. Bears and elephants entered dancing and drinking from bottles, while two goats played seesaw, one continually bouncing the other off the end of the board, "after the manners of naughty little boys." The domino-playing dogs suddenly seemed to drop dead, only to be revived into a dancing frenzy by a monkey playing a fiddle.[35] In a lithograph from the same circus program, an eager monkey pupil wore a sign reading "Pinch Me" on its tail while a crying pig wore a dunce cap (fig. 28). The literary critics Peter Stallybrass and Allon White write that hogs occupied an uneasy position in urbanizing societies because their fertile dung, once so useful in the rural world, was now a major source of pollution. The pig's appearance and habits were eerily human: their pink skin resembled European pigmentation, they ate a similarly omnivorous diet, and they lived in close proximity to human beings, as they rooted around for refuse near farmhouses and in fetid city streets.[36] (Indeed, the Maori and Polynesian term "long pig" refers to human flesh as food for cannibals.)[37] In 1842 the *New York Daily Tribune* counted some ten thousand pigs ranging in the city. Susan Strasser points out that these urban hogs ate such great quantities of trash, and provided such a good food source for the impoverished, that attempts to remove them were met with organized resistance. Even into the late nineteenth century, after local sanitation laws had banished pigs from many streets, small pigsties were allowed in New York City, where some tenement residents still kept pigs in their basements and apartments.[38]

Circus acts revealed this uneasy, liminal proximity of pigs to human beings. In addition to performing as an "educated" porcine or naughty schoolboy, the circus pig played a human baby in front of unsuspecting audiences. In one enduring act, a clown tenderly nursed a fully swathed "baby" with a bottle. The gentle scene suddenly ended when the baby, now squealing, wriggled out of its swaddling blankets, urinated all over the clown, and promptly revealed its true identity.[39] Such staged encounters perhaps had even greater resonance in an age when the spatial bifurcation of human beings and animals rendered by trains, cars, electric trolleys, bicycles, and other artifacts of modernity was seemingly growing at breakneck pace. Even as early animal welfare activists in the Gilded Age affirmed the shared sentient nature of human and animal consciousness by protesting acts of cruelty toward animals, they sought to compartmentalize human-animal encounters in the urban environment.[40] Their efforts to retire exhausted

Figure 28. "Sweet Bye & Bye," P. T. Barnum's Circus Museum and Menagerie, 1887. Walking upright, riding bicycles, singing, and studying, these mostly dressed animals confounded the boundaries between human and animal. (Interior program illustration courtesy of Buffalo Bill Historical Center, Cody, Wyo., MS6, Davidson.10)

carriage horses and chafed, bleeding cart dogs, and their push to ameliorate abominable conditions for cows and pigs traveling in unventilated railroad cars, were part of this larger project.[41] As such, they attempted to rid the streets of strays, sickly, and dying animals by establishing new humane societies that practiced a "noiseless process" of euthanasia, thus removing the troubling specter of visible suffering from the streets.[42] They also fought to shut down unsanitary dairies and miasmal urban slaughterhouses.[43] Edward Bellamy described a futuristic utopia set in the year 2000 wrought by centralization and mechanization in his best-selling novel *Looking Backward* (1888): "Ceasing to be predatory in their habits, [people] became co-workers, and found in fraternity, at once, the science of wealth and happiness." Likewise, animal welfare activists such as the New York shipping heir Henry Bergh (founder of the American Society for the Prevention of Cruelty to Animals in 1866) imagined a time when technology would render animal labor, and hence animal cruelty, obsolete.[44] The humane movement was therefore predicated upon a decoupling of older, everyday, utilitarian human-animal relationships, a curious "dehumanizing" of sorts, as activists fought to remove animals from the public sphere. But the pervasive presence of wandering dogs, hogs, and chickens on city streets at the turn of the century spoke to the continued presence of an older preindustrial order. The circus, with humanized animals and animalized humans, highlighted this ambiguity of modern people's position within the natural world.

The popular images of various "human" animals have their intellectual underpinnings in the Enlightenment. During this period, artists, poets, and naturalists began to speculate about the common origins of human beings and other animals. The seventeenth-century French artist Charles LeBrun suggested that painters could learn to depict human emotions better by studying animals.[45] Seventeenth-century anatomists discovered that animal physiology was remarkably close to that of humans.[46] Voltaire and David Hume departed from Descartes's assumption that animals were simply living machines by arguing that animals, like humans, were sentient beings.[47] The theory of evolution, corroborated by anatomical science, cemented the physiological link between the human and the animal. Although Charles Darwin's *Origin of Species* (1859) made no mention of humanity's position in the natural order, his later works, *The Descent of Man, and Selection in Relation to Sex* (1871) and *The Expression of the Emotions in Man and Animals* (1872), asserted that humans had evolved directly from primates.

In the 1890s showmen exhibited racial difference in explicit, evolutionary terms. A new act, the "Race of the Races," pitted various human "races" against each other in footraces, and on horseback on the hippodrome track. In 1891 Adam Forepaugh's circus included an "Indian Foot Race, a genuine contest for supremacy between the Brule and Assinnibone Indians."[48] In Buffalo Bill's quarter-mile "Race of the Races" (1896), an international spectrum of "cowboys" competed against each other: a Cossack, a Mexican, an Arab, and a Gaucho, all riding horses of different "races."[49] A footrace in 1900 featured women of different races and places like the newly annexed Philippines, thereby mirroring the political rhetoric of Theodore Roosevelt and Alfred Beveridge, who applied Darwin's theory of natural selection to human society as a "scientific" justification for war and empire building.[50]

As a popular forum on evolutionary theory, the circus also playfully destabilized the distinction between elite intellectual culture and mass entertainment by transmogrifying science into burlesque. A Ringling Bros. route book from 1894 contained an engraving of a grinning, gap-toothed monkey, entitled, "Man, Previous to His Degeneration." A circus program cartoon, "Johanna's Soliloquy," depicted Johanna the chimpanzee atop a pile of books, one clearly labeled "Darwin," with a finger in her mouth while pondering a human skull. In 1909 a newspaper article, bearing the headline"Circus Almost Looses Missing Link Again," described how one of the Ringling Bros. monkeys, "Darwin," was rescued from a fire in his cage, taken to a luxury hotel, and then transported to Madison Square Garden in a taxi: "Darwin looked so much like a human being that the cab man wanted to charge him the regular fare."[51]

The nineteenth-century animal welfare movement proliferated in this intellectual environment, in which the relationship between human beings and animals was under heated reconsideration. Because evolutionary theory linked human beings and animals to shared biological origins, some circus audience members saw the caged animal as an enslaved human being. Consequently, animal acts remained virtually the only aspect of the circus that received consistent public protest in the late nineteenth century and the early twentieth.[52] Animal welfare organizations such as local Jack London Clubs contended that captive animals stood as a symbol of "threatened" manhood which had been newly imprisoned by urbanization, technology, bureaucratic institutions, and shrinking wilderness. Just as men's site of masculine renewal, the frontier, had been pronounced "closed" by the U.S. Census Bureau in 1890, circus animals were forcibly taken from a vanishing state of nature. Animal welfare organizations charged that wild animals' "natural" dignity was violated because they were made to dance in pink tutus and pose on their haunches, among other seemingly ridiculous human postures.[53] In an attempt to thwart such criticism, Buffalo Bill's Wild West claimed that its horses were "natural": "there is absolutely no artificiality in the entire programme—the horses are as they were intended to be, and the men ride as horses should be ridden. They are not trained to act before the limelight, neither are any of them subjected to tortures in order to serve the whims of their masters."[54]

Some defenders of the circus and Wild West show argued that animal labor was no different than human labor. In contrast to animal welfare advocates, William T. Hornaday asserted that labor for all creatures was critical for survival:

[T]here is no sound reasoning or logic assuming that the persons of animals, tame or wild, are any more sacred than those of men, women and children. We hold that it is no more "cruelty" for an ape or a dog to work in training quarters or on the stage than it is for men, women and young people to work as acrobats, or actors, or to engage in honest toil eight hours per day. Who gave to any warm-blooded animal that consumes food and requires shelter the right to live without work? *No one.* I am sure that no trained bear of my acquaintance ever had to work as hard for his food and shelter as does the average bear out in the wilds. . . . Now has he anything on the performing bear? Decidedly not.[55]

Despite accusations from animal welfare advocates decrying unnatural circus animal labor, the beasts offered audience members opportunities for imaginative rejuvenation: spectators could feel the rumble of a herd of elephants thundering through the big top, or smell the dung of camels and polar bears; one could actually see a tiger's muscles ripple as it jumped through a flaming hoop. With this physical proximity to potentially violent animals came the exciting specter of accidental animal escapes when the big top was blown down by a storm, or when a deadly railway accident forced cars packed with animals to lurch off the tracks. During the 1890s the Walter L. Main circus published a pamphlet which detailed its fatal railroad wreck near Tyrone, Pennsylvania, on May 30, 1893. At least five men and forty-nine horses died, and a panther, lion, tiger, zebra, yak, hyena, and scores of monkeys and elephants escaped into the countryside. The pamphlet vividly described how the loose tiger shattered the bucolic peace: "Then the untamed monster started out in the country looking for new fields. He came to the farmyard of Alfred Thomas, where a woman was milking a cow. The woman left suddenly and the tiger sprang upon the cow and killed her. He was devouring his quivering meal when the farmer appeared with his rifle and shot the tiger. Pleased with his royal sport, Farmer Thomas shouldered his rifle and started in pursuit of a panther that he knew was cavorting on the mountainside." [56]

Although such opportunities for hunting circus animals on the loose were rare, the circus offered its audience members the imagined pleasures of the hunt. The boy hunter at the circus was a common character in children's circus fiction. Plucky, daring orphan boys single-handedly wrangled a loose circus lion or bear before it destroyed a community.[57] P. T. Barnum wrote several children's books featuring boy protagonists who tracked, captured, or killed wild beasts. He dedicated one of his earliest efforts, *Lion Jack* (1876), to "the many boys of America, who have gazed with round-eyed wonder and admiration at the wild beasts which, for their amusement and instruction, I have gathered together from all parts of the world into my menageries[.] I dedicate this story of a good and brave American boy, who fought with lions in their lairs and other wild animals in African jungles and Asiatic deserts, and gained much glory and wealth."[58]

Show programs and published manuscripts of animal agents described harrowing accounts of capturing wild animals in distant lands, an activity made all the more dangerous because animals were valuable to the circus only if captured alive.[59] The animal dealer Charles Mayer remembered that when he captured sixty elephants running amok on a Malay sultan's territory, three Malay men—who remained nameless in his account—were killed in the process.[60] As part of their quest to showcase male power, proprietors consciously purchased "superlative" animal specimens that were the largest, fastest, strongest, fiercest exemplars of their "race."[61] An animal's size magnified its appeal and danger, for the animal might injure its handler, or it might escape, causing pandemonium in a peaceful town. Strength, musculature, size, and ferocity were all signs of superlative animal manhood. In the winter of 1881–82, P. T. Barnum created a public uproar in England when he purchased Jumbo, a towering African elephant living at the London Zoo. When Jumbo initially refused to board the ship on which he was to be taken to the United States, the British press wrote of Jumbo's "desire" to remain in London, consequently sparking an unsuccessful nationwide letter-writing campaign for Jumbo to stay. Captured as a baby by Arab hunters in Abyssinia in 1861, Jumbo was still growing when he reached the United States in 1882 (plate 4).

Jumbo, the "Lord of All the Beasts," was portrayed as a model of manly kindness who—despite his power and potential fury—allowed children to ride him. Show programs chronicled his altruistic character, especially after his sudden death in 1885. Just hours after a performance at St. Thomas, Ontario, Jumbo sacrificed his own life when he allegedly threw his best friend, Tom Thumb, a dwarf clown elephant, and his keeper, Mathew Scott, out of the path of an oncoming train. Jumbo was crushed and dead within minutes. P. T. Barnum recounted Jumbo's bravery in a melodramatic children's story in which he took three small children (Tom, Trixie, and Gay) to his circus. "'Who was Jumbo?' asked Trixie. 'Oh, a tremendous elephant, as big as six of these rolled into one! He went to Canada, and there a locomotive smashed into his brain, and he turned over and died. But first he wrapped his trunk around the baby elephant and flung him safe off the track,' [said Tom]. 'Good Jumbo!' said Gay with a smile; but there were tears in Trixie's eyes. 'Yes, baby; and that's the way we would jump for you in any danger,' added Tom."[62]

Depicting the fatal collision between Jumbo and the train as the inevitable triumph of industrial technology over nature, program engravings portrayed the elephant as many times larger than the train: "The leviathan of the rail and the mountain of bone and brawn came together with a crash that made the solid road-bed quake. The heavy iron bars of the engine's pilot were broken and twisted as if they had been but grape vines."[63] After hiring the scientists Henry A. Ward and Carl Akeley to prepare the elephant's corpse, Barnum profitably exhibited Jumbo's massive hide and "majestic" skeleton in the name of scientific uplift along his transcontinental route.[64]

Perhaps best known for his "marriage" to Johanna the ersatz gorilla in 1893, Chiko the chimpanzee was marketed as insatiable, athletic, and racially "black." One scientist, R. L. Garner, stated that Chiko was "several inches taller than the best" he had ever seen and the "finest specimen" of his "race"—although official scientific opinion was divided regarding Chiko's actual "race": was he a chimpanzee, a "black" orangutan, a gorilla, or of mixed descent? Or might he be human? One thing is certain in hindsight: Chiko was no gorilla, because circuses did not acquire genuine gorillas until 1921.[65] Zoologists noted that Chiko's thumbs—unlike those of other apes—were virtually human, and that he could probably learn to play the piano and bass.[66] Garner tried to prompt Chiko to speak in order to record his voice. Chiko, "a perfect gentleman unless otherwise provoked," roared in response.[67] Agents further identified Chiko as a "missing link" by conjuring racially charged stories about his political "career" in Africa before capture. They based these tales upon the observations of a fictitious traveler in the Congo who noted that Chiko had been an alderman in the wild. The traveler also reported that Chiko possessed prodigious strength and appetite, and could eat dozens of apples followed by a swallow of coffee. Press agents also noted impishly that Chiko's high social rank in the Congo made him a "society favorite" in the United States.[68]

Speculation regarding Chiko's racial "purity" became especially intense when Chiko was "married" to Johanna—whose "race" was also in question. The wedded chimps were a metaphor for contemporary discourses about race and the permissible bounds of racial "mixing," given the frequent references to couple's plans to procreate.[69] After Chiko's untimely death in 1894 (from "a surfeit of apples"), he was stuffed and displayed at the American Museum of Natural History. Posed in a standing position with his "strong" right arm extended, Chiko wore a wistful expression. His bones had been replaced by wood, and his skeleton was on exhibit nearby.[70]

Because Chiko was the "finest specimen of his race," showmen promoted him as an archetypal chimpanzee (or gorilla or orangutan, depending upon one's scientific opinion). Donna Haraway observes that turn-of-the-century zoologists commonly hunted for adult male animals as "typical" representatives of their species.[71] As such, Chiko symbolized virile manhood, capable of bending the bars of his cage and terrorizing his keeper. Advertising Chiko as an African "statesman," circus proprietors endowed the chimp with the same physical qualities that contemporary Euroamerican racial theorists bestowed upon black men. One lithograph from 1893 (the same year that Chiko was added to the menagerie) depicts an unnamed gorilla, incorrectly listed as an orangutan, walking upright, wielding a rock in one hand, and holding a frightened white woman under his other arm: "Just secured and now added. A Giant Black Orang . . . more closely resembling man than any other creature known to exist using knives, forks, cups and other articles in precisely the same manner as a human being. The veritable missing link" (fig. 29). Chiko, like Johanna, performed next to Africans at the Ethnological Congress. Haraway argues that in death, animals could be manipulated, through the process of taxidermy, into imagined natural perfection, offering museum and circus patrons "a peephole into the jungle."[72] In addition, she postulates that "taxidermy fulfills the fatal desire to represent, to be whole; it is a politics of reproduction."[73] At museums and circuses, showmen and taxidermists "reproduced" dioramas of dead animals to fashion an idealized natural world where males were physically superior to females, even though androgynous living animals like Chiko's strong "wife" Johanna confounded such normative notions.

Early-twentieth-century animal photographers also used the dangerous

*Figure 29. "A Giant Black Orang," Barnum & Bailey, 1893. Billed as an orangutan,
a gorilla, and correctly as a chimpanzee, this circus primate used silverware, sat in a
rocker to the astonishment of the crowd, got "married," and hauled off a white woman in a
racialized depiction of a humanized animal. (Lithograph courtesy of Circus World
Museum, Baraboo, Wis., with permission from Ringling Bros. and Barnum & Bailey,®
The Greatest Show on Earth,® B+B-NL38-93-25)*

large male as their primary "representative" subject (just as deer hunters
have always prized the buck with the biggest antler spread, using standards
created by the Boone and Crockett Club). Martin and Osa Johnson, wildlife
photographers and filmmakers from Kansas who chronicled "wild" human
and animal subjects from Africa and Asia in their memoirs and in circus
advertisements, described a lone male African rhinoceros that was ready to
charge: "Then one day we saw a beautiful specimen, perfectly posed, with
both background and lighting exactly right for a picture. His Roman nose
and splendid horns were clearly outlined, his heavy shoulders and muscles

rippled magnificently in the sun. Martin set up his camera and turned it over to me; at a signal I was to start grinding." [74] The circus, like the wildlife photographer, captured and commodified the vanishing animal wilderness as a portable, snapshot spectacle.

ANIMAL TRAINERS

Armed only with a thin chair, a whip, a starter pistol, and sugar or slivers of meat, the animal trainer calmly commanded cats, giant Asian and African quadrupeds, and pachyderms to dance in formation, leap through flaming hoops, or ride atop one another in unfathomable configurations of predator and prey. Trainers and press agents actively promoted their use of a recent late-nineteenth-century training method called the "kindness method" in order to rebut accusations of animal cruelty. [75] As its name implies, the kindness method used a reward system in which the animal was given treats and praise for learning new skills; unruly behavior in cats, for instance, was usually punished with little more than a short rap to the nose. Circus proprietors and trainers reasoned that beating an animal made little sense because animals were an expensive investment which yielded high returns only with good treatment. They asserted that circus animals were healthier and lived longer than zoo animals. [76] In his autobiography, Carl Hagenbeck took credit for developing the kindness method when he began training animals for circuses in the mid-1880s. [77] Other famous trainers, including Frank Bostock and Mabel Stark, also advocated kindness principles. Katherine Grier connects the rhetoric of "kindness" to the growing cultural authority of the American middle class. In the social context of the "kind" republic in the early nineteenth century, these Americans began to idealize a "domestic ethic of kindness" toward animals. Currier and Ives lithographs and sentimental novels portrayed animals as loyal guardians who formed monogamous, self-contained families comparable to those of their owners. [78]

At the beginning of the twentieth century, the domestic focus of the kindness method ideology also complemented contemporary attitudes about the "white man's burden." Trainers following kindness principles prided themselves for being reserved, stoic, and infinitely patient with their unpredictable charges. Dressed in paramilitary garb, such trainers saw themselves as stern father figures to their unruly animal "children." Trainers likened animals from tropical zones to people of color from nonindustrial societies over which Europe and the United States held financial, military, and strategic control. Frank Bostock, a member of an illustrious English family of animal trainers, referred to his circus animals as "untamed men and women" and contended: "The training of my dumb companions is never cruel—less so . . . than the firmness exercised occasionally in the correction of an evilly disposed child."[79] Similarly, Carl Hagenbeck spoke of people from countries where he captured untamed animals as "uncivilized . . . no less wild than the beasts" with which they worked. This juxtaposition of the human and animal made the trope of the white man's burden visually complete, as people of color and beasts were "trained" together for profit and ostensible edification.[80]

Although performances with animals highlighted the biological interconnectedness of the animal and the human, trainers took pride in their "mastery" over beasts.[81] As models of disciplined manliness, animal trainers argued that absolute sobriety was an essential part of their craft.[82] Bostock emphasized that the trainer had to be on the job around the clock. Carnivores, for one, had to be trained at night, because these nocturnal predators were dull and lazy during the day. Bostock noted that successful trainers— even when injured—were always calm, adding that the first principle of training a wild animal was "never let an animal know his [own] power."[83] Bostock added that a trainer must never lie down when working with an animal, that only the upright trainer was master. On the ground he became fair game for attack.[84]

Showmen juxtaposed the trainer's detached calm with the omnipresence of animal dangers. In an interview for the *Detroit Free Press* in 1890, George Conklin, a lion tamer, nonchalantly introduced the reporter to the menagerie: "These are the man-eating tigers of India. . . . They are full grown and in the prime of life. The one with a chain on his neck has lately killed his man. . . . Yes, indeed. These are the genuine killers. Do not go too near the cages. They have a fearfully long and sudden reach."[85] Newspaper articles focused upon the male trainer's stoicism when he was injured in front of an audience. Still, women and men both obeyed the circus's ethos of "the show must go on," regardless of the severity of one's wounds. When Jack Bonavita was mauled by a group of lions in Indianapolis in 1900, he coolly shoved the handle of his whip down one cat's throat and shouted commands at the others, and all continued their tricks. The audience roared approvingly, but after four stunts Bonavita bowed gracefully, staggered off stage, and was whisked off to the hospital. The journalist Cleveland Moffett noted that "Bonavita's steady nerve saved him."[86]

Whereas the circus cultivated the erotic image of women animal trainers, male trainers were marketed as models of manly stoicism. Usually dressed in khaki or formal wear, male trainers did not encounter the same kinds of problems as women trainers, who were often forced to wear impractical, fluffy dress. In the second decade of the twentieth century, Mabel Stark, for one, was required to don an awkward feather headdress during a lion act with the Al G. Barnes circus. Attracted to the birdlike movement of the headdress, the lion pounced on Stark and cut a five-inch gash on her head.[87]

By contrast, prominent trainers who worked exclusively with domestic animals were not marketed as quintessential "manly" men. Their acts were intended to be humorous, not death defying. Alf Loyal, a dog trainer with Ringling Bros. and Barnum & Bailey, was never advertised as "fearless" or "masterful." Instead, his billing focused on the funny, human exploits of his dogs: "Dogs that actually think and reason. Introducing 'Toque' who rides, leaps and juggles like a man; and 'Chiquita,' the clown-dog whose real sense of humor will merit your closest attention."[88] Emil Pallenberg, who trained brown bears, was similarly invisible as he worked with "[w]onderful acrobatic cycle-riding, rope-walking bruins" seemingly devoid of virile ferocity, just as the cuddly "teddy bear" (a national craze modeled after Theodore Roosevelt beginning in 1906) rendered the potentially lethal bear into a child's toy.[89]

As an arbiter of nature, the trainer commanded predator and prey animals to perform together as friends in the biblically evocative "happy family" act (plate 5). In reality, though, the trainer's skill had little to do with the animals' docility: they had been fed so thoroughly that they were little danger to each other. At the turn of the century, Frank Bostock trained lions, tigers, hyenas, sloth bears, polar bears, and Tibetan bears to work together with their respective enemies. Bostock proclaimed mixed-group training "wonderful," because "[the animals] have been subjected to this gross indignity by the superiority of man."[90] Captain Jack Bonavita, a pupil of Bostock, exhibited twenty-seven lions at once in 1900. Reportedly, Theodore Roosevelt admired Bonavita's simultaneous mastery over "twenty-seven kings of the forest."[91]

The trainer became master of life and death when animals became violent or, in circus parlance, "went bad." The trainer either banished the animal through permanent caging (the commonest punishment) or arranged the offender's execution. Bostock observed that "going bad" was an occasional and inexplicable part of the aging process that only struck a few species, most commonly lions, tigers, and elephants. Advertisements lured spectators to see these rogue animals with lurid stories about man-eaters. One headline roared: "Fierce Battle for Life of Boy Crushed in Assassin Tiger's

Jaws," after Frank Bostock's tiger, Rajah, broke loose in New York City in 1901, then killed and partially ate a sixteen-year-old boy. The article continued, "[Rajah the tiger] cannot understand why the dainty morsels which pass and repass the cage in the form of exclamatory, admiring women cannot be thrown to him as choice tidbits. But Rajah is doomed to a life of enforced abstinence, for even the bloodless pleasures of other days are to be denied him."[92]

The execution of an intractable circus elephant, or "bull," was the least common but most spectacular instance of the animal trainer's ultimate power. Pachyderms "gone bad" were put to death by firing squad, poison, or strangulation. On October 8, 1888, the Adam Forepaugh circus publicly strangled Chief Forepaugh, a forty-year-old elephant, after he killed seven men during several rampages in Philadelphia, Grand Rapids, Michigan, Topeka, Kansas, and Akron, Ohio, where he temporarily "took possession" of the town. Circus workers strangled Chief by tying a noose around his neck and attaching the ropes to elephants on either side of him who then moved in opposite directions. Like other dead circus animals, Chief was taken to a scientific institution, in this case the University of Pennsylvania, where he was skinned, stuffed, and studied.[93] P. T. Barnum recounted that Albert, "a very large and treacherous Asiatic elephant," was sentenced to death after he killed one of his keepers in 1885. On July 20, Albert was publicly executed on the outskirts of Keene, New Hampshire. The elephant was chained to four trees; the location of his heart and brain were marked with chalk. Thirty-three members of the Keene Light Guard stood in line at sixteen paces, and at the word of their commanding officer, fired at Albert, who collapsed without a struggle. His remains were donated to the Smithsonian Institution.[94] After Adam Forepaugh donated Tip the elephant to the Central Park Zoo in 1889, Tip continued to terrorize (and kill) several keepers. Finally deemed "bad" by his last keeper, William Snyder, Tip was publicly poisoned in May 1894. The *New York Times* covered the graphic scene: "In his paroxysm he whirled about the little limits of his cage, reared his great body against the heavy timbers, and charged upon [the crowd] with his blunted tusks. He raised his trunk high into the air and trumpeted in agony. From his mouth he spouted big drops of blood, and then, gathering himself with all his might, he made one dash toward the rear of his pen, beyond which lay the green lawn of the Park. Chains that bound him broke like springs, and he was almost free. But the poison was doing its work, and when the monster seemed sure to dash himself against the outer cell wall, upon the chain which still held fast about his foot he tripped. It stopped him in his rush."[95]

Ted Ownby has written that the elephant execution represents a ghoulish metaphor for lynching—itself a form of violent community spectacle.[96] Proprietors often depicted the elephant as racially "black." Exported from Africa or India, elephants portrayed "savage" masculinity in its largest land-mammal form. Osa and Martin Johnson blamed the African elephant's intractability on the African man, who they claimed was too "uncivilized" to domesticate the elephant, unlike the "more advanced" Asian races: "Is it any wonder that the African elephant has also remained a savage, when the members of the human race that reside near him fall so low in the scale of man?"[97]

Showmen attributed the elephant's crazy behavior to "must," a frenzied sexual state that made the elephant combustible. Euroamerican discourses about black men similarly articulated the imagined dangers of black male sexuality in an explosion of racist scholarship at the turn of the century: Charles Carroll, *The Negro a Beast* (1900); William B. Smith, *The Color Line: A Brief in Behalf of the Unborn* (1905); and Robert Shufeldt, *The Negro, a Menace to American Civilization* (1907). According to Bederman, white southern men saw the powerful fiction of the "Negro rapist" as an enormous threat to their manhood, their communities, and "civilization."[98] A record number of 161 recorded lynchings occurred in 1892, and actual numbers were undoubtedly higher.[99] Nevertheless, most African American victims of the lynch mob were never accused of rape. Many were murdered because they were an economic threat to local white businesses. The circus elephant sexually run amok, as well as the financially successful African American man and the fiction of black sexual prowess, symbolized a threat to Euroamerican community order.

MANLINESS UNDER THE BIG TOP

Wearing thin tights and a leotard, the male big-top player's body was always in plain view. As P. T. Barnum suggested, the circus's visual exhibition of the disciplined, athletic male body provided an excellent lesson in virtue for the young boy. "[T]he [circus] athlete demonstrates the perfection of training of which the human body is capable. His feats of strength and graceful agility pleases the understanding as well as the eye, and if the average small boy does stand on his head and practice turning 'hand-springs' and 'flip-flaps' with exasperating persistence for three weeks running after going to the circus his physique will be all the better for it."[100]

The flying trapeze replicated contemporary thrills such as balloon flight, mountain climbing, and the nascent sport of flying an airplane. Originally popularized by the French physical-education teacher and impresario Jean Léotard in the 1850s (whose tight costuming still bears his name), the flying trapeze transformed its practitioners into somersaulting, muscular missiles through space. Programs described Charles Siegrist's stunts as "daring displays of unrivaled accomplishments," "startling feats of skill and sureness . . . defy[ing] the laws of gravitation."[101] But the program left it at that: in stark contrast to their treatment of big-top women, whose family-centered origins in show business were highly touted, press releases and programs paid little attention to Siegrist's life outside the ring. His life story mimicked the individualistic, manly "rags to riches" trope of several circus owners. Born in a covered wagon on the Oregon Trail in 1880, Siegrist (whose original surname was Patterson) was orphaned as a young child. Supporting himself as a newspaper seller, young Charles, who had a severe speech impediment, attracted his customers' attention by doing acrobatic tricks on street corners. As a nine-year-old he was recruited to become a blackface minstrel performer with the O'Brien Brothers, and in 1898 James A. Bailey contracted him to work for the Barnum & Bailey circus, where he stayed until 1931. Charles adopted the surname of his mentor, the trapeze artist Toto Siegrist, shortly after he joined the circus.[102] Until his death in 1953, Siegrist was an active part of the outfit, even after suffering a broken neck.

Euroamerican big-top men (like their lady colleagues) incorporated new machines into their thrill acts. In 1903 Nick Howard worked as "Cyclo, the demon wheel man, Davis." Howard's act consisted of riding a bicycle twenty-one times around a sixty-four-foot track containing a vertical wall. In an interview, "Cyclo" explained that fear would only kill him, because if he stopped moving when ascending the wall, he would fall and crush his head. The circus promised audiences "two minutes of sheer terror," stating that "only such a trick . . . comes from marvelous skill and undaunted intrepidy."[103] In contrast to the marketing of Mauricia de Tiers and other female riders in the "L'Auto Bolide" act, there was no mention of Howard's "good breeding" or "refined" education to temper the physicality of his act, because the powerful performance of the "Kinetic Demon" confirmed rather than subverted male norms.

The looming possibility of a mishap enhanced a player's brave, manly persona. Headlines blared: "Fall from Trapeze: George Dunbar Nearly

Killed"; or, "Women Scream as Rider Falls: Ancillotti Fails to Clear Gap in His Circus Feat and Is Badly Hurt."[104] When Otto Kline, a Wild West trick rider with Barnum & Bailey's circus, was killed after he fell off his horse at Madison Square Garden in 1916, all local papers covered the tragedy, featuring headlines like "Cowboy Rider Killed before Circus Crowd," and offered extensive coverage of the funeral.[105] In most instances, nonetheless, circus men survived; press releases cited this sort of unflinching "show-must-go-on" comportment as proof of quintessential manliness, because these well-proportioned male bodies betrayed little sign of pain, even when it was excruciating. By contrast, showmen depicted clowns and African American men to be outrageously sensitive to minute discomfort. Dramatizations of stoicism and pain were tied to racial stereotypes.

Asian bodily stunts tapped into stereotypic "oriental" practices, notably fire walking and sleeping on the proverbial bed of nails. Japanese, Arab, Chinese, and South Asian acrobats commonly played scenes of imagined bodily torture. Japanese acrobats frequently exhibited the perch act, a harrowing stunt involving several people, often whole families: one artist balanced a pole or ladder upon his shoulders or forehead, and smaller players—usually children—climbed the object in balance and performed acrobatics and balancing stunts in midair. In 1894 Okeo Akimota ascended a ladder of swords with bare feet.[106] Arab troupes—often simply called Bedouins—executed a series of impressive ground acts that included leaping, somersaulting, and the aptly named reversed pyramid building, a stunt which required one person to hold up several others in an inverted triangle: a world turned upside down (plate 6).

At Carl Hagenbeck's Greater Shows in 1907, press agents unveiled Poline, the famous Hindoo Fakir, who could lift sixty-pound weights with his eyelid. A picture depicted Poline with a rope attached to his "muscular" eyelid, which was connected to a small child positioned at his feet, whom he would later lift.[107] A Barnum & Bailey lithograph from 1916 depicts the members of a Chinese troupe of acrobats calmly sipping tea while being suspended by their hair (fig. 30). Tiny Kline watched Chinese aerial acrobats tumble in "tortuous" poses wearing "kaleidoscopic" costumes, while juggling plates on strawlike sticks.[108] The wives of the Chinese acrobats participated in the act's finale.

[E]ach couple sat down to a small table to which the chairs were attached at the base, facing each other and a tea service before them. Two ropes were lowered from a crane-bar overhead as the assistant passed the

Figure 30. *"A Chinese Tea Party,"* Barnum & Bailey, 1916. *Echoing missionary representations, Asian performers frequently appeared in ascetic scenes of bodily contortionism at the circus. (Lithograph courtesy of Circus World Museum, Baraboo, Wis., with permission from Ringling Bros. and Barnum & Bailey,® The Greatest Show on Earth,® B+B-NL44-16-1U-3)*

hook attached to the rope, through the tightly braided and twisted knob of hair of each, while man and wife inter-locked their legs under the table. At the signal of the equestrian director's whistle, all three couples were hoisted in the air; joined by their legs they appeared to be sitting relaxed on the chairs, though actually being suspended by their hair,—supporting the weight of the table as well. . . . With the pulling of the scalp affecting the facial expression, it took on a look of fright, as if they had suddenly seen a ghost. Sitting there in the air, they poured and sipped their tea as though they enjoyed the party . . . meanwhile, four of their comrades were being hoisted up at various points along the track to the ropes, stretched from the top of quarter poles at far ends of the big top, down toward the center, and as the tea-party was lowered to the ground, down came these four flying Chinamen in a "slide for life"—suspended by their queues; each carrying the American flag in one hand, the Chinese flag in the other, thus assuring themselves of applause which I am sure they would have earned even without "presenting colors."[109]

The majority of circus contortionists were Euroamerican men, but they were often racially masked, labeled as "indiarubber men" or "klischniggers" (after a famous European, Edward Klischnigg, who played an ape in English theaters and circuses in the 1830s). J. H. Walter, an Englishman also known as the "Serpent-man," was an internationally renowned contortionist in the 1880s whose writhing flexibility enabled him to bend backward with his head peering out between his feet; in this position (known as the Marinelli bend), Walter clenched a mouth grip and lifted his body into the air without the use of his hands. An interviewer once pressed Walter to reveal whether women found him irresistible given his ability to bend his body into such unusual positions; Walter sadly replied that his severe bodily regimen had rendered him weak and impotent: "Sir, the chastity which monks do not always observe is forced upon an artist of my class. . . . I have all the appearance of a strong man; my chest is wider than your own, but beneath it I conceal the lungs of a child; they are stunted by the daily pressure of my thoracic cage."[110]

As a site for remarkable bodily contortions, the big top was also home to gender ambiguity and play. Even though showmen valorized brawny male performers, big-top acrobats—both men and women—were most successful if they were petite. As a result, their virtually identical dress and similar degrees of muscularity made male and female big-top athletes androgynous. In England, the voyeuristic barrister Arthur Munby was mesmerized by watching "feminized" men such as Jean Léotard work the single trapeze.[111] Moreover, explicit drag acts had long been part of the American circus. In the 1840s, Robert Stickney introduced "The Frolics of My Granny," in which an old woman wearing a bonnet and skirt rode into the arena with a bouncy and fidgety young boy on her shoulders. Suddenly, the old woman and the boy were cast aside—revealed to be a wicker frame—and the human mass was transformed into a single male rider in tights.[112] In 1859 G. A. Farini, a wire walker and also a circus manager, walked across Niagara Falls on a cable while dressed as Biddy O'Flaherty, an Irish washerwoman who laundered clothes as she made the treacherous crossing.[113] In the 1870s Farini's adopted son, El Nino, wore drag as Lulu, a popular leaper who catapulted twenty-five feet into the air off a hidden springboard. Wearing drag, Fred Biggs opened the big-top program for Sells-Floto in 1913 as "The Initial Laugh. . . . He is the only man on the American continent who can by his solitary efforts, entertain such a vast gathering as daily visits the Sells-Floto Circus. Watch for Biggs. You'll laugh at him. You can't help it. Fun is his middle name and he'd bring a chuckle to a man with the mumps."[114]

Duping an audience of gullible rubes was an integral part of drag's pleasures. Route books and newspaper articles documented male riders and acrobats who enjoyed standing in for female colleagues suddenly taken ill. At the Ringling Bros. circus at Watertown, South Dakota, June 1892, the route book noted: "Blanche Reed having measles, Mike Rooney [a fellow rider] substitutes. With curly wig and cheeks like two blush roses; with corset upside down and dainty dress of taffy-candy pink, Mike looked 'too sweet for anything.' Made a hit with his bow. Did splendid."[115] At the Barnum & Bailey production at Chicago in October 1904, Jeremy Silbon substituted for an ill female trapeze artist. According to the *Chicago Daily Journal,* "When the big audiences at the Barnum & Bailey show are applauding the daring mid-air performances of Mademoiselle Cleveland and voting that she is the prettiest girl in the show, that 'young woman' is enjoying the joke . . . . [She is] in reality, a fine healthy boy."[116]

As an itinerant amusement that attracted outsiders, the circus generally accepted drag artists. Georgie Lake, a transvestite, worked comfortably at the cooch show in the years surrounding World War I. Players like Berta "Slats" Beeson, a wire walker with the Ringling Bros. and Barnum & Bailey circus, slipped into male and female personas with ease, inside and outside the canvas big top. As "the world's only aerial danseuse, the madcap of the wire-running, dancing, leaping, swinging, pirouetting on a slender thread of steel," Beeson in his act did not betray his "real" gender. Tom Barron, "the World's Tallest Clown," recalls that "[Beeson] was a female impersonator, but he wasn't feminine acting at all, in fact, he was captain of the baseball team. A terrific guy, he was really funny." [117]

But Beeson's female impersonator act, along with Albert Hodgini's role as "Miss Daisy" (described in chapter 4), do speak to the existence of a larger, vibrant drag culture in the early twentieth century. The female impersonator Julian Eltinge advanced his vaudeville and movie career by publishing the *Julian Eltinge Magazine and Beauty Hints*, which provided eager female consumers with advice on cosmetics and clothing. Drag played an important role in the saloons, resorts, live sex shows, dance halls, and dime museums of the Bowery in New York City. George Chauncey suggests that drag was but one part of a complex "gay world" in the early twentieth century. He argues that "fairies," "inverts," "female impersonators," and "traders" were more broadly accepted in the first third of the twentieth century than in subsequent decades, in part because contemporary scientific and popular constructions of an intermediate "third sex" made homosexuals less threatening to standard male codes. "[The fairy] was so obviously a 'third-sexer,' a different species of human being, that his very effeminacy served to confirm rather than threaten the masculinity of other men. . . . Their representation of themselves as 'intermediate types' made it easier for men to interact with them (and even have sex with them) by making it clear who would play the 'man's part' in the interaction." [118]

In this environment, Eltinge, Hodgini, and Beeson enjoyed a wide audience. The breadth of circus drag stars was especially remarkable because they brought aspects of this urban gay world to isolated rural areas. Chauncey suggests that popular notions of the "third sex" allowed a man to participate in this rich and fluid gay world without being stigmatized as "deviant," "so long as he abided by masculine gender conventions." [119] The theater historian Laurence Senelick notes that Eltinge "normalized" his drag act by behaving "like a man" offstage. [120] Hodgini and Beeson remained at the center of the circus community, in large part because they conformed to social norms—married with children, captain of the baseball team—once they left the ring.

CLOWNS

Clad in pancake white and sleek jester garb, or in big, baggy clothes, huge shoes, a tiny hat, and a gigantic nose, the clown toddled around the big top, childishly teasing the ringmaster. An essential part of the circus since its American arrival in the late eighteenth century, the clown "interfered" with the serious acts and "covered" for workers setting up another act or players who were unexpectedly injured. Unlike those male players whose manliness was augmented by staged encounters with fierce animals, the clown usually worked with relatively harmless creatures: pigs, mules, geese, and pigeons. Jules Turnour and Emmett Kelly, among others, took up clowning because their bodies could no longer handle the physical strain of acrobatics, contortionism, and riding; in many cases the clown, in contrast to other, more athletic big-top performers, represented the male body in physical decline.

With historical roots in traveling medieval troupes and the European tradition of the court jester, clowns worked a variety of raucous acts. They played in big, off-key clown bands during the concert, did solo acts under the big top, danced in drag at the cooch show, and played in sprawling big-top clown congresses. In addition, several accomplished riders and acrobats played clownish roles in floppy garb and occasional pancake. Like the "genuine" clown, these comic riders mocked conventional norms, staging spastic and undignified acrobatics on a horse's rump.

No one particular look characterized the clown; each clown's costuming and makeup style served as his distinctive trademark (fig. 31). Al Miaco, for example, was a traditional "whiteface" clown and dressed as a Shakespearean jester, wearing caps and bells. Although covered in clown white, his facial features essentially remained his own. He performed over-the-top caricatures of bodily containment and cultural refinement. By contrast, the auguste clown, also known as the "proper clown," who appeared in or out of whiteface, burlesqued his body with an exaggerated nose, mouth, eyes, and ears, shaggy hair, and decrepit, oddly sized, zany, too-bright clothes. Whitefaced clowns often performed with their auguste counterparts in what Paul Bouissac calls a dichotomy of "culture" versus "nature": the suave whitefaced clown playing the violin, for instance, while the dirty auguste clown frolics around the big top making music with a rubber glove that he has just pretended was a cow's udder.[121] The presence of animals augmented the centrality of displacement and liminality to clowning. Physically masked, the clown worked with the stubborn, humorous mule, neither horse nor donkey, and the pig, a fully liminal, fully humanized animal at the circus. He also drove ungainly teams of ostriches instead of horses. Clowning often blended with the sideshow as giants, midgets, and other players with physical abnormalities often played clowns. In the 1920s Ernie Burch dressed in drag, complete with enormous rubbery breasts, gowns, gaudy

Figure 31. Clown group, ca. 1905. Note the presence of whiteface and auguste clowns, a tramp, a pot-bellied police officer, a drag clown, a Hebrew clown (standing far left), and mules for the January Act. (Photograph courtesy of Circus World Museum, Baraboo, Wis., CL-N81-GRP-3)

pancake makeup, and a big, blonde wig, à la Mae West; in the auguste tradition, Lou Jacobs also worked in drag, wielding an overgrown baby carriage in his act, sometimes with dachshund in tow. The oddly shaped "wiener dog" enhanced the act's subversive look.

By the turn of the century, the clown's act had become almost purely visual. His voice could no longer be heard amid the din of three rings and two stages of activity under the cavernous big top.[122] This silent "joey" (a generic term derived from the name of the eighteenth-century English clown Joey Grimaldi) was a throwback to clowning's antecedents in European pantomime. The "talking clown" had been an integral part of the American circus until Barnum & Bailey instituted the three-ring standard for the largest circuses in 1881. At the one-ring show, the talking clown "guyed" (teased) the ringmaster, made jokes about local politics and current national news, sang comic tunes, and often spoke several languages to satisfy polyglot immigrant communities. In the cozy atmosphere of the forty-two-foot circle, the talking clown maintained an intimate relationship with his audience. A French immigrant, Jules Turnour, clowned for small wagon shows before joining the Ringling Bros. in 1889. He recalled his earliest days at the one-ring circus and his constant, mad-dash preparations to tailor each performance to its specific location: "The tents were not nearly so large as they are now and you could talk to your audience and be readily understood. Accordingly, I made haste, as soon as I reached a town, to get a local newspaper, find out what was going on, and then I made a reference to it in my clowning. It never failed to please the spectators."[123] The most famous talking clown in American circus history was the whiteface clown Dan Rice (1823–1900), who started out as a puppeteer in Reading, Pennsylvania and then moved on to a trained pig act before becoming a clown. During his heyday in the 1860s and 1870s, Rice reportedly earned $1,000 a week.[124] He also worked in blackface and was an accomplished rider.[125] Wearing colorful tights, puffy shorts, and a leotard top for his clown act, Rice exuded dexterity and excellent comic timing. Although a few publications (*Porter's Spirit of the Times*, for one) found Rice's poor grammar and general lack of education objectionable, he generally remained popular with the American public during the antebellum and Civil War eras—some, however, called Rice a traitor for playing both sides of the Mason-Dixon Line during the war.[126]

Rice's connection to blackface was his strongest connection to turn-of-the-century clowning. Stuart Thayer suggests that minstrelsy had its beginnings at the circus, and that the two amusements overlapped considerably during the nineteenth century. As part of his repertoire of comic songs, George Nichols gave what the program listed as a "minstrel scene, titled Jim Crow," at J. Purdy Brown's circus in 1831. The plantation slave, "Jim Crow," and the urban dandy, "Zip Coon," were stock circus minstrelsy characters and soon became a central part of the new antebellum minstrel show. Drag minstrel characters also had an early home at the circus: Daniel Gardner played a "wench dancer" named "Miss Dinah Crow" at the Green & Waring's Eagle Circus in 1836. Not surprisingly, many antebellum minstrel players began their careers as circus clowns.[127] Eric Lott writes that the American clown—particularly the auguste clown—borrowed heavily from the slave trickster, an integral figure in the African American folk narrative tradition. The clown, along with the trickster, depicted "lovable butts of humor and devious producers of humor." Both stood as champions of the weak who slyly defeated the strong through sheer wit.[128]

As the clown harmlessly tottered around the big top, he deflated the ringmaster's pretentious presentation. He seemingly sabotaged the musically orchestrated precision of the three-ring whirl just as the blackface minstrel performers constantly "guyed" the highfalutin interlocutor. Jules Turnour described a typical clown stunt, the "January Act," in which a lowly clown character named January trades a mule for the ringmaster's fine horse. When the mule refuses to budge for the exasperated ringmaster, he pays the clown to take the obdurate mule away. The obstinate, sneaky clown dupes the ringmaster into giving him the horse, the mule, and money.[129]

In the "Peter Jenkins Act," a star female bareback rider was suddenly unable to appear in the ring because a horse had kicked her backstage. Meanwhile, her beautiful "rosin back"[130] pranced riderless around the arena. Upon hearing the announcement that Mademoiselle La Blanche would not appear, a drunk rose from the crowd and stumbled down to the ringmaster with booze in hand, loudly condemning the program a bust. The ringmaster challenged the drunk to ride the horse. The drunk readily accepted and mounted clumsily. As he lurched forward, his ragged clothes suddenly fell off, revealing a graceful body in tights and spangles that proceeded to perform complicated bareback acrobatics. The audience roared with approval, greatly enjoying their having been duped. Mark Twain's Huck Finn also eagerly witnessed this act: "[The acrobat] just stood up there a-sailing around as easy and comfortable as if he warn't ever drunk in his life—and then he begun to pull off his clothes and sling them. . . . And, then, there he was, slim and handsome, and dressed the gaudiest and prettiest you ever saw, and he lit into that horse with his whip and made him fairly hum—and finally skipped off, and made his bow and danced off to the dressing-room, and everybody just a-howling with pleasure and astonishment."[131]

Riding mules instead of horses, clowns dressed as policemen in ratty clothes, with a big, lopsided badge, sooty black face, and pendulous abdomen. At Trenton, New Jersey, in 1907, a press release disguised as a newspaper article further poked fun at law enforcement. The "article" announced: "Scene at the Barnum & Bailey Circus Not on Real Program." While watching the clown number, "Clarence the Cop Chasing a Tramp," two Trenton officers reportedly moved quickly in to help their "colleague" arrest the "criminal."[132] Clowns also worked in a large group for the big-top "fire number," in which they crazily pretended to douse a fire while impersonating fire chiefs and lieutenants. On one level, it would seem that such impersonations were harmlessly gentle celebrations of a world turned upside down, as childlike clowns tottered around playing big in oversized "grownup" clothes on Circus Day. But these acts could also be read as carnivalesque inversions. Dating back to medieval and early modern Europe, clowns lewdly impersonated the clergy and nobility at annual feasts and fairs.[133] Circus clowns underscored local authorities' very real ineptitude in controlling large crowds; their stunts, then, can also be interpreted as a critique of social surveillance. Such was the case with a melee at Buffalo Bill's Wild West in Brooklyn, New York, in 1892. The *Brooklyn Citizen* reported that the lone police officer on the scene cautiously poked his head out from behind a canvas tent, but then "made up his mind that he was not wanted, and withdrew his head. When a bystander went to call him, he had disappeared."[134] Still, local police commonly targeted circus workers as the sole perpetrators of community disorder rather than focus on members of the crowd.[135]

Because the clown's precise identity was often ambiguous, he represented what Raymond Williams has called "structures of feeling." In other words, the clown embodied "a kind of feeling and thinking which is indeed social and material, but each in an embryonic phase before it can become fully articulate and defined exchange."[136] By tapping into "structures of feeling," clowns often played unconscious racial stereotypes that helped reinforce social norms. Some circus programs contained portraits of clowns in literal blackface, with huge red mouths and bulging eyes, strumming energetically on a banjo, but often the auguste clown's blackface was metaphorical. He created his racial identity through the act of "whitening up" with thick pancake. His greasy whiteness and exaggerated bodily zones—huge red mouth, lolling, paint-encircled eyes, big fake nose, ears, and feet—made his look strikingly similar to blackface. Showmen played upon this visual connection by arguing that African American men literally *were* clowns because of their supposed affinity for clowning and the circus. The Ringling Bros.' route book from 1895 and 1896 contained a section, "The Plantation Darkey at the Circus," which imagined—in almost orgasmic language—black men as minstrel characters.

> [T]he great American "nigger" has a laugh not only his own but one that owns him as well. In the presence of the clown he and the laugh are firmly bound together. They can't get away from one another; not the nigger and the laugh! Oh, no! It is with him under every inch of his black

skin, in every nerve, muscle, sinew, even in his bones. Every atom in his body responds to it. It wiggles his toes, bends his knees, puts a double-action, spring-hinge in his back and electrifies his whole being with the most exquisite emotions of a tickling which, as his burnt-cork counterfeit would remark, "Can't be scratched." It does everything to him but to take the kinks out of his hair. At times it leaps out of his capacious mouth, like a flame of fire . . . and just as you think it is getting away from him for good and all, back it darts through the white archway of his dental orifice into his interior regions way out of sight, but not out of evidence, for you can see it bulge out his ribs and you wonder why he doesn't explode.[137]

Proprietors further conflated the African American man and the clown by arguing that both were completely controlled by their emotions, not reason. Superlative examples of white manhood—the big cat tamer, the wire walker, and so forth—demonstrated little emotion during life-threatening acts. The clown, by contrast, howled in mock fear when he saw a mouse, or shrieked in pain at a mosquito bite. Showmen characterized male African American spectators in a similar vein as giddy and superstitious. The Ringling route book stated: "The animals in the menagerie, usually considered an instructive feature of a circus, are to the negro weird beings from a world rather more remote than the sun or moon . . . his particular horror is the snake. . . . [If a lion roars] [o]ne negro shouts, 'He's loose,' and instantly the thousands of assembled black people take up the cry of 'He's loose,' and stampede. It makes no difference who 'he' is that is 'loose'; they run like mad, men, women, children, shouting, 'He's loose, he's loose, he's loose, he's loose.' In every direction they scatter."[138]

Actual big-top acts made this rhetorical relationship between the clown and the African American complete. In 1888 Eph Thompson trained the elephant John L. Sullivan at the Adam Forepaugh circus. Wearing a boxing glove at the end of his trunk, the elephant sparred with Thompson in the ring and frequently "punched" him so hard that Thompson went flying over the ring bank.[139] Unlike the white trainer who dominated powerful animals, Thompson played a clownish coward—constantly vanquished by the boxing pachyderm—and consequently remained unthreatening to Euroamerican audiences. Yet Thompson still had a difficult time finding employment with American shows. As a result, he moved to Europe where his career flourished.

In line with the tenets of nineteenth-century romantic racialism, showmen's portrayals of black men and clowns reflected contemporary representations of white women: late-nineteenth-century scientists argued that "excessive" emotionalism defined women, racial "savages," and children of all races. The German Darwinist Ernst Haeckel and the Americans Edward Drinker Cope and G. Stanley Hall were all proponents of recapitulation theory, positing that every organism repeats the life history of its "race" within its own lifetime, evolving through the less developed forms of its ancestors on its path to maturity. They contended that Euroamerican women and "primitives" remained mentally and emotionally fixed in lower ancestral stages of evolution. Accordingly, only white boys were physiologically and mentally capable of reaching the highest stages of racial and gender development as fully evolved men. This line of thought used pseudoempirical phrenological evidence to claim that African American men were perpetually emotional and juvenile, just like the clown.[140]

The painted clown acted out childish behaviors and infantile pleasures. He reveled in dirt, cried freely, openly adored the serious "adult" acts, and played physical pranks on everybody, from ringmaster to the audience. If playing a hobo (popularized most fully by Emmett Kelly's "Willie" tramp character during the Depression, when at times nearly one-quarter of the American workforce was unemployed), the auguste clown's persona was defined by dirt. Laughing loudly at the clown's antics perhaps transported audiences back to the unrestrained pleasures of their own collective infancy and childhood. More than a "low Other" who simply represented a tantalizing version of what they were not, the unfettered clown symbolized what clock-bound, alienated adult Euroamerican men perhaps felt they had lost.

Playing European ethnic characters, clowns portrayed preindustrial dupes, perpetually bumbling with the fast, sophisticated pace of urban American life. The idea of racial whiteness itself was hotly contested at the turn of the century, an era when white skin color by itself still did not confer automatic white privilege. Racial theorists, politicians, and circus showmen alike ascribed "primitive" qualities to European immigrants from Ireland, southern Europe, and eastern Europe . Matthew Frye Jacobson shows that contemporary magazine descriptions of Italian immigrant communities were strikingly similar to travel writing about Africa and the Levant. He suggests that such imaginings had violent consequences: in 1892 eleven Italian Americans in New Orleans were lynched after being convicted of murdering the local police chief.[141] In 1924 a revised Immigrant Act enacted quotas that virtually halted immigration to the United States from

Figure 32. "Immigrants Just Arrived from Ellis Island," Kassino Midgets, 1927.
Clowns worked at the sideshow, in addition to the big top. Stock clown characters are
represented here: two rustics on the left, one of whom appears to play a Romany drag
character, Italians, and Hasidic Jews (last two on the right). (Photograph courtesy of
Circus World Museum, Baraboo, Wis., CL-N45-MGT-51)

southern and eastern Europe and Asia (fig. 32). Three years later, the Kassino Midgets played a rag-tag group of "Immigrants Just Arrived From
Ellis Island" at the circus. Costumed as Hasidic Jews with oversized beards
and dangling side curls, Italian rustics, and a Romany drag character, these
miniature clowns provided a comical exaggeration of unassimilable racial
difference.

Clinging to mules, geese, and plows, native-born farmers were also a
common clown subject. In 1893 Adam Forepaugh's burlesque after-show
concert featured a "Yankee Farmer" in addition to an "Irish Knock-About,"

a "Black Face Comedian," and "Black Face Sketch Artists."[142] As part of its clown constellation, Barnum & Bailey's program in 1906 included a "Funny Rustic," a "Fat Boy," and an "Odd Zany," among more specific ethnic types including a "German Broad Face" and an "Austrian Looby."[143] The Hebrew clown was another turn-of-the-century staple, a visible part of a lively Jewish American entertainment culture that humorously chronicled travails of immigrant life in the United States.[144] Like the symbolic uses of blackface through whiteface, agrarian and Old World characters nostalgically reminded male circus audiences of childish pleasures from an era when labor was seemingly more leisurely, tied to the seasons rather than the industrial clock.

The emasculation of the childish rural bumpkin clown was so complete that this character was commonly played in drag: clowns wore enormous fake dresses, wigs, and breasts, in which domestic geese frequently nestled as the clowns played the fiddle or recorder. Tied to other acrobatic and riding traditions, drag has been an important element of circus clowning since its genesis in eighteenth-century Europe. In 1786 the English clown Baptiste Dubois performed the "Metamorphosis of a Sack," in which he changed costumes—and gender—while inside a bag. This stunt later became a standard riding act at the American circus. In the folk tradition of the English comic yokel, Dubois also dressed as a rustic booby, complete with red wig and ruddy face.[145]

In some respects, the clown's gender was indeterminate, obscured by thick pancake white, eye pencil, lipstick, and loud, floppy dress. On one level, the clown's emasculated masquerade rendered him harmless, a friend of children. But on another level, the drag clown explicitly challenged gender norms, because he demonstrated the shifting, socially constructed ground on which "natural" norms were based. Hence, the stereotype of the "scary" clown as sexual predator or mental derelict lives on, exemplified by the actor Tim Curry's sociopathic clown character in the film version of Stephen King's *It*, and punk bands such as the Insane Clown Posse.[146]

The drag clown was most startling at the cooch show, where he often wore tight, plunging gowns, garters, and voluminous inflatable breasts. There he and other drag clowns danced and twisted suggestively, playing gender-bending pranks on dumbfounded men who expected to see nude women. But in some cases, the "men only" cooch show audience did expect to gaze at men in drag. In fact, some gay clowns had sexual encounters with male audience members during and after anonymously crowded scenes. The profession of clowning itself attracted gay men, who found circus life—with its spectrum of human diversity—to be a haven where they could work and live in relative peace.[147]

THE SIDESHOW AND ETHNOLOGICAL CONGRESS

As an interactive performance site, the sideshow and ethnological congress featured varied men who talked to their audiences, stretched and twisted their bodies, danced, and sold postcards of themselves. As in other areas of the circus, their gender performances were shaped by race. The "born" Euroamerican freak, for one, performed as a pillar of domestic virtue. Based on the model of the successful nineteenth-century showman Tom Thumb, midgets were humorously marketed as men of high military rank, like "Major" Burdett, "the world's smallest man."[148] Charles Tripp, a "Legless Wonder" with Barnum & Bailey's sideshow, was advertised as "well educated," "intelligent, level-headed, and well informed," "a very sociable man" from a solid family.[149] Bourgeois patriarchy was another common element in the construction of the "born" Euroamerican freak. Born in 1844, Eli Bowen, another "Legless Wonder" with the Barnum & Bailey sideshow, was frequently photographed with his wife Mattie and their four sons. Staged against a parlor setting, Bowen and his well-dressed family appeared financially successful and dignified.[150] Still, the presence of the family could also be sexually evocative. Audiences perhaps found the domestic scenes especially appealing, because they offered "proof" that the freak still had functional genitalia, practiced sexual intercourse, and produced children. Audiences surely imagined the logistics of sexual activity in the case of Eli Bowen, who had flipperlike protuberances instead of legs.

Euroamerican-made freaks and "novelty" acts were often in racial disguise. For instance, in 1896 the Forepaugh & Sells Brothers circus listed W. H. McFarland and Wife as a "Mexican Knife Throwing" act.[151] Dressed in turbans and quasi-"oriental" garb, Euroamerican sword swallowers were variously called Arabs and South Sea Islanders.[152] The tattooed man achieved his masculine primitiveness once his body became colored with injected ink. Whereas tattooed women were marketed as victims of forcible abduction, the tattooed man's "color" was the mark of his travels around the globe, like those of the sailor or soldier. His tattoos were a permanent record of his rites of passage into manhood, a living memento of his physical contact with faraway people. The first tattooed Euroamericans typically were eighteenth-century seamen who had sailed to South America, Asia, and the Pacific Islands, many of whom reportedly had had sexual relationships with indigenous women.[153] Charles Tyng (1801–79), a sea captain based in Boston, received several tattoos from a ship's mate during his early career in the second decade of the nineteenth century: "I had letters, anchors, hearts, on my hands and arms, and a fancy double heart with C. T. in one, and S. H. [Sally Hickling, an unrequited sweetheart from Boston] in the other, red roses between, each heart pierced with cupid's dart, the red showing the drops of blood dropping from the wound. This was on the left arm."[154] Several tattooed artists, or "Living Picture-Galleries," claimed to have been tattooed forcibly, or willingly during indigenous marriage rituals, but virtually all intimated that the procedure was sensual. In reality of course, virtually all had been tattooed in the United States in order to enter show business.[155]

The tattooed man first became widely popular at the American circus in 1876. That year, P. T. Barnum hired Constantine, or Captain Costentenus, in all likelihood an ordinary Italian immigrant who got tattooed as a way to enter show business. But his manager, G. A. Farini (who also managed Krao, "the Missing Link"), told a different, clearly more dramatic story. Supposedly the child of Greek royalty in Albania, Constantine was captured by a Turkish despot and raised in harems both in Turkey and in Egypt, where scores of women fawned over him and occasionally dressed him as a girl. As an adult in the 1860s, Captain Costentenus participated in a French expedition to Cochin China (Vietnam) and Burma, where he was reportedly kidnapped and tattooed on nearly his entire body, including his eyelids and ears. Press agents suggestively noted that the only parts of his body remaining unmarked were his palms and the soles of his feet. The famous nineteenth-century Indologist Max Müller supposedly examined Costentenus, noted that he spoke six languages, and concluded that his copious tattoos were primarily depictions of the Hindu goddess Durga, wife of Siva. Costentenus heightened his authenticity by wearing a skimpy loincloth (amply illustrating that his tattoos continued beneath the cloth) and his long hair in braids atop his head. The talker tantalized his audiences by declaring, "And this wild tattooed man is always much admired by all the ladies."[156] With the publication of Cesare Lombroso's work linking tattooing and criminality in the 1880s, Constantine—himself a consummate showman—began to market himself as a murderer.[157]

The tattooed man willfully disfigured his body, which made him seem impervious to pain. While showmen praised the big-top athlete's manly ability to hide his pain, the tattooed artist seemed to feel no pain at all—an attribute that the new field of criminal anthropology linked to "dangerous classes" and to "savages" like the Hindu ascetic. Furthermore, tattooing itself was physical evidence that the "Living Picture Gallery" had been intimately touched by another man. Like the act of blackening up, the tattooing process conferred a sort of erotic license upon its European and American practitioners, allowing them to script themselves in sensual terms without official censure. The contemporary social scientist Albert Parry explained the act of watching the tattooed man in recapitulation terminology: "An American circus-goer, gazing at the tattooed man in the sideshow, relives his own past of untold centuries back. Moreover, he can now imitate the freak. He can get a tattooed design or two onto his own skin—and thus blissfully revert to his own distant, primitive type, incidentally experiencing a certain erotic pleasure in the process of being tattooed."[158]

Other forms of willful bodily mutilation further transmogrified the male body. Tiny Kline remembered watching men carefully prepare their bodies to become marketable "rubber-men" and "human ostrich" glass-eaters.[159] Kline knew several "human ostriches" and recalled how they ate drinking glasses and light bulbs, cutting their mouths with great frequency. Usually, the glass-eater would chew bits of glass, spit them into a glass of water, and then "drink" the prickly mixture.[160] One "human ostrich" died in the middle of another act at an outdoor performance. Kline notes that his autopsy revealed the severe internal damage done by his glass-eating career: "And so it came to light: the tumorous growth attached to his stomach when dissected, revealed bits of glass, nuts and bolts and other small hardware imbedded in that semblance of a gizzard; as if mother nature, when seeing this human trying to imitate the fowl, went right along with him in his unhealthy pursuit, aiding him."[161] By publicly eating harmful substances for pay, the glass-eater methodically subverted the limits of the body itself by gradually committing, in Kline's words, "practically retarded suicide."[162]

The ingestor's practice of internal bodily disfigurement linked him to the traditions of ritual bodily disfigurement found in preindustrial societies. His demonstrations of bodily punishment tapped into popular depictions of South Asian sadhus (celibate religious ascetics) reclining on beds of nails with withered, perpetually upright arms, which Americans saw in missionary tracts or in the pages of *National Geographic*.[163] The anthropologist Kirin Narayan demonstrates that the sadhu's bodily disfigurement—on the bed of nails in particular—has been commodified and emptied of its original meaning in American advertising and language.[164] In our time, this connection between bodily mutilation, racial transmogrification, and desire has continued with "Mr. Lifto," a former member of the Jim Rose circus sideshow, whose act is inspired by South Asian ascetic practices. Mr. Lifto infuses ascetic South Asian dress and religious imagery into his act as he lifts irons and bricks with his long, flaccid, pierced penis.[165]

As part of a culture that embraced racial hierarchy, the range of characters that "genuine" men of color were hired to play at the turn-of-the-century sideshow was limited to the "wild man," royal "savage," "missing link," and childlike "primitive." Black players in particular were often staged as actual apes, "undeveloped" men, or exemplars of masculine "savagery." William Henry Johnson, an African American from Bound Brook, New Jersey, staged many variations on the "primitive" stereotype throughout his sixty-six-year career as a freak. Billed at different times in his career as a "wild boy," a "missing link," a "Siamese tree-dweller," a Martian, an Aztec, a "nondescript," and most famously as "Zip . . . What Is It?," Johnson always remained mute on stage, from his early days with P. T. Barnum at the American Museum in 1860 until his final years at Coney Island, where he silently worked until his death in 1926 at eighty-four.[166] Accounts of Johnson's early life are sketchy: perhaps he was sold to a sideshow at the age of four by his destitute parents; other reports state that in 1854 P. T. Barnum rescued him from slavery.[167] Clad in fur, a grass skirt, and posed next to a spear, he also strummed the ukulele during the sideshow "blow off" in later years. He shaved his head, except for a small tuft of hair, a few inches above his occipital bone, that was teased into a stiff triangle to exaggerate his head's sharp point and small size. Johnson's domineering manager described "Zip" to the press as a perpetual juvenile, racially "limited" from reaching full manhood. "In private life [Zip] is a noisy and irrepressible child. . . . In the sideshow tent when the crowds are barred, Zip casts aside his reserve and frolics with [Princess] Wee Wee, the midget. He plays with his choicest possessions—a broken watch, and frequently becomes mightily upset when he thinks his dignity has been hurt." [168]

Johnson's long career of utter silence, coupled with his odd head shape, has given many people cause to assume that he was mentally retarded and afflicted with microcephaly.[169] However, those who knew him in private remembered him differently. Tiny Kline worked with Johnson about the time of World War I and recalled that he was a "normal colored man." Although Johnson's contract stipulated that he remain silent on stage, he mixed freely with circus workers behind the scenes. Kline remarked: "Should any folks have dropped into the side-show during the 'off' hours, between five and seven P.M., however, they would have found [Johnson] down on his knees in the circle with the other men, before a blanket spread out on the ground, shooting 'crap,' and with typical Southern accent, repeating the magic words used in the game: 'Come on, seven!' 'Come, 'leven!' or 'Baby needs new shoes!' 'bet I'll make it!' 'Here I come!'"[170]

It was a tribute to Johnson's skills that his act effectively eclipsed any notions of normality witnessed by his coworkers behind the scenes. As a mute "savage," Johnson had a public persona that complemented contemporary ideologies concerning race and manhood. Recapitulation theorists would argue that Johnson's race kept him "fixed" in an earlier stage of development—similar to the "noisy and irrepressible child" that he was supposed to be. Unlike the Euroamerican freak, Johnson, a bachelor, was never staged in a fancy parlor setting with a wife and children—even though he earned a good income as "Zip." In line with scientific constructions of racial difference, Johnson's race kept manliness, and all its associations with whiteness and the respectable family, out of his "developmental" grasp. But among the closely knit, interracial traveling town of circus workers, Johnson was an ordinary man who participated freely in circus community life.

At the ethnological congress, the absence of northern Europeans and the native-born implicitly placed them at the top of the racial hierarchy of human beings and animals. In 1894 Barnum & Bailey announced that "The Australian Bushmen, The Lowest in the Human Scale of All the Peoples of the Earth are also to be Seen in the Congress."[171] Indeed, this racial-animal juxtaposition took its most dramatic and cruelest popular form in 1906 at the Bronx Zoo, where Ota Benga, a Batwa pygmy, lived against his will in the Monkey House with a parrot and an orangutan. African Americans successfully petitioned for his transfer to a Colored Orphan Asylum, but Benga committed suicide ten years later.[172]

William T. Hornaday, the director of the Bronx Zoo who arranged and managed Ota Benga's captivity, turned to the animal world to explain racial difference. He proclaimed that the most "primitive" people on earth were the "canoe Indians" of Tierra del Fuego, "the lowest rung of the human ladder." Hornaday stated: "Their only clothing consists of skins of the guanacos loosely hung from the neck, and flapping over the naked and repulsive body. They make no houses, and on shore their only shelters from the wind and snow and chilling rains are rabbit-like forms of brush, broken off by hand."[173] Hornaday concluded that orioles, caciques, and weaver birds were more intelligent than the "canoe Indians" of Greenland, or the "Poonans" of Central Borneo, because the birds at least demonstrated elaborate nest making (i.e. home making) skills. By projecting middle-class ideals about privacy and property ownership onto a taxonomy of humans and animals, Hornaday justified the racial politics of the day.

The evolutionary notion of "vanishing races" was another facet of masculine racial representation at the circus and Wild West. G. Stanley Hall proclaimed: "Never, perhaps, were lower races being extirpated as weeds in the human garden, both by conscious and organic processes, so rapidly as to-day."[174] Showmen claimed that people of color were vanishing alongside the landscape to which their racial identity was ostensibly tied. Press agents characterized Native Americans as "fine specimens of a race of people doomed . . . to . . . extinction, like the buffalo they once hunted."[175] "Step by step with the departing buffaloes he has kept an always backward pace. There seems to be no power on earth to save the departing Red Man. His doom seems to be fixed, his day on earth is apparently short."[176] Advertisements exhorted audiences to see the Wild West show soon, to catch a last, live glimpse of the nearly "extinct" American Indian. Impresarios marketed the ethnological congress as a conservation project of sorts, where modern Euroamericans could still witness the unrestrained masculinity of the "natural" man in a world fast becoming culturally homogenized by the industrial revolution and western imperialism.[177] Yet the eventual face-to-face meetings often shook up audiences' nostalgic expectations of the Other as a static relic. One newspaper reported that two boys were shocked to hear Wild West Indians speak in perfectly clear English. "The boys gasped in astonishment and looked ready to cry to think that a 'red devil' should talk in that modern way."[178]

Native Americans were hired at the Wild West and circus to dramatize hegemonic, sweeping declension narratives about vanishing people, animals, and habitat. But paradoxically, such employment also gave them an opportunity to cement cultural ties among themselves—in direct contradiction to the normative tropes of decline that they were hired to play. Native American players often met with fellow Indians in the audience after the performance. At Ashland, Wisconsin, in 1896, the Wild West was an occasion for peacemaking between some historical enemies, the Lakota Indians and the 500 Ojibwa attending the production. Cody and the federal agent in charge of the Ojibwa helped arrange a meeting at which the Indians held a powwow and smoked the peace pipe. The route book observed, "This is the first time in nearly forty years that these two old enemies have met on friendly terms. . . . The meeting was all that could be desired, and the 'hatchet' is forever buried between these two tribes, who have been enemies for so many years."[179] L. G. Moses argues that Native American performers often found the circus and Wild West show to be places in which to affirm cultural traditions, much to the dismay of assimilation-minded reformers who saw these seemingly unassimilated Indians "playing themselves" as a threat: reformers thought that Cody's traveling outfit celebrated traditional cultural forms and, worst of all, encouraged Plains Indians (in particular) to remain nomadic.[180]

At the ethnological congress, showmen portrayed men of color living lazily in "sun-kissed" lands, while their female counterparts labored. Just as impresarios depicted female labor as evidence of racial "primitiveness," the supposed "ease" of these performers stood in distinct contrast to an industrious Victorian manly ideal. Barnum & Bailey's ethnological congress in 1894 promised audiences "an entire family of intelligent Javanese, the women busily occupied in deftly weaving vari-colored straws into beautiful mats, *while the men sit in front of their huts smoking*, and the children are at play" (emphasis mine).[181] Regarding Barnum & Bailey's "delegates from the East," one article in the same year observed that "The Women Wear 'Bloomers' and the Men Petticoats: With Rings in their Noses and Diamonds in their Feet."[182] Another announced that the "Wild Men of New Guinea . . . [although] Unaccustomed to Clothes . . . are Fond of Ornaments."[183] Thus displays of male "laziness" produced spectacles in drag, where women did "men's work" and men sat ornamented, idle.[184] Still, exhibits of nonlabor could also undermine showmen's goals of edifying their audiences. Bluford Adams argues that ethnological congresses "offered a glimpse of a world where labor was not alienated," thus providing working-class audiences with a less regimented alternative to modern industrial life.[185]

Press releases, programs, and newspaper articles were obsessed with the scantily clad, nonwhite male body as a model of sensual, premodern masculinity, as the following headlines from 1894 suggest: "South Pacific Savages: The Men Are Models of Robust Vigor" (*New York Sun*); "[T]he Muscles of the Men Are as Hard as Those of Trained Athletes: And Their Countenances Scarred . . . in Battle Denote the Presence of a Brute Courage Such as Only They Possess" (*New York Advertiser*).[186] Press releases beckoned Euroamerican audiences with descriptions of a "Symphony in Coffee . . . A Coffee-Colored Congress"; or, "Black and brown skinned, copper colored, white, olive. Every shade, color and kind of savage people from mountain, valley, forest, jungle or cave."[187] The "coffee-colored congress" was filled with nearly nude men, rustling in skimpy grass skirts and leather loincloths, wearing brightly colored war paint and tattoos, while Euroamerican families looked on. Proprietors further teased their audiences with liberal mention of polygamous practices, providing a direct contrast to their promotions of Euroamerican players as monogamous patriarchs.[188] But these marketing strategies could have concrete consequences. At an ethnological production in London in 1899, "Savage South Africa," the local media complained that British women and the barely clad South African performers (who later worked at the Louisiana Purchase Exposition in St. Louis in 1904) had become too friendly with each other, observing that "grown women not only shake hands with them but stroke their limbs admiringly. . . . These raw, hulking and untamed men-animals are being unwillingly and utterly corrupted by unseemly attention from English girls."[189] Although power relations were seemingly framed and codified by the distinction between performer and audience, the jammed environs of the circus and Wild West provided opportunities for interracial contact among audiences and show workers in an era when such encounters were forbidden.

As part of their efforts to heighten the differences between the preindustrial and the modern, press agents freely documented the initial reactions to urban society of male Native American members of the ethnological congress. Their stories always juxtaposed the physicality of the Native American man—complete with waist-length hair and great height—with the dense, "effete" urban landscape of the eastern United States.[190] Nostalgia animated such imaginings. William F. Cody mournfully observed that settlement had created creaky, settled men and posited that masculine renewal came with expansion into the wilderness:[191] "[P]ioneers fought their way westward into desert and jungle. . . . From the mouth of the Hudson River to the shores of the Pacific, men and women and children have conquered the wilderness by going to the frontier and staying there—not by crowding into cities and living as do worms, by crawling through each other and devouring the leavings."[192] The Wild West, though, reportedly helped stifle the effeminizing influence of modern civilization: a courier in 1907 announced that Buffalo Bill's Wild West would "stimulate not only the manliest, but the most heroic qualities of both mind and body. . . . [O]ut and out and through and through the manliest exhibition of our day . . . without exception, the hundreds of representatives from the various enlightened, civilized, semi-barbarous and savage nations included in its anthropological, military equestrian and tribal divisions are nature's *noblemen* in physique, fearless audacity, consummate skill. . . . Both as individuals and as a whole, wherever they appear they command respect and admiration, and their manly leader can truthfully say of everyone of them: '*This is a man.*'" (emphasis in original).[193]

Male audience members often read these exhibitions of male athleticism on America's continental and overseas frontiers as liberating—just as many found the itinerant culture of the circus itself so unfettered and attractive that they "ran away" with it. Mark Twain praised the Wild West show in a letter to William F. Cody, stating that he saw the show twice and "enjoyed it thoroughly," and that "it brought vividly back the breezy, wild life of the great plains and the Rocky Mountains, and stirred me like a war song."[194] After attending "Pawnee Bill's" Wild West in Montclair, New Jersey, Army Captain C. W. Briggs expressed melancholic desire for the vanishing "wild" West in a letter to *Billboard* in 1900 (in which, tellingly, Briggs notes that he must take a train—the most pervasive symbol of the new age—to reach the "frontier"): "Nearly half of my life has been spent on the great western frontier, and it is no exaggeration to say that the panorama, as enacted, carried me back to the old days and filled my heart with a feeling of homesickness for the wide and boundless prairie, which, even as I write of it now, comes back to me in a manner so strong that I feel like smashing the oak desk at which I sit writing, packing my camp outfit and taking the first train for God's open wilderness, where fresh air and cool spring water at least are free, and the four walls of brick and mortar, the city man's world, can no longer encompass me."[195]

UNSCRIPTED SPECTACLES

In many respects, the volatile, freewheeling scene outside the circus tents was more of a masculine space than the scripted gender displays from within. Outside, laborers shouted orders at each other, drove stakes, moved heavy equipment, wagons, canvas, and trunks, loaded the trains, all the while wearing tight, short-sleeved shirts—revealing dirty, "blackened up" bodies drenched with sweat. Male spectators milled around the tents and watched the laborers, or gambled in one of the many "grift" joints that followed most circuses. Just as male bodies under canvas were superlative examples of manly athleticism or racial "primitiveness," the workingmen outside the tents provided audiences with a masculine show of physical labor. Work crews were generally divided by race, but nearly all roustabouts had two things in common: youth and strength. Just as the acrobats and riders were athletes, so were these workers, whose bodies were a source of constant fascination for spectators. Proprietors further fetishized the bodies of African American roustabouts by using them in specs set in Asia or Africa. Black workingmen strode around the big top dressed in robes and head-pieces. During the course of the spec, these men went to a separate tent several times, where they gradually took off pieces of their costuming until at the end of the spec, their bodies were nearly bare.[196]

Workingmen's labor was also exciting to watch because it was just as dangerous as the athletic stunts under the big top. Roustabouts were frequently injured and sometimes killed by heavy equipment. During their public "performance" of efficiently setting up and tearing down the circus, workingmen were occasionally crushed by a loose wagon, a tent pole, or a train car rolling away. Route books were peppered with such entries. On July 31, 1894, in Willimantic, Connecticut, "while sleeping on a flat car, [Dennis Kearns, a razor back,] falls off and rolls beneath the wheels of the fastly moving trains and has both legs severed from his body . . . after a short struggle he died, speaking of his dear mother."[197] On July 16, 1892, at Beaver Dam, Wisconsin, "In unloading Frank Tuttle, a trainman, was run over by the big tiger den. One wheel passed over his jaw and another over his breast, crushing him terribly. Blood ran from his mouth, ears and nose, and formed in a pool around him. 'Good-bye, boys, I am dying,' he said. That night he passed over the dark river which all must cross in time. His brother, summoned from Oshkosh by a telegram, took the body home."[198] On July 14, 1898, "'Slivers' Holland, assistant boss canvasman, [was] severely burned by a flame from a beacon, which was upset by the rear wheel of pole wagon and then exploded."[199]

Roustabouts often derailed the disciplined production of labor in which they played such a critical part. They swore, fought, and occasionally killed one another. One newspaper reporter was especially surprised when he heard no "blue" language spoken among the workingmen. Route books also noted that roustabouts were occasionally arrested for verbal profanity.[200] Because these anonymous men were generally transient and their working conditions were rough, they had little personal stake in the traveling community which bound other circus workers so tightly together; as a result, they were more prone to violence. For example, in Dubuque, Iowa, a canvasman, Lewis Hart, was hit over the head with a stake by W. Johnson after an argument. Hart was taken to Mercy Hospital, where he died. Johnson was arrested and charged with first-degree murder.[201]

But audience members were more dangerous. Despite the presence of local police or Pinkerton agents on the large railroad outfits, crowds of men gathered on Circus Day to posture, throw stones, and pull punches. During Barnum & Bailey's annual spring parade in New York City in 1892, local hoodlums suddenly pelted players with snowballs and rocks. A woman riding atop the Mother Goose wagon was struck in the mouth with a snowball laden with heavy chunks of coal. Her mouth bled and she lost several teeth. Angry Native American workers dove into the crowd and brawled with the young punks. "Pistols Fired and Women Scream for Help," proclaimed one headline.[202]

In general, the evening show and its aftermath were the time for men to fight—after women and children had gone home. In darkness, town "toughs" confronted circus workers and fellow spectators alike. At Bowling Green, Kentucky, one circus employee, J. H. Lewis prevented a "shooting affray between town guys" and secured "one of their revolvers as a memento."[203] During a rough night in Anamosa, Iowa: "After show, Anamosa and Rock City toughs, full of fighting whiskey, had a grand battle royal. One Irishman got his face pounded off, but denied that he hollered 'enough.'"[204] In some isolated rural areas, the annual arrival of a circus became a predictable, ritual stage for violence, an opportunity to "settle scores" or vindicate one's manly honor in a public setting. Circus workers looked at Kentucky's hill country as a predictable venue for such confrontations. Emmett Kelly recalled: "[B]y nightfall, I could see hundreds of lanterns lights bobbing on the mountain paths, and before the show our tent was jammed to capacity. This was rough country with plenty of moonshine and the fighting that goes with it. An odd thing, though, was that the toughs who got loaded and picked fights never bothered the circus people, but always fought each other."[205] The Ringling Bros.' route book in 1893 chronicled the following incident: "Monday, September 18th. Williamstown, Ky. The show has the effect of bringing them in from the hills. All have an opportunity to see high life among the natives. The little town is in holiday attire, and to fittingly commemorate the happiness of the day, and incidentally to vindicate some very urgent cases of 'personal honor, sah,' that have been neglected for some time, 'a bit' of cutting and shooting is done. Only two were killed. The wounded have not yet been counted."[206]

These public spectacles of violence complicated the relationship between performer and spectator because the roles were often reversed, as circus workers—normally construed as primitive Others in the eyes of the audience—judged the audience to be unruly and dangerous. Covering a spectrum of planned and spontaneous acts, the circus shakes up the idea that the "gaze" is unidirectional and hegemonic. At the circus, male workers and audiences both played active subjects and objects of the male gaze—therefore complicating studies that analyze the gaze exclusively as an expression of male power over women.[207] At the circus, no one possessed exclusive ownership of the gaze because it was a site of multiple surveillance, a three-ring "theater in the round" which enabled people to watch each other from many vantages.

The circus was a ritualized gathering site for men across the United States, for "strangers" pouring in from miles around. In this setting of a world-town seemingly turned upside down, men freely engaged in bad behavior. But outbursts of male violence were not simply random. Rather, the commotion addressed many aspects of the social order. Although Euroamerican men engaged most commonly in recreational fighting with each other, they also created scenarios of displaced abjection because they attempted to raise their own precarious social status by picking fights with men of color.[208] In addition fights could express familial or class-based divisions within a community. Eruptions could also articulate the frustrations of bored, alienated men who led increasingly regimented lives. Circus Day, after all, represented a day away from work, and provided consumers with live images of the world beyond their borders.

At the turn of the century, normative circus images of male gender were a part of a burgeoning consumer culture targeted at young boys. Dime novels, such as "Tom Throttle, The Boy Engineer of the Midnight Express," placed vigorous Euroamerican boys squarely in the thick of dangerous, wholesome athletic adventures, often in foreign locales.[209] But as this chapter has argued, the circus's promise of an exciting world beyond provincial boundaries offered its consumers something even more startling. Alongside its manly white big-top athletes and animal trainers, the circus contained androgynous acrobats, drag clowns, "wild" animals, and animalized men—all of whom provided flexible exhibitions of male gender identity which challenged contemporary gender norms but still reflected the racialist standards of the day.

As national popular-culture forms, the circus and Wild West had a lack of local ties that compelled showmen to promote their spectacular gender performances on the national and international stage. The ubiquitous figure of the powerful Euroamerican male circus athlete in boy's fiction and show programs also resonated with theorists who linked physical fitness with nationalism. G. Stanley Hall exhorted his fellow educators to include physical exercise in their curricula because there were "many new reasons to believe that the best nations of the future will be those which give most intelligent care to the body."[210] At the turn of the century, many Americans were already confident that their country, having thoroughly trounced Spain in 1898, was fast becoming one of "the best nations." In this context, the circus's vigorous celebration of the "strenuous life" had national, indeed global, ramifications at the dawning of the "American Century."

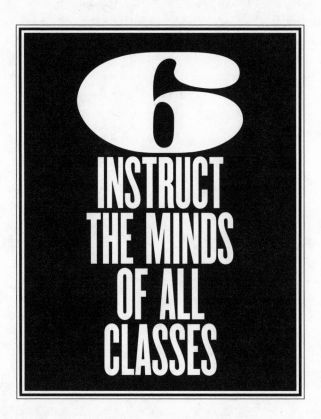

6

INSTRUCT THE MINDS OF ALL CLASSES

After the United States handily won the Spanish-American War in 1898, the circus proprietor Peter Sells was jubilant:[1] "We have placed an object lesson before the world that will cause tyrants tremble and inspire the oppressed with hope. We have taken our place at the very head in the front rank of nations. We have taught the whole world a lesson that has started every nation on earth to meditating upon the figure America is to cut in the world's politics of the future. . . . We have succeeded in bringing the English-speaking peoples of the whole world together into a moral alliance that will be of greater value to mankind than all the wars of all the nations for the past thousand years."[2]

Sells's bold pronouncement points to a unique moment in the history of the American circus. The turn of the twentieth century marked the first time that railroad showmen stressed the moral, political, and economic dominance of the United States in world affairs. It was also the only time that they produced an extensive array of extravagant, facsimile reenactments of American foreign relations mirroring actual overseas expansion.

To be sure, circuses had staged occasional dramatizations of foreign events throughout the nineteenth century, focusing on India and China in particular.[3] Yet these earlier, relatively puny displays made no larger claims about the capacity of the United States for world leadership. At the turn of the century, railroad proprietors crafted a range of U.S. foreign relations spectacles, thereby bolstering their assertions that their outfits were educational, "uplifting," and, according to a Barnum & Bailey program from 1894, able to "instruct the minds of all classes."[4]

In addition to adding didactic heft to their programs, the new empire was a rich source of new subject matter that helped the circus and Wild West to remain novel and salable. A press release for Buffalo Bill's new season in 1900 noted, "The military features have risen into greater prominence because they illustrate the things which are now in everybody's mind."[5] Indeed, the United States was involved in a flurry of military and economic activity overseas at the time. Although political and business interests crept into noncontiguous areas after the Civil War when the United States acquired Alaska and the Midway Islands in 1867, the nation's reach overseas accelerated in the 1880s and 1890s. The United States agreed to share Samoa with England and Germany in 1889. Four years later in Hawaii, American sugar growers helped overthrow the reigning monarch, Queen Liliuokalani, to secure a strategic gateway to Asian markets. In 1898 the United States annexed the islands by a simple congressional majority. After the Spanish-American War, the United States gained direct control over huge amounts of noncontiguous territory previously belonging to Spain and virtual sovereignty over Cuba through the Platt Amendment (1901). American military activity had become permanently far-flung, with troops stationed around the globe.

The circus and Wild West gave their vast audiences an immediate, intimate look at America's new position in world affairs with live translations of abstract foreign relations ideologies. A program for Buffalo Bill's Wild West in 1899 beckoned American audiences to meet "Strange People from Our New Possessions."[6] Tightly packed together by the thousands, spectators could share a cosmopolitan experience of the new empire that offered more than the purely visual sensation of reading about foreign affairs in a newspaper: they could smell the acrid odor of gunpowder during a battle re-enactment, or hear the soft sounds of Hawaiian grass skirts rustling during a hula demonstration.[7] Tody Hamilton, a press agent for Barnum & Bailey, recognized the circus's power to make America's military prowess intimate to its audiences when he advertised Barnum & Bailey's exhibition of model battleships from the Spanish-American War in 1903: "Many in the interior of America, who have their money voted yearly in vast appropriations for naval defense, have never even seen a battleship. It will be a pleasure to give them their first sight of the ships that defend them in times of war."[8] Providing people with their "first sight," the circus, as this chapter will contend, had tremendous power to help shape audiences' ideas about the expanding nation-state and its changing position in world affairs. Because the circus and Wild West also performed profitably across Europe, these popular imports were especially powerful promulgators of the nation's rising place in the world.

Circus and Wild West spectacles framed the new empire within the American exceptionalist tradition. However inaccurately, these amusements defined U.S. expansion as a distinct counterpoint to European formulations of formal empire solely characterized by colonization and military domination, because the nation's acquisition of noncontiguous territory was predicated on an abiding sense of moral "uplift" through economic intervention. As the historian William Appleman Williams put it, "In the realm of ideas and ideals, American policy is guided by three concerns. One is the warm, generous, humanitarian impulse to help other people solve their problems. The second is the principle of self-determination applied at the international level. . . . But the third idea . . . is one which insists that other people cannot *really* solve their problems and improve their lives unless they go about it in the same way as the United States" (emphasis in original)[9] As moral cheerleaders of expansionism, circus and Wild West owners echoed these paradoxical convictions in their staged spectacles. They also participated in the new empire in a complementary way, through the physical procurement of people and animals from other countries. American circus proprietors engaged in a form of diplomacy that should be characterized as informal because these showmen were nongovernmental actors. As sites of fun making, the circus and Wild West show had even greater ideological significance because their instructive messages were blanketed and thus naturalized in hair-raising fun.[10]

THE INTERNATIONAL POLITICS OF THE CIRCUS TRADE

Throughout its history, the circus was an international business—its extraordinary content comprised people and animals from faraway countries. But the turn-of-the-century circus partially owed its size and scope to European imperialism. As European countries began scrambling for parts of Africa in the 1860s, animal traders purchased a greater number of elephants, giraffes, hippopotamuses, zebras, and rhinoceroses for the American circus business and for allied amusements such as the zoological garden and world's fair, which also expanded during the second half of the nineteenth century. German firms dominated the market. The Hagenbeck family, in particular, was the chief supplier to Barnum & Bailey and to Adam Forepaugh's circus during the 1880s and 1890s. Concurrent technological developments like canals and railroads made exotic acts increasingly accessible throughout the world. As a result, turn-of-the-century showmen justly claimed that their programs were more authentic than ever before. One article, bluntly titled "Freak Hunting in India," contended that foreign freaks were no longer "made" but instead were "born," because railroad lines and telegraph networks in colonized countries had given impresarios easy access to a real "Wild Man of Borneo," whereas in the past, he would have been played—in the words of the press agent—by a "Virginia Darky." [11]

The international circus animal trade was a lucrative yet risky business. Traders contended with unfriendly local authorities, tropical diseases, strange food, and volatile weather. Carl Hagenbeck's son, Lorenz, remembered a particularly perilous thirty-seven-day sea journey off the eastern coast of Africa in 1906 while securing 2,000 dromedaries for the German government. Aboard the huge, rickety SS *Hans Menzel* with hundreds of dromedaries, eighty Arab dromedary handlers, European traders, and a full shipping crew, Hagenbeck and the others experienced constant, stifling heat, three fires on board, and violent storms which swept two dromedaries overboard. In his memoirs, Lorenz remembered the malodorous scene: "I was always fearing the worst. Below deck the heat was unbelievable. The combination of equatorial heat, bunker fires and the accumulated warmth of the crowded transport of dromedaries, produced a temperature which glued one's shirt to one's back. It was a glad moment every time I was at last able to clamber up, through the open hatch, away from it. If ever I had needed any additional proof of what a dromedary can stand, it was to hand now. They had been trained by centuries to hard conditions and chronic thirst." [12]

The animal trade was most treacherous for the local men hired to capture big game. Carl Hagenbeck described in detail how his traders worked with provincial African rulers to gain permission to hunt and to hire hunters. Hagenbeck's traders brought delicacies and cash to the authorities, who in turn fed them and provided entertainment: war dances, female dances, camel and horse races, and sham fights. The following day, crowds of village men came seeking employment as hunters. Hagenbeck and his German hunters paid local leaders for the privilege of using their land, and gave indigenous hunters a small wage for the dangerous task of ambushing wild game. Hagenbeck observed: "But for the natives big-game hunting is a very different matter. Then the fray is far from one-sided; the weapons of the man are little, if it all, superior to those of the brute; and the 'hunting' is more of the nature of the human combatant. Should a horse stumble—an accident which, on that uneven ground, intersected by underground streams, is only too likely to happen—death either to the animal or its rider is the probable result. We need not be surprised that the Sudanese assert that the professional elephant hunter never dies at home, but ends sooner or later under the tusks and feet of a hunted elephant." [13]

W. C. Coup, a veteran circus owner, recorded similar observations from an elephant hunt in the Sudan by Paul Ruhe, a "master-hunter" with the Reiche Brothers: "First we try to distract the attention of the female from her young. Then a native creeps cautiously in from behind and with one cut of a heavy broad-bladed knife severs the tendons of her hind legs. She is then disabled and falls to the ground. We promptly kill her, secure the ivory and capture the little one. Of course we sometimes have a native or two killed in this kind of a hunt; but they don't cost much—only five to six dollars apiece. The sheiks are paid in advance, and do not care whether the poor huntsmen get out of the chase alive or not." [14]

Coup's chronicle of this indifference toward indigenous hunters was common among showmen and traders. In line with the marketing of the rogue tiger or killer elephant, circus owners heavily advertised the attendant risks of the hunt and the numbers of local people who assisted in the capture of wild animals. As early as 1851 P. T. Barnum's Asiatic Caravan claimed that a "drove of elephants was captured in the jungles of Central Ceylon, by Messers. Stebbins, June and George Nutter, accompanied by 160 natives, after a pursuit of three weeks and four days in the jungles." [15] The racial politics of the hunt for circus animals mirrored the racial division of labor in colonial societies across the globe. From South Africa to Vietnam, imperial authorities hired colonial subjects at scant wages to perform dangerous jobs: mining, ditch digging, and railroad construction.

The hunt was also an omnipresent part of imperial popular culture. [16] In British India, colonial administrators spent their leisure time in tiger-rich places like Rajasthan and the Sunderbans, stalking and killing Bengal tigers. During the hunt (*shikar*), some British administrators and Indian royals (*rajas* and *nawabs*) mixed freely in a social setting.[17] Other Indians participated in the hunt in roles that reified colonial hierarchies: as servants hired to carry water, food, and supplies. In the late nineteenth century and the early twentieth, upper-class American men also took part in colonial hunts. In addition to his forays into the Dakotas and other parts of North America, Theodore Roosevelt shot wild animals in South America and Africa. The hunter and taxidermist Carl Akeley, who created the Hall of African Mammals at the American Museum of Natural History, logged five safaris to Africa beginning in 1896. Armed with cameras and guns, he treated the black African personnel who accompanied him as children. Akeley reprimanded his assistant, "Bill" (whose real name was Wimbia Gikungu), for shooting a gun, an act that represented, in Donna Haraway's words, a "usurpation of maturity." Black Africans were typically prohibited from hunting independently with guns in the presence of white men, thus illustrating how imperial governments used the hunt as an instrument of colonial domination.[18] Earlier, Akeley had a direct tie to the circus: in 1885 he helped stuff P. T. Barnum's elephant Jumbo for a traveling display.

The reach of the American circus was immense, from small hamlets in the United States to the Kalahari Desert and beyond. But the scope of this international business was dependent upon colonial stability. During periods of rebellion, the animal trade shrank. The African trade in camels, elephants, black rhinoceroses, and hippopotamuses fell sharply after 1879, when the Mahdi's rebellion in Egypt and the Sudan began.[19] In other instances, circus animal traders helped solidify colonial authority. Lorenz Hagenbeck's perilous journey aboard the SS *Hans Menzel* in 1906 was sponsored by the German government, which needed dromedaries for its army in German South-West Africa to quash the Herero tribal rebellion on the Kalahari Desert.[20]

In addition to securing animal performers, showmen from the United States conducted informal foreign relations to recruit human circus workers. Acrobats, for instance, helped inaugurate diplomatic relations between the United States and Japan in the mid-nineteenth century. The first American consul general in Japan, Townsend Harris, saw remarkable Japanese jugglers performing for the emperor at Yeddo in 1858, just five years after Commodore Matthew C. Perry first landed there. One juggler could catch a spinning top, make it land on the edge of a sword near the hilt, and send it skittering to the tip.[21] Several years later, the Japanese tycoon finally allowed the performers to tour the United States. Their ship landed at San Francisco in 1867, thereby making this troupe of twenty male and female acrobats, top-spinners, butterflyers, and magicians the first Japanese to receive visas in the United States, as members of "Professor Risley's Imperial Troupe of Wonderful Japanese."[22] Attracting crowds of several thousand, the troupe gave audiences a powerful "first sight" of Japanese people, an image heightened by their spectacular, seemingly ascetic bodily stunts of revolving pyramids, acrobatics, and contortionism. President Andrew Johnson received the performers at the White House and proclaimed their visit the beginning of a profitable relationship between the United States and Japan.[23]

Carl Hagenbeck's diplomacy with the British colonial government reveals the extensive international networks that showmen relied on during the hiring process. Initially, Hagenbeck's request to hire South Asian players for his "Triple Circus East India Exposition" in 1906 was shuttled back and forth between British army officers and a provincial secretary. Finally, influential friends and his maze of animal agents, many of whom had ties with local *rajas* and *nawabs*, intervened on Hagenbeck's behalf and got him credentials to see the viceroy of India, who grudgingly gave his consent. But the viceroy stipulated that Hagenbeck deposit a bond of £5,000 into the royal treasury to insure proper care of the circus performers. Hagenbeck's agents helped him hire two men, two women, and two children from the various "Hindoo races," including "fighting" Rajputs and "weak" Bengalis, all of whom were "physically perfect representatives of their races." Hagenbeck employed one hundred South Asians, who in early 1906 traveled from Calcutta to London and then on to New York.[24]

On a few occasions, proprietors did not even bother getting permission from foreign governments, or from the entertainers themselves. Perhaps the most horrific example of a documented kidnapping involved a group of Aborigines from Palm Island, Australia, who were abducted by P. T. Barnum's agents J. B. Gaylord and R. A. Cunningham. In January 1883 the *Spirit of the Times* reported, "One of our Australian correspondents encountered at Queensland, bargaining for specimens of the Moira tribe, the irrepressible J. B. Gaylord, who is touring the world in search of curiosities for Barnum."[25] Soon after, Cunningham took the Aborigines on tour in the United States with Barnum, Bailey and Hutchinson's circus. Standing four foot eleven inches tall, Jimmy Tambo played Tambo Tambo, a "Ranting Man-Eater" at the sideshow. After performing in Cleveland, Ohio, Tambo died and was left behind in the basement of a local funeral home until 1993, when his descendants and a British anthropologist located his embalmed body and returned it to Palm Island for proper burial with full rites.[26]

FROM BRIC A BRAC TO SPECTACLE

From its inception in 1793, the American circus and related amusements like the museum and menagerie capitalized on exotic animals, people, and artifacts from colonized areas. Early examples of the outlandish included "Iranistan," the mansion in Bridgeport, Connecticut, of the future circus proprietor P. T. Barnum, erected in 1846. Costing over $150,000 at the time,[27] the palace was based upon the architecture of the Brighton Pavilion (home of the British prince regent), which in turn was inspired by the colonial Indo-Saracenic architecture of British India.[28] In New York City, roving sea captains and other collectors provided Barnum's profitable American Museum with an enormous international hodgepodge of extraordinary living and dead creatures, and historical relics. One could find the following items in Case 794, at the Sixth Saloon: "Ball of Hair found in the Stomach of a Sow; Indian Collar, composed of grizzly bear claws; Sword of a Swordfish penetrating through the side of a ship; Algerine Cartouche Box; Algerine boarding pike; African pocket-book; Chinese pillow; Horses foot, injected; a petrified piece of Pork, which was recovered from the water after being immersed sixty years; Fragments of the first canal-boat which reached New York through the Canals; African Sandals; Turkish Shoes; Sultan Slippers; Turkish Slippers; Ancient Iron Breast-plate, found in Wall Street, 1816; Arabian Bridle; Wrought metal Mexican Stirrup; Turkish Ladies' Boots."[29]

However, these early images paled in comparison to the railroad circus's full-blown imperial scenes at the end of the nineteenth century. By the 1890s showmen had transmogrified happenstance foreign objects into teleological exhibitions of American nationalism.[30] Huge circus specs transformed what the British historian John Springhall has called "little wars of empire" into entertaining proof of the growing influence of the United States on the world stage.[31] These widely circulating performances promoted broad support for American military activity, using selective scenes from generally "popular" wars like the Spanish-American War and the Indian Wars. Collectively, these specs portrayed the United States as a democratic republic whose style of government, economic system, and "way of life" spread worldwide would herald a utopian age of unprecedented prosperity. These specs reiterated the theme in different ways: as narratives of national origins, as frontier stories, as au courant reenactments of contemporary events, and as crusading, missionary-style narratives of Christian progress.

The wars that the circus and Wild West chose to ignore were as significant as the ones they reenacted. Scenes from the Mexican War and the American Civil War—wars that reminded the nation of its slavery past— were virtually absent. Although individual amusement proprietors toured urban areas during the Mexican War with bulky moving panoramas (vast paintings on canvas panels sewn together and wound from one huge roller to another to simulate a moving landscape), circus owners did not incorporate these scenes into their exhibitions given the escalating sectional crisis that followed.[32] The Civil War itself, moreover, could not be easily watered down so as to be palatable on both sides of the Mason-Dixon Line. The circus reenacted Civil War battles only sporadically: in 1862 as part of Yankee Robinson's production "Three Great Epochs," in 1863, in his "Battlefield of Antietam," and again in 1887 when the Sells Brothers circus executed a brief reenactment from the battle of Shiloh, a "soul-stirring artillery race." But the outfit's star feature that year was "Pawnee Bill's Historical Wild West."[33] At the turn of the century, only a couple of circuses made even brief reference to the Civil War. In 1905 the Adam Forepaugh & Sells Brothers circus opened with a military pageant of men (including Theodore Roosevelt's Rough Riders) dressed in uniform from every American war. The show program did not specify whether the Civil War troops would be Union or Confederate. The pageant's effect was largely ceremonial, as hundreds of soldiers dressed in the appropriate uniforms simply marched around the big top to the accompaniment of a brass band. Lumped together with other American wars, the Civil War, now neutralized, became an undifferentiated part of the nation's historical march to the present.[34]

Just as world's fairs commemorated seminal historical occasions like the Louisiana Purchase of 1803, the circus and Wild West used the past to substantiate contemporary nationalist celebrations of the new empire. The Hungarian immigrant brothers Imre and Bolossy Kiralfy created several spectacles, notably "Columbus And the Discovery of America," an expensive production that linked the arrival of Columbus on North American soil to later nation building. The spec reportedly cost $500,000, an expenditure which the circus freely advertised.[35] In Keokuk, Iowa, the *Constitution-Democrat* stated that Barnum & Bailey had spent $75,000 for scenery,

Figure 33. "Columbus Takes Possession of a New World," Barnum & Bailey, 1892.
In this painterly romantic landscape poster, Columbus stakes his claim to the New World
before an obedient crew and obeisant Native Americans. (Lithograph courtesy of Circus
World Museum, Baraboo, Wis., with permission from Ringling Bros. and Barnum &
Bailey,® The Greatest Show on Earth,® B+B-NL38-92-1F-16)

$250,000 for costumes, $50,000 for armor and weapons,[36] and the rest for incidental expenses. The spectacle announced that Columbus had been the first to ignite the "engine" of American progress. A poster in 1892 grandly proclaimed that the production told the "Whole History of a Great Nation." As evidence of the circus's fusion of the instructive and the titillating, hundreds of dancing girls writhed under the moonlit sky at the Alhambra palace, upon Columbus's departure from Spain.[37]

At the circus, Columbus "takes possession" of the New World, as obsequious shipmates and Native Americans bow at his feet (fig. 33). While Columbus meditates, "a *vision of progress and civilization* appears before his ecstatic eyes, revealing to him the wonderful results in invention, science and art which future generations will glorify as the result of his stupendous discovery" (emphasis in original).[38] Commemorating the four hundredth anniversary of Columbus's travels across the Atlantic, the spectacle was an "invented tradition," as it used Columbus's mistaken journey for South Asian spices to validate contemporary notions of Euroamerican progress and empire with the heft of ostensible historical precedent.[39]

The American Revolution was another part of this origins narrative of contemporary nationalism. Showmen characterized the Revolution as a contest between a despotic, colonial authority and "freedom-loving" "indigenous" people, a victory of republicanism over monarchy, and a blueprint for democratic self-determination throughout the world. Even in the 1790s, circuses already reenacted Revolutionary battle scenes. Performing at Philadelphia in May 1798, a circus contained a political play: "The Death of Major André, and Arnold's Treachery: Or, West Point Preserved, Three Acts."[40] But in the late nineteenth century, spectacles used the Revolution to make broader claims for America's global authority. In 1893 the Adam Forepaugh circus exhibited "1776, Historic Scenes and Battles of the American Revolution" in nine scenes, declaring, "The Revolution made us a nation destined to be the very greatest of all the earth, consequently it is the principal and primal event in our history."[41] The cast of the spec was composed of fifty-one military officers, actors portraying all of the signers of the Declaration of Independence, and hundreds of chorus members, playing the Green Mountain Boys and the Pioneers. Proclaiming the Declaration of Independence "The Greatest Document ever written being signed by its Brave Upholders," the spec breathlessly described "The Waiting Mob without. The Glad Tones of Old Liberty Bell. 'Freedom to all men throughout the Land.'" The production ended with President Washington's inauguration, as chorus members cheered, "Long Live George Washington, President of the United States."[42]

In "1776" the "mob" was passionate, yet peaceful. Unlike French firebrands who massacred members of the aristocracy during the Reign of Terror in 1793, these American revolutionaries merely destroyed a statue of George III, not the king himself, nor did they force a wholesale transformation of the social fabric. They eagerly celebrated the restoration of order through the democratic election of a "natural" aristocrat, George Washington (plate 7). The French Revolution, which caused the destruction of the ancien régime, was never reenacted at the circus. Circus spectacles depicted the restrained, not radical, transfer of power as an "exceptional," uniquely American impulse, to be followed throughout the world.

Circus and Wild West shows linked the Indian Wars to this larger origins narrative—as part of the nation's "inevitable" movement from "savagery" to civilization. These reenactments reflected contemporary foreign relations ideologies for two additional reasons: at the turn of the century, most Native Americans were literal foreigners in their own country, because they were not uniformly recognized by federal law as U.S. citizens until 1924. Secondly, the Indian Wars—indeed, the whole history of Euroamerican relations with Native Americans—was a colonizing project, an ideological and strategic blueprint for subsequent U.S. entanglements overseas.

Buffalo Bill Cody's dramatization of the Battle of the Little Big Horn absolved General Custer of any personal responsibility for the ambush, at which a concert of Plains warriors killed every soldier under his command in 1876. At the Wild West, Custer (a fearless, yet dim warrior who graduated last in his class at West Point) became a heroic martyr whose 365 men stood no chance against 6,000 Indians. First produced in London during a European tour (1887–92) and in the United States at the Columbian Exposition in 1893, "Custer's Last Fight; or, The Battle of the 'Little Big Horn,'" characterized the actual battle as "savage warfare, in which the foe was absolutely merciless, where capture meant torture, added to death; where no quarter was given, and no prisoners taken."[43] In the epilogue, Cody recast a minor confrontation between federal troops and a small band of Cheyenne riders (who had not even participated in the Battle of the Little Bighorn) as retaliation for Custer's loss. This incident on which the coda was based happened twenty-two days after the battle itself, when Cody killed and scalped the Cheyenne chief Yellow Hand.[44] For theatrical effectiveness, Cody wrote in the show program that his decision to join federal troops had been based upon his desire for revenge—but, as the historian Richard Slotkin points out—Cody in fact enlisted some three weeks *before* the Battle of the Little Big Horn took place.[45] Transforming a bumbling gunshot death into an exciting hand-to-hand knife fight, the illustrated show program depicts Cody standing over Yellow Hand's body at the edge of a cliff, poised with a knife in his right hand, and the bloody scalp and headdress clenched in his outstretched left hand. From below, scores of federal troops look up solemnly at Buffalo Bill. The caption reads, "Death of Yellow Hand—Cody's first scalp for Custer."[46]

By concluding the exhibition of defeat with a declaration of federal power, Cody negated the unsettling ideological implications of the Native Americans' victory at the Little Big Horn.[47] Custer's widow, Elizabeth, for one, greatly appreciated Cody's seamless interpretation of the past. "I have been rejoicing for nearly three weeks in the success of your exhibition for your sake and also that you are teaching the youth the history of our country where the noble officers, soldiers and scouts sacrificed so much for the sake of our nation's land. I always thank you from my heart for all that you have done to keep my husband's memory green. You have done so much to make him an idol among the children and young people."[48]

As a positive exhibition of recent events, Cody's production did little to acknowledge the genesis of the Battle of the Little Bighorn: show programs did not mention the federal government's decision to open sacred reservation lands in the Black Hills to gold prospectors in 1875, when in fact this was a catalyst for war. Programs did prominently display a photograph of Buffalo Bill (with rifle in hand) shaking hands with the Hunkpapa Sioux chief Sitting Bull: "Foes in '76 — Friends in '85."[49] Sitting Bull had been a brilliant military strategist at the Battle of the Little Big Horn in 1876. The next year, he led a band of almost 1,000 people to Canada to escape federal troops, but found little to eat during nearly four years of difficult exile. Returning with his band in 1881, Sitting Bull surrendered to federal authorities and was imprisoned at Fort Randall in the Dakota Territory; eventually he was forced to farm an allotment and live in a log cabin at the Standing Rock Agency, where in 1885 he met Cody. He toured with Cody's Wild West show for one highly successful season. After returning to Standing Rock, Sitting Bull remained a vocal critic of federal Indian policy, particularly the Dawes Allotment Act of 1887, which in exchange for dubious promises of citizenship imposed individual land allotments on the Indians, who had always practiced communal land ownership. This devastating piece of federal legislation opened up approximately 15 million acres for Euroamerican settlement. Sitting Bull became a participant in the Ghost Dance Movement, initiated by the Paiute holy man Wovoka in 1889: this promised the imminent arrival of a messiah who would remove the whites and return the buffalo if followers danced a Ghost Dance and wore magical clothing to ward off Euroamerican violence. Alarmed at the spread of this resistance movement, U.S. military leaders followed a familiar policy of extermination. In mid-December 1890, Cody and Sitting Bull met for a final time at the Standing Rock Reservation in South Dakota, after General Miles had asked Cody to convince Sitting Bull to meet with U.S. Army officials; the meeting ended horribly, however, when Sitting Bull was accidentally shot and killed by zealous U.S. marshals.[50] On December 29 the military slaughtered Lakota families at Wounded Knee, South Dakota. Shortly thereafter, Adam Forepaugh's circus translated tragedy into spectacle with a series of Ghost Dance scenes, "recalling the Weird Scenes of the late outbreak and the terrible frenzy of the Savages seeking the advent of the Red Messiah."[51]

The Indian Wars provided a precedent and guide for future overseas military conduct by the United States during the Spanish-American War and its subsequent occupation of the Philippines (1898–1946).[52] Walter Williams asserts that the U.S. government's relations with Native Americans during the nineteenth century, and especially the Supreme Court's ruling in *Cherokee Nation v. Georgia* in 1831, provided the United States with a model for governing colonial subjects.[53] Richard Drinnon traces the interconnectedness of continental and overseas empire building even further back, to colonial efforts to exterminate Indians after the Powhatan chief Opechancanaugh's attack on Jamestown in 1622, and the 1637 Pequot War fought in the Massachusetts Bay colony.[54] More than two and a half centuries later, most U.S. regiments fighting in the Philippines were from western states, and a full 87 percent of all U.S. generals in the Philippines had fought in the Indian Wars. American soldiers referred to the Filipino nationalist Emilio Aguinaldo as "Tecumseh" or "Sitting Bull," and in describing Filipino military tactics used the phrase "injun up."[55] In declaring Filipinos "unfit" for self-government, Vice President Theodore Roosevelt claimed, "To grant self-government to Luzon under [the Filipino leader] Aguinaldo would be like granting self-government to an Apache reservation under some local chief."[56] Circus proprietors offered identical representational strategies. They juxtaposed scenes from the Spanish-American War with earlier battle scenes from the Indian Wars and featured Native American actors playing the Spanish and Filipinos as living symbols of the continuity of U.S. expansionism.

Cody's vivid, "authentic" Wild West also attracted thousands of European immigrants to the American West. During Cody's many European tours from the late 1880s to the early years of the twentieth century, millions of Europeans formed their first impressions of the trans-Mississippi West based on his live reenactments.[57] As further testimony to the pervasiveness of Buffalo Bill's frontier among European spectators, Wild West dime novels sprang up throughout Europe at the turn of the century, including Germany (*Texas Jack "Der große Kundschafter"*), France (*Texas Jack la Terreur des Indiens*), and Italy (*Buffalo Bill L'Eroe del Wild West*).[58] The Wild

West promoted the trans-Mississippi West as a vast empire of liberty for white cultivators who chose to "improve" or "civilize" western lands, while it simultaneously claimed that this boundless continental frontier (and its Indian inhabitants) was rapidly disappearing in the face of such settlement.

As the political and economic interests of the United States became increasingly globalized during the 1890s, the focus of the circus and Wild West also expanded. Impresarios enlarged their exhibitions of the American "frontier" to include overseas frontiers. These interpretations were strikingly similar to the ways in which contemporary American intellectuals and policymakers now conceptualized the frontier: the Far East had become part of the "Wild West" as a site for new American markets and cultural commerce. In 1893 Buffalo Bill's Wild West became Buffalo Bill's Wild West and Congress of Rough Riders of the World. L. G. Moses emphasizes that this change was prompted initially by the fears of Cody's co-owner Nate Salsbury that assimilation-minded reformers might convince the Interior Department to ban Indian employment altogether at the Wild West show.[59] The new name also capitalized on the growing visibility of America's global frontier. "Cowboys" from around the world congregated at the Wild West: Cossack, Mexican, Arab, and Syrian cowboys performed feats of horsemanship in colorful native costumes.[60] After the Spanish-American War, Buffalo Bill's Congress of Rough Riders of the World now included Cuban, Hawaiian, Puerto Rican, and Filipino "cowboys."[61] In 1909 Buffalo Bill merged his operations with those of another Wild West owner, Major Gordon W. Lillie ("Pawnee Bill," who had worked as an interpreter with the Pawnee) to form Buffalo Bill's Wild West combined with Pawnee Bill's Great Far East, or, as it was known to show workers, the Two Bills show. In the initial grand review, "The Red Men of two hemispheres ride side by side and Many Nations contribute Man and Beast to a Triumphal March of the Ethnological Congress."[62]

"A SPLENDID LITTLE WAR . . ."

At the onset of the Spanish-American War, William F. Cody publicly expressed his desire to send "30,000 braves" to Cuba.[63] His Wild West opened the 1898 season with a Color Guard of Cuban Veterans who marched around the arena. The program noted that all were "on leave of absence in order to give their various wounds time to heal, all have fought for the flag of Cuba and will soon return to that country to act as scouts and guides, for which their familiarity with the topography of the island especially commends them."[64] When the war ended on August 12, audience members at the Forepaugh & Sells Brothers circus in Beloit, Kansas, erupted after a worker interrupted the performance to read a telegram announcing the big news. The circus route book noted: "[This] was the occasion for a mighty ovation, the entire audience rising to their feet and cheering the stars and stripes. Merrick's Military band struck up the air of 'Star Spangled Banner,' and again the audience burst into an uproar. It was surely a memorable occasion."[65] The national mood was jubilant after a four-month war that resulted in few U.S. casualties (385 battle deaths, 2,061 deaths from other causes, and 1,662 nonmortal wounds), and the acquisition of overseas territories that remapped the nation's position in world affairs.[66]

The circus and Wild West portrayed the Spanish-American War not in terms of colonial conquest but as evidence of liberal progress and democratic equality. Subsequent U.S. actions, however, clearly belied such lofty aims. Relations with the Philippines quickly spiraled into a protracted guerrilla war after McKinley annexed the islands with the Treaty of Paris in February 1899. More than 200,000 Filipinos died from battle wounds, famine, and disease from 1899 to 1902, and sporadic fighting continued thereafter. In Cuba, U.S. intervention soon turned into a military occupation that transformed the newly liberated nation into a virtual U.S. colony under the auspices of the Platt Amendment of 1901. In Puerto Rico, U.S. financial policies prompted local elites to consolidate landholdings so that they could grow crops primarily for export.[67] After the U.S. military began its occupation of Haiti (1915–34), the U.S. government drained Haiti's treasury, controlled its customs houses, introduced racial segregation into hotels, clubs, and restaurants, reinstituted a corvee system of forced labor for road building, and enforced a "shoot on sight" curfew in the wake of the mob killing of President Vilbrun G. Sam. Brenda Gayle Plummer writes that U.S. policies in Haiti during the occupation constituted formal, rather than informal, imperialism.[68] But even the U.S. ideal of economic development and free markets, William Appleman Williams contends, contained the seeds of formal empire. "When an advanced industrial nation plays, or tries to play, a controlling and one-sided role in the development of a weaker economy, then the policy of the more powerful country can with accuracy and candor only be described as imperial."[69] Despite the contradictions between the rhetoric of self-determination and actual military domination, many prominent Americans like William F. Cody supported expansion by the United States. Cody, for instance, voiced his support for its rule of the Philippines in his correspondence with his friend Theodore Roosevelt.[70]

Figure 34. "American Fleet of Battleships," Barnum & Bailey, 1899. Although touring
Europe during the Spanish–American War, the circus immediately fashioned a facsimile
reenactment of selected triumphal war scenes on an ocean of "real water" to promulgate
America's rising position in the world. (Lithograph courtesy of Circus World Museum,
Baraboo, Wis., with permission from Ringling Bros. and Barnum & Bailey,®
The Greatest Show on Earth,® B+B-NL200-99-1F-2)

Artifacts of war saturated the circus and Wild West to such an extent
that one newspaper writer was prompted to compare the war in the Philippines to the Wild West show: "The theory of the Administration is that the
trouble in the Philippines is like the Wild West show. It isn't war, but it looks
a good deal like it."[71] In 1899 Barnum & Bailey fashioned "America's Great
Naval Victory at Santiago" while touring Europe (fig. 34). The spectacle
was "Presented on a Miniature Ocean of Real Water, with real War Ships,
Guns, and Explosives," and ended with the annihilation of the Spanish fleet
while "The Star Spangled Banner" played in the background.[72] Viewed by

countless thousands of Europeans, the circus announced to British, French, and German rivals America's new position on the world stage and the power of its expanding navy.

Upon returning to the United States in 1903, Barnum & Bailey exhibited official models of U.S. warships (based on plans from Secretary of the Navy John D. Long),[73] ranging in length from three to nine feet.[74] The 1904 Barnum & Bailey program announced: "These models were built as an expression of the appreciation of the management of this exhibition of the power and glory of the American navy as so magnificently manifested during the late Spanish-American war. [Because the circus was traveling in England] . . . the people with the . . . show realize, perhaps, more fully than their fellow-Americans who have remained at home, the magic potency of the name 'American' that has been given to it by its splendid navy."[75] The presence of miniature battleships, government-sanctioned objects, enhanced Barnum & Bailey's credibility as a respectable source of information about current events. As the press agent Tody Hamilton had stated earlier, "It will be a pleasure to give them their first sight of the ships that defend them in times of war."[76] The battleships were powerful fetish objects, linking an abstract, faraway war to an intimate material reality, a riveting "first sight" that could be inspected and touched by curious circus-goers, thereby giving Americans a concrete sense of how the government was spending their excise tax dollars.

A year after the Spanish-American War, Buffalo Bill's production "Battle of San Juan Hill" exhibited the multiracial Congress of Rough Riders doing battle with the Spanish. The artists playing Theodore Roosevelt's regiment of Rough Riders included Euroamericans, Cubans, African Americans, and, in a twist of intentional racial disguise, Native Americans as the Spanish villains.[77] In the second act, "The Rough Riders' Immortal Charge," Roosevelt's "virile" regiment defeated the "wine-soaked" Spanish through "manly" courage and discipline: "There is a frantic yell of admiration and approval as the soldiers—*white, red and black*—spring from their cowering position of utter helplessness and follow Roosevelt and the flag. On and ever onward they leap, struggle and crawl. . . . The Spaniards cannot believe that so small a force would dare an assault so forlorn of all hope. They erroneously infer that an army is charging close behind it, and as it breathlessly comes closely on for a hand-to-hand death grapple, they pale, they flinch, and at last they turn and fly in panic. Their gold and crimson emblem of ruthless oppression is torn from the ramparts, and Old Glory streams on the breeze, triumphant in its place" (emphasis in original)[78]

Showmen's seamless spectacles of the Spanish-American War dramatized a racially and economically diverse American military destroying the
"decrepit" and "effete" Spanish colonial empire and (ostensibly) replacing it
with a democratic American empire of liberty and free markets.[79] African
American soldiers played a crucial role in the Rough Riders' victory, and
show programs acknowledged their presence. In 1900 Buffalo Bill featured
Negro Cavalrymen in a new act.[80] Programs for the "Battle of San Juan
Hill" celebrated the presence of multiracial American troops in the Spanish-
American War as evidence of the readiness of the United States to "guide"
the rest of the world by its own multicultural model.

After such hair-raising recreations of dramatic Spanish-American War
battle scenes, Panama Canal diplomacy was a seemingly bland, curious entertainment subject. The acquisition of the canal zone by the United States
involved no formal declaration of war or exciting military battles. In 1903
the U.S. government gave strong financial and military support to the Pana-
manian nationalist movement which sought independence from Colombia.
After immediately recognizing Panama, the United States paid $10 million
plus $250,000 a year[81] for rights to the Canal Zone, a swath of land ten miles
wide that divided the country in two.[82] The Panamanians immediately realized that they had only partial control of their new country, especially after
the United States began building the canal. Some members of Roosevelt's
cabinet opposed acquisition of the Canal Zone, as did several newspapers.
The Hearst papers proclaimed that the "Panama foray is nefarious . . . a quite
unexampled instance of foul play in American politics."[83] Yet most Americans fully supported Roosevelt's efforts to control the Canal, and cared little
about how he got it.[84]

The circus transformed the questionable acquisition of the Canal Zone
into an unquestioning public celebration of American might. Adam Forepaugh & Sells Brothers' spectacle "Panama; or, The Portals of the Sea; or,
The Stars and Stripes" vigorously publicized the canal's construction. The
spec opened the program and featured "The Grand, Imposing, Majestic,
Military, Ideal."[85] Couriers stated that the production "Idealiz[ed] in He-
raldic and Military Magnificence America's Opening of the Panama Canal
to the Nations of the World. Historical in its Magnificent Lessons, Educational in its Elevating Nature, Moral in its Imposing Theme, Patriotic in its
Noble Tendencies, Unparalleled in its Tremendous Magnitude."[86] Forecast-
ing the ideological tenor of the Panama-Pacific Exposition in San Francisco
(1915–16), this spec predicted that the canal would be an oceanic bridge,
bringing prosperity and democracy to the rest of the world.

The circus also remade into spectacle the tedious course of actual diplomacy. Like the "Canal" production, Barnum & Bailey's spec "Peace, America's Immortal Triumph" (1906) contained no thrilling battle reenactments. Written by Bolossy Kiralfy, the production reconstituted an event that seemingly had little impact upon American life—Theodore Roosevelt's mediation in 1905 of the Portsmouth Treaty, which ended the Russo-Japanese War—into a colorful paean to rising U.S. globalism. In the spec, several European countries (represented by ponderous floats) try to reconcile Japan and Russia, but fail. Finally, the U.S. "Columbia" float appears. Represented by a towering, beneficent female statue named "Peace," the United States quickly resolves the war, without firing a single shot. According to Barnum & Bailey's route book: "At this moment the portrait of President Roosevelt is unfolded and the Angels of Peace sound their trumpets proclaiming to the world that the great deed has been accomplished and that Peace reigns throughout the world."[87] "Peace" ended when the American float left the stage, flanked by jubilant Japanese and Russians.

The spec's "happily-ever-after" conclusion—not surprisingly—belied actual events. Although the Portsmouth Peace Treaty ostensibly honored the open door by declaring that no power (Japan or Russia) should be allowed to establish its own colonial sphere of influence in China, the agreement hardly guaranteed peace. European monarchs had asked Roosevelt to mediate the peace, because they were concerned about the dangerous implications of the antitsarist revolution of 1905 in Russia, which coincided with the Russo-Japanese War. The resulting treaty favored the Japanese by granting Japan controlling interests in Korea, the southern half of Sakhalin Island, and several Chinese ports.[88] The Portsmouth Treaty thus helped sow the seeds of Russian instability and growing Japanese domination on the world stage. In *Collier's* magazine, W. E. B. Du Bois predicted that Japan's victory marked the beginning of the end of "white supremacy" throughout the world. Likewise, Haitian intellectuals called for a "Meiji Restoration" for Haiti.[89]

As a whole, showmen predicted that the benefits of an American empire of liberty would soon be felt worldwide. Peter Sells, proprietor of the Adam Forepaugh & Sells Brothers circus, proclaimed: "The building of the Nicaragua canal, enlargement of our standing army, a navy that will equal, if not surpass that of any other nation on earth, a wise and just supervision over Cuba and Porto Rico, the acquisition of ports at several points in Asia for coaling and resting stations for our commercial and war ships, are all essential to the future welfare of this nation."[90] Captain Alfred Thayer

Mahan also called for an enlarged U.S. navy to protect America's increasingly far-flung commercial interests: "The ships that thus sail to and fro must have secure ports to which to return, and must, as far as possible, be followed by the protection of their country throughout their country throughout the voyage. . . . The necessity of a navy . . . springs, therefore, from the existence of a peaceful shipping."[91] In an address to the American Historical Association at the 1893 Columbian Exposition in Chicago, Frederick Jackson Turner set forth his frontier thesis, which linked the nation's future prosperity to its ability to solve the crisis of overproduction by establishing (and protecting) new overseas markets. But as the co-owner of a vastly popular, live, nomadic show, Peter Sells advanced expansionist ideology in a far more accessible medium than print. The circus, a national entertainment, projected an ostensibly unifying patriotic consensus to a diverse audiences of immigrants, African Americans, Native Americans, and native-born whites — although those same audiences often used Circus Day and Buffalo Bill Day as opportunities to engage in fractious, racially charged altercations with each other.

CIRCUS CRUSADES

Some spectacles focused on countries where the United States had few economic, political, or strategic interests. Portraying the United States as a Christian nation led by chaste, manly Euroamerican capitalists, circus expositions of Christian "industry" and "civilization" complemented the goals of missionaries at the turn of the century. Admiral George Dewey and Alfred Thayer Mahan, among other well-known imperialists, supported the YMCA's missionary work overseas and Dwight Moody's Student Volunteer Movement, because these private religious groups helped establish American capitalist culture abroad and new markets for American goods.[92] Woodrow Wilson and William Howard Taft both believed that the spread of Christianity abroad formed "the only basis for the hope of modern civilization."[93]

Yet in 1899, members of the Chinese group, the Harmonious Righteous Fists, challenged this missionary-capitalist nexus. Sanctioned by the Empress Dowager, leader of the disintegrating Manchu dynasty, the xenophobic "Boxers" (as they were called in the West) terrorized foreign missionaries and private citizens as part of their plan to remove outsiders from Chinese soil. In response, President William McKinley ordered 5,000 troops to the Chinese mainland. Other countries with spheres of influence

[*Instruct the Minds of All Classes*]

Figure 35. "The Rescue at Pekin," Buffalo Bill's Wild West and Congress of Rough Riders of the World, 1901. Depicting the United States as the leader of an international consortium preserving the "Open Door" to China, this spectacle underscored the continuity of continental and overseas expansion by hiring Native Americans to play the Chinese Boxers. (Lithograph courtesy of Circus World Museum, Baraboo, Wis., BBWW-NL4-01-1F-2)

in China also sent troops to protect their own citizens and investments. Within a few months, the rebellion was quashed and foreign powers agreed to obey the "open door" notes. Written by Secretary of State John Hay in 1899 and 1900, the notes asked all powers to preserve "Chinese territorial and administrative integrity."[94]

In 1901 Buffalo Bill's Wild West reenacted the uprising and its immediate aftermath in a brand-new spec, the "Battle of Tien-Tsin" (or the "Rescue at Pekin") (fig. 35). Colorful regiments of American, German, Russian, British, Japanese, French, and South Asian Sikh "cowboys" thundered around a sawdust arena, shooting at queue-clad Native Americans playing the Boxers perched atop a stone wall.[95] The Wild West show depicted the U.S. military as leading a cooperative international charge—although in reality, each power had scrambled to protect its own assets in China and some attempted to cut land deals with the Boxers along the way. The program argued that the Boxers were "savage . . . uncompromising [and] indifferent to the civilized world. . . . The Royal Standard of Paganism floats proudly defiant of the Christian world. [Later] the Royal Standard comes down and the banners of civilization take its place."[96]

Historical productions valorized the Christian soldier. In 1903 Ringling Bros. produced "Jerusalem and the Crusades," celebrating the first Crusade (which began in 1095).[97] Departing from France, the Knights thundered off to save the city of Jerusalem from the "fanaticism of the Mohammedans." Meanwhile, the Saracenic Emir in Jerusalem "held high feast and made merry while the Crusaders besieged the Holy City." The Emir was surrounded by "ladies of the harem," omnipresent, diaphanously dressed "oriental" dancing girls in racial disguise. His opulence and polygamy stood him in stark contrast to the celibate Christian Knights, who vowed "to lead sweet lives in purest chastity, To [eventually] love one maiden only, cleave to her, And worship her by years of noble deeds, Until they won her."[98] The monogamous Crusaders defeated the polygamous Muslim "infidels" and the spectacle came to a triumphant Christian finish. The accompanying program book reminded the modern circus audience: "The centuries have witnessed many changes since knights wandered over the earth in search of adventure but the virtues of the ideal knight of the Middle Ages remains the ideal of the Christian Gentleman of the Twentieth Century."[99] "Jerusalem and the Crusades" also tapped into current American nostalgia for the medieval knight—reinvented as an antidote for the "over-civilized," neurasthenic, urban man. Consumers commonly saw the knight's image on commercial products and in literature. Charles Major published *When Knighthood Was in Flower* in 1898; the next year, the novel was already in its twenty-third printing. Winston Churchill's *Richard Carvel* (1899) was also a best-seller.[100] Likewise, "Jerusalem and the Crusades" cast the medieval knight-cum-"Christian Gentleman" as a natural world leader, a progressive counterpoint to Islamic "decadence" and "ease."

During the Spanish-American War, Peter Sells characterized the Spanish in a similar vein, as a feeble contrast to the sober, virile American soldier: "[The war] has proven forever that brains are superior to booze for the stimulation of bravery. It proves that the 'Yankee pigs,' as the pompous, wine-soaked, effete Castillian has termed us, are their superiors as statesmen, as soldiers and as sailors. That while they were feeding their people upon lies and attempted to bolster up their tottering and rotten government with pompous brag, our people were resting their cause upon truth, and in the contest between truth and falsehood 'truth beareth victory.'"[101] As an enlisted man during the war, Sherwood Anderson and his fellow soldiers anticipated the circus's effeminate Spanish stereotype: "To the soldiers the Spaniards were something like performers in a circus to which the American boys had been invited. It was said that they had bells on their hats, wore swords and played guitars under the windows of ladies' bedrooms at night."[102] Collectively, these florid images helped give the United States' victory in the Spanish-American War a gendered face, as a triumph of "sober," republican manliness over a mawkish, inept military.

Even though the United States did not have direct colonial interests in India, the subcontinent was a frequent circus subject. American Protestant missionaries had targeted South Asia as a major evangelical site since 1813, when the British government's Charter Act lifted a ban on missionaries in its Indian provinces.[103] Missionaries from the United States returned home with spectacular, circuslike representations of animal gods, royal "Hindoo" ceremonies atop elephants, ferocious animals, and body-contorting sadhus. Showmen soon transmogrified these startling reports into live performances.[104] Some of the most familiar circus animals, particularly Asian elephants and Bengal tigers, came from India. At the circus, several South Asian animals were named after South Asian rulers. Arriving in America in 1821, the ship *Bengal* brought the "fighting" elephant "Tippoo Sultan"—named after Tipu Sultan, a courageous potentate in southern India who attempted unsuccessfully to stop British expansion into southern India during the late eighteenth century.[105] Elephants were also occasionally named "Mameluke," an Arabic word for a military slave who was a member of the Turkish-speaking cavalry ruling Egypt and Syria as part of the Mamluk dynasty during the twelfth and thirteenth centuries.

As a site of American missionary activity, South Asia was a moral frontier rather than an economic or political one. More broadly, circus proprietors' treatment of Indian culture reflected the state of affairs between the United States and Great Britain. As the United States consolidated its own regional empire in the late nineteenth century, the British invested more capital in American foreign ventures than any other European power. Consequently, the United States and Britain enjoyed an informal alliance, based on strong cultural and economic ties. And these ties were racial. As Reginald Horsman has written, "Anglo-Saxonism" was an English and American intellectual invention that helped unify white people of different classes and ethnicity under a shared racial heritage.[106] The racial ideology of Anglo-Saxonism was founded upon nineteenth-century Teutonic germ theory, which posited that the seeds of democracy traveled westward with the Teutonic conquerors to Britain, and then North America. Stuart Anderson contends that Anglo-Saxonism provided a major reason for diplomatic rapprochement between the United States and Britain at the beginning of the twentieth century.[107]

In supporting British colonial rule, circus exhibitions also legitimized the fledgling American empire. The circus depicted colonized India as a "queer," "heathen" culture, in need of "guidance" from mature, industrial, Christian England. Such representations had long roots. In 1858 four American circuses reenacted scenes from the Sepoy Rebellion of the preceding year, an indigenous South Asian revolt against British colonial authority.[108] All circus renderings transformed this South Asian challenge to British rule into a harmless, ceremonial feast of colorful costumes and athletic feats atop horses and elephants that concluded with an affirmation of British colonial authority. At Wilkes-Barre, Pennsylvania, in 1858, a newspaper advertisement for the Rivers & Derious circus proclaimed: "Among the novel performances will be presented the thrilling dramatic spectacle entitled the war in indea; or, The Siege of Lucknow! With all the beautiful effects of character, music, costume, properties, &c. To commence with a Grand Cavalcade A la Turk, by a full corps of lady and gentleman Equestrians."[109]

The scale of turn-of-the-century South Asian specs was gargantuan compared to such mid-century displays. Barnum & Bailey's spec "Oriental India" (1896) claimed to depict daily life in India in a manner comparable to the ethnological congress along the midway at contemporaneous world's fairs. One colorful lithograph pronounced: "Eastern Home Life & Occupations Revealed to Christian Eyes in Vivid Pictures by Genuine Natives of India and Ceylon." Another containing a "series and views of living groups of strange and curious people" crammed onto a single stage a plethora of stereotypical Indian characters—"the very people themselves, whole families and groups of them": the sacred cow, the male snake charmer, a man smoking *bhang* (marijuana) from a water pipe, scantily clad men climbing trees, Hindus praying, the "famous dancing girls of Madras," "Silver and Devil Mask dancers from Kandy [Sri Lanka]," and women making textiles by hand while caring for seminude children (fig. 36).[110] Barnum & Bailey staged "Oriental India" on a raised platform surrounded by a long oblong band of Indian animals—elephants, tigers, sacred cattle—and others that were not from the region like Johanna the ersatz gorilla, now billed as

Figure 36. "Oriental India," Barnum & Bailey, 1896. Containing snake charmers and temple dancers among other South Asian stereotypes, the spectacle was exhibited alongside the ethnological congress in the menagerie tent. (Lithograph courtesy of Circus World Museum, Baraboo, Wis., with permission from Ringling Bros. and Barnum & Bailey,® The Greatest Show on Earth,® B+B-NL38-96-1F-2)

"Chiko's Widow." A narrow walkway for audience members separated the human exhibits from the animals.[111]

In general, the circus depicted India as an immutable cultural landscape fixed in an ancient "Asiatic" mode of production. "Oriental India" was exhibited in conjunction with a series of "chaste, refined and elegant representations of groups of ancient classic and modern" statuary and paintings, enhancing the impression that Indian culture itself was a static work of art, composed of immobile figments, easily exchanged in any preindustrial cultural context.[112] But showmen occasionally alluded to transformations in the colonial economy. Barnum & Bailey's ethnological congress in 1895, which employed "Trinidad Coolies" and "Hindu Creoles from the Equator," displayed members of the South Asian diaspora as royalty, although these Indians virtually all worked as manual laborers.[113] After the British abolished slavery in 1833 in British Africa and the West Indies (but not in India until 1843), a diasporic South Asian labor pool served as a replacement for black slave labor. During the late nineteenth century the British employed vast numbers of South Asians in the imperial factory system, which stretched throughout the Empire: from huge new textile mills in Bombay and Ahmedabad to rubber plantations in Malaysia. In Uganda, Kenya, and other British colonies in Africa, Indian work gangs built railroads under brutal conditions.

Proprietors further effaced the bleak reality of British colonialism with showy spectacles of Indian political pageantry. One event in particular, the *Durbar* (royal court) of Delhi, became an enduring circus subject. In 1903 Lord George Curzon, British viceroy of India, sat atop a bejeweled elephant as he crowned King Edward VII the new Emperor of India. Hundreds of parading elephants, camels, horses, and Indian *rajas* and *nawabs* attended the ceremony. Lady Curzon and the Duke and Duchess of Connaught also participated in this celebration of British imperialism. Delhi was the old Mughal capital of India: by donning the accouterments of Indian rulers in an imperial setting, Lord Curzon implicitly invented a historical link between his own authority and the legacies of past emperors, notably Akbar, the sixteenth-century Mughal.

In 1904 in New York City, the Durbar of Delhi became an eighteen-minute sawdust show under Barnum & Bailey's three-ring big top.[114] In a lavish scene of elephants and camels caparisoned in gold and silk, Euroamerican actors depicted South Asians and the English. A program souvenir described the Durbar with jumbled, discrete orientalist references: "[T]here is a troop of native soldiers riding upon lofty, swaying camels and preceded by the mystic priests of Buddha, leading the sacred zebus and the sacrificial cattle; there is a prince of Siam with his retinue of warriors and shapely oriental dancing girls. . . . There is a brief halt while the Potentates of the Indian kingdoms pay their tribute to the Imperial power. Then once more the procession moves on; the royal elephants join the pageant and the long line of splendour disappears through the parted curtains of that unknown land of mystery where the artists prepare for the feats of the arena." [115] As hundreds of Indians bowed to British authority, the "Durbar of Delhi" ballyhooed the colonialist stereotype that only a tiny minority of British officials were needed to rule millions of Indians, even though an organized Indian nationalist movement was already in full swing after the formation in 1885 of the Indian National Congress.[116] Because Americans received scant information about India from other contemporary media, the circus's ritualized, ceremonial images of the subcontinent were potentially all the more potent.[117]

India's popularity as a circus subject underscores what T. J. Jackson Lears has termed an "antimodern" response to modernity at the turn of the century.[118] Proponents of the Arts and Crafts Movement (taking their cue from William Morris and John Ruskin, who led the nineteenth-century English craft revival movement) saw the resuscitation of artisan craft production as a way to ameliorate class conflict.[119] The circus's sentimental depiction of India as stagnant and yet spiritual reflected this intellectual trend. The circus idealized India as a closely knit agrarian and artisan society—a stark contrast to the reality of contemporary labor struggle in mining towns and urban centers across America. More generally, the popularity of the Indian Swami Vivekananda in America after he addressed the World Parliament of Religions in 1893 at the Columbian Exposition in Chicago,[120] the rise of Theosophy, and the growing popularity of vegetarianism in the United States all mirrored this "antimodern" American fascination with India.[121]

Despite its nostalgic representations of preindustrial India, the circus still depicted overseas expansion as an essential component of a powerful, modern nation-state. Its exceptionalist messages reached an even wider audience when it toured Canada and Europe. American specs abroad contained the same sort of seamless, consensus-driven vision of current events that shaped show programs at home. During the Boer War (1899–1902), American officials and private investors favored the British (even though the U.S. government officially declared neutrality). Euroamerican public opinion, however, supported the Boers, whose independence movement reverberated with their own revolution against England.[122] During its 1901 season in the United States, Buffalo Bill's Wild West program highlighted its "Rough Riders of the Transvaal. Representative Boers Fresh from the South African War," remarking that "the American is prone to sympathize 'with the underdog,' especially if he is putting up a particularly good fight."[123] But shortly before Buffalo Bill's Wild West toured England and the Continent from 1902 to 1906, the show abandoned its popular Boer War spec. Perhaps Cody wanted to avoid offending British spectators, or perhaps he simply deemed the spec "old hat." However, the Ringling Bros. showed similar tact in Canada, a member of the British Commonwealth. A circus performer named Wiser had played a Boer scout for show dates in the United States, but in Canada he became the "Australian boomerang thrower" in the opening pageant. Al Conlon, who operated the show's "picture machine" on the grounds, reversed the film during Canadian show dates, thus "reversing" the course of events therein: the Boers now "ran" from the British and "in the end" were "totally annihilated."[124]

While touring England in 1898, Barnum & Bailey dramatized Britain's first Sudan Campaign (1884–85) against the Mahdi of Egypt's Islamic nationalist uprising (plate 8). Written by the English journalist Bennet Burleigh, war correspondent for the *Daily Telegraph* (London), "The Mahdi; or, For the Victorian Cross" was a tale of romantic colonial chivalry set on the Sudanese frontier, even though the struggle was ongoing. Burleigh freely acknowledged that "time, place and grouping" had "been slightly changed, for the purposes of dramatic representations." As a result, the Mahdi's revolt became the basis for melodrama in which the young Sergeant McLean sacrificed his own life to save British and American ladies.[125] At the circus, the Sudanese rebellion became dramatic fodder for high adventure and romance, an undifferentiated "little war of empire" now drained of its original revolutionary meaning and its current urgency in favor of jingoistic entertainment.

Circus and Wild West scenes of triumphal empire building flourished in a society that largely supported overseas expansion: the pro-imperialist presidential candidate William McKinley, for one, handily defeated his anti-imperialist opponent William Jennings Bryan in 1896 and 1900. Theodore Roosevelt, an ardent expansionist, easily won the presidency in 1904, and William Howard Taft, an architect of "dollar diplomacy," beat Bryan in 1908. A wide range of ordinary Americans—from farmers and industrial workers to urban businessmen—supported expansion, which particularly in the wake of the devastating panics of 1873 and 1893 was seen as a necessary antidote to overproduction and a failing economy.[126]

But a small group of citizens rejected the new empire's utopian promise. Discrete, dystopic groups of anti-imperialists argued that the dawning of U.S. globalism signaled the nation's constitutional and cultural demise. Charles Towne, a former congressman from Michigan, gloomily forecast the rapid decline of the United States within a year of the Spanish-American War, "from the moral leadership of mankind into the common brigandage of the robber nations of the world."[127] In the small town of Bonham, Texas, the editor of the *Bonham News* posed his readers a set of somber rhetorical questions in an endorsement of the anti-imperialist Democratic candidate,

William Jennings Bryan in the presidential election of 1900: "Do you favor Imperialism, Trusts, Militarism, War taxes, Foreign Alliances, subsidies to favorites, and extravagant expenditures? Do you think the time has come for Americans to put aside the Constitution and the teachings of the Declaration of Independence? If not, Tuesday, November 6 is the time to show your disapproval."[128]

Within a year after it was founded in 1898, the Boston-based Anti-Imperialist League claimed 70,000 members, the highest number in its history.[129] Although virtually all anti-imperialists supported participation by the United States in the global economy, they were mightily concerned with the boundaries of its expansion. As Richard Drinnon writes, "[Expansion] was about whether the U.S. empire should be hemispheric or global, and secondarily about the nature of the Constitution: did that document follow the flag?"[130] Still, anti-imperialist ideology was often contradictory. While some anti-imperialists, including Mark Twain, W. E. B. Du Bois, and Jane Addams, disavowed the establishment of a formal empire on humanitarian antiracist and constitutional grounds, others like Senator "Pitchfork" Ben Tillman of South Carolina, a vice-president of the Anti-Imperialist League, maintained that imperialism would inevitably "mongrelize" American racial identity.[131] Accordingly, anti-imperialist debates were interconnected with the explosive racial tensions of the period. White supremacists expressed their outrage when President Theodore Roosevelt in 1901 invited the black leader Booker T. Washington to the White House.[132] In some respects, Roosevelt's position as a vigorous expansionist and advocate of domestic civil rights (a position similar to that of his friend William F. Cody) might seem paradoxical, but like other politicians and showmen of the Progressive Era, Roosevelt was a paternalistic believer in the "white man's burden." He asserted that people of color throughout the world could become "civilized" through "proper" education and "moral uplift," of the sort provided by Washington's Tuskegee program of "Thrift, Patience, and Industrial Training for the masses."[133]

Anti-imperialist white supremacists dreaded the specter of racial amalgamation. After the United States ratified the Treaty of Paris, which sanctioned its annexation of the Philippines, the U.S. Senate passed the McEnery Resolution, which stipulated that the Filipinos would never become U.S. citizens. However, many anti-imperialists argued that inevitably, people of color from "America's possessions" *would* become U.S. citizens—just as African Americans had done so as a result of the Thirteenth, Fourteenth, and Fifteenth Amendments. Linking the racial identity of African Americans to that of so-called mongrel people of color throughout the world, the editor of the *Jackson* (Miss.) *Clarion-Ledger* announced that only racial segregation—at home and abroad—would preserve the social order: "The surest step to joint degradation and deterioration is amalgamation."[134] In the immediate aftermath of the Spanish-American War, the *New Orleans Daily Picayune* warned its readers: "The American people must understand that the acquisition of Cuba, Porto Rico and the Philippine Islands will bring in not less than a dozen million peoples of alien races utterly unfit to understand, much less to appreciate, the constitutional government and free institutions of this Republic, and, therefore, to admit such peoples to citizenship and a participation in its public affairs would be a most pernicious and ruinous policy."[135] Senator Donelson Caffrey of Louisiana went even further: "The 12,000,000 negroes in this country should be deported to the Philippines and never be allowed to set foot in this country again. If the South could be rid of the negro problem it would prosper beyond the dreams of man."[136]

In this social context, it would seem that the presence of Cuban "brown-skinned American revolutionaries" and of Filipino, Hawaiian, and black Rough Riders would unsettle certain Euroamerican audiences. Those from the southern United States, where racially based anti-imperialism was fierce, might have found these exhibits particularly disturbing. "The Battle of San Juan Hill," for one, freely displayed armed African Americans and Native Americans fighting for the independence of another people of color.[137]

Furthermore, armed African American cavalry drills at the Wild West sharply contradicted contemporary depictions of black soldiers by the mainstream American press. Newspapers frequently carried lurid stories about drunk, lazy, volatile, violent black troops (one described a soldier who was "drunk and threateningly displayed his revolver into a crowd"). The *Daily Picayune* wrote: "The negro troops in Cuba are proving a disgrace to the United States, and a serious menace to the lives, property and public order of the Cubans themselves. . . . Murder, robbery and rape are said to be a common business with them, and the people seem to have no redress." Another headline claimed: "Negro Regiment from Virginia Has Proved a Failure."[138] Military officials took special notice of African American soldiers who deserted in the Philippines, drawn to the Filipinos' calls for a united movement of people of color against white supremacy. General Frederick Funston, who supervised the capture of the nationalist leader Emilio Aguinaldo,

put a bounty on the black deserter David Fagen's head and was greatly satisfied to hear that the head had been delivered—in a wicker basket.[139] Such
reportage undermined the decisive role that African Americans played in
winning the war in the Philippines, in particular that of the 24th Negro In-
fantry in capturing Aguinaldo in 1901. Black newspapers, however, noted
these achievements in exacting detail.[140]

Still, southern newspapers made no connection between multiracial circus acts, armed people of color, and anti-imperialist racial ideologies. In
fact, articles praising circus scenes of multiracial athletic wizardry existed
alongside the above-mentioned diatribes against the supposed "intractability" of African American soldiers in Cuba and Filipinos' "inability" to become "civilized." What accounts for this disjuncture? What made the circus
and Wild West a seemingly safe social space for exhibiting anti-imperialists'
worst fears concerning racial amalgamation and claims to citizenship? This
absence of protest is especially striking because the circus and Wild West
were traveling communities whose nomadic multiracial members lived and
performed in close proximity to one another. Given these logistical reali-
ties of a traveling outfit, certain audiences might have been even more likely
to read pro-imperialist exhibits of U.S. expansionism as socially ominous.
But they did not make such explicit connections. For that matter, African
American newspapers did not comment on these displays either.[141]

Perhaps the fanfare of the racial Other at the freak show helped offset
the subversive implications of the pro-imperialist displays. The range of
professional African and Asian "savages," "missing links," and legless, armless, conjoined, and hirsute people of color collectively reified racial distinctions through bodily exhibition, similar to the evolutionary juxtaposition of
humans and animals at the "Ethnological Congress of Strange and Savage
Tribes."[142] Perhaps the performance of racial "savagery" helped diminish
the potentially transgressive claims to equality found in other parts of the
exhibition. To some audiences, the players might have served as examples
of inassimilable racial difference, living proof of the prevailing "wisdom" of
racial segregation. The literary scholar Amy Kaplan points out that Roosevelt himself diffused the disturbing implications of his multiracial Rough
Riders by rewriting the course of the battle shortly after it took place: in
several accounts, he minimized African American accomplishments, instead
characterizing black soldiers as comical, lazy, shiftless, freakish, and impotent without white commanders—despite the critical role that African
Americans actually played in storming Kettle Hill.[143]

Burlesque also may have played a role in diminishing these potentially jarring displays. As much as circuses and Wild West proprietors earnestly took pride in their shows' verisimilitude, they also poked fun at international affairs. As Emilio Aguinaldo triumphantly returned from exile to his native Philippines on the U.S. gunboat *McCulloch* to draft a new constitution fashioned along American lines, the Ringling Bros. circus menagerie displayed its big Philippine boa constrictor named "Emilio Aguinaldo." In September 1898 Aguinaldo supposedly swallowed himself.[144] During the Russo-Japanese War (1904–5), Barnum & Bailey press releases hinted that the sideshow's conjoined "Corean Twins" might have to be separated so that they could serve in the Japanese military.[145] The circus's constant subversion of bodily boundaries perhaps helped lessen its claims of realism among its racist anti-imperialist audiences.

Yet as other parts of this book have suggested, Circus Day and "Buffalo Bill Day" *were* sites of real racial and ethnic anxiety. One can turn to the behavior of the enormous multiethnic crowds—sometimes more than 20,000 milled around the show grounds—as another barometer of the ways in which this amusement served to "instruct the minds of all classes."[146] Circus workers observed countless examples of racial violence on Circus Day. For example, on October 25, 1890, in Navasota, Texas, a white man named James Whitfield shot and killed an African American man at the Sells Brothers' sideshow door after supposedly being insulted by him.[147] After a white man and a black man scuffled at the circus grounds at Falls City, Nebraska, in 1898, the Adam Forepaugh & Sells Brothers route book reported that "there was talk of lynching the negro."[148] Despite its fleeting presence, the circus and Wild West provided a moment of community articulation, bringing dystopian racial anxieties into sharp relief just as much as it highlighted unifying national narratives of modernization, global power, and prosperity.

With U.S. participation in World War I looming in 1916, the circus and Wild West plugged the readiness of U.S. troops.[149] In 1916 Buffalo Bill and Miller Brothers 101 Ranch Wild West Combined produced a new military pageant, "Preparedness," designed not only to entertain with its colorful pageantry and athleticism, "but also to arouse public interest in the enlargement of the army and in 'Preparedness' for defense in case of possible attack."[150] On the eve of U.S. entry into the war, the War Department granted furloughs to scores of soldiers to allow them to participate in "Preparedness." The soldiers demonstrated the dangers of trench warfare and the work of the scout and sharpshooter in an effort to arouse public support for the war effort. Circus posters during the war exhorted Americans to buy Liberty Bonds, and the aerialist Bird Millman personally sold war bonds to fellow workers and audience members; other circus members sewed socks for soldiers and volunteered for the Red Cross.[151] Working with the federal government, the circus and Wild West actively supported the war.

Yet circus spectacles of World War I were limited to military exercises *before* the advent of actual American fighting. In the context of contemporary anti-German hysteria, xenophobia, and bloodshed, battles from World War I—unlike those of the Spanish-American War—were never recreated at the circus and Wild West. In contrast to Barnum & Bailey's enthusiastic reenactment of the Portsmouth Treaty of 1905, no showmen chronicled the establishment of the League of Nations, because an increasingly enfeebled President Wilson failed to gain Senate ratification for U.S. membership.

The age of the great circus and Wild West foreign affairs spectacle, then, ended with World War I. Although the victory of the United States marked the start of its clear-cut economic (and later military) domination on the world stage, this same period was also marked by extraordinary global unrest. Amid the Bolshevik Revolution and its aftermath, a devastating influenza epidemic, massive strikes, the Palmer Raids, and heated debates about isolationism versus internationalism, circus owners avoided controversy by choosing fictive, erotic, ahistorical pageants set in the Middle East, Africa, and Asia, devoid of any references to the concurrent demise of the Ottoman empire. The geographic interchangeability of such orientalist programs was constant: in 1919 Al G. Barnes's circus performed "Alice in Jungleland"; Hagenbeck-Wallace produced "A Night in Persia" in 1923, which resurfaced as "Geisha" in 1928–29 and finally as "An Oriental Fantasy" in 1935. These circus pageants contained lecherous "oriental" despots, scores of "oriental dancing girls," and military stunts on elephants and camels. Furthermore, with the growing popularity of movies and radio (media that also chronicled up-to-date foreign affairs), circus proprietors no longer emphasized their specs' verisimilitude as a way to draw the crowds. However, the ideological thrust of the turn-of-the-century circus and Wild West spec—characterized by a triumphant, Disneylike emphasis on America's moral and economic stewardship—remained a critical component of the nation's foreign relations "mission" during the rest of what Henry Luce later termed the "American Century." According to William Appleman Williams, this ideology manifested itself as "a firm conviction, even dogmatic belief, that America's domestic well-being depends upon such sustained, ever increasing overseas economic expansion."[152]

After World War II, live circus exhibitions of the racial Other and the politics of performer procurement became increasingly problematic for U.S. officials. The French Congolese "Ubangi" women had been a great hit with the American public when they toured with Ringling Bros. and Barnum & Bailey in 1930. Yet in 1954 the social climate had changed when McCormick Steele, the Ringling Bros.' "foreign rarities scout," tried to bring the Ubangis back to the United States during the Cold War. The French colonial government refused to issue passports to the Ubangis, because the French had banned all native practices of bodily disfigurement—in line with colonial attempts to reconfigure and reform the "native" body along western lines. Furthermore, the vice consul at the American embassy would not help Steele procure the Ubangis because of "possible racial antagonism in U.S.A."[153]

Here, the circus provides a fascinating perspective into the racial politics of the Cold War: amid the growing civil rights movement in the United States and rising nationalist movements in newly decolonized Third World nations, the U.S. government was highly sensitive to Soviet propaganda in developing countries, which asserted that capitalism caused American racism. Consequently, the decision of U.S. officials to support the French ban complemented larger U.S. foreign policy objectives.[154] In the age of anticolonialist movements and the Cold War, circus displays of the Other had become politically hazardous. But this is not to say that racially exoticized circus images disappeared. Rather, they endured in other ways. As a potent location for the roots of American global popular culture, the human and animal Others of the fin-de-siècle railroad circus continued to shape America's vision of the world through the interconnected realm of popular culture and politics.

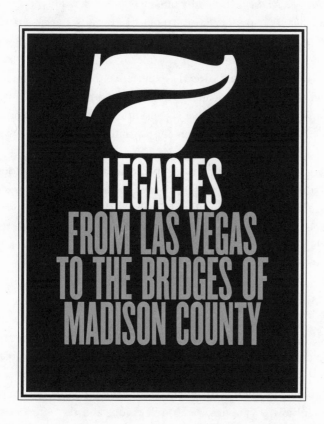

7

LEGACIES
FROM LAS VEGAS
TO THE BRIDGES OF
MADISON COUNTY

At the turn of the century, the gigantic railroad circus descended upon a community, shut it down, and then moved on. Yet the impact of the circus was far from ephemeral. As a powerful agent of global representation in an age before electronic media, the railroad circus collapsed the world under canvas—right at home—for urban and rural consumers across the United States. As a corporation on wheels, the circus's labor performances of the new industrial order, its variegated exhibitions of human and animal relationships, and its spectacles of America's growing power in world affairs heralded the arrival of a new modern age, framed around an unsettling matrix of bodily celebration and prudery, social conformity and marginality, jingoism and internationalism, racial hierarchy and racial fascination.

The world of the railroad circus suggests, however, that the nation's departure from the Victorian era was hardly wholesale. Rather, the circus articulated the tensions of a culture at the crossroads of Victorian and modern. The circus's rhetorical conventions and work rules (emphasizing separate spheres, propriety, and racial hierarchy) embraced a nineteenth-century *Weltanschauung* defined by binary opposites, while its performances presented a playful, modernistic, stream-of-consciousness sensory flux in which the distinction between men and women, and between people and animals, became indeterminate.[1] By "running away" with the circus's subterranean rovers, one could directly experience the world in the immediate present, on the edges of society.[2] The circus's simultaneous contradictory impulses of nostalgic, normative representation and subversion of established social hierarchies made it an appropriate emblem of an age of transition.

The gargantuan railroad circus helped catapult a "nation of loosely connected islands" into a modern nation-state with an increasingly shared national culture.[3] Blanketing its far-flung markets months in advance of its coming, the circus abetted the rise of modern advertising with its totalizing tactics: thousands of colorful posters featuring lithe bodies beckoned audiences with images of eroticism and danger, while press releases personalized the upcoming show with tantalizing stories (à la *True Story*) about the lives and loves of various human and animal stars. The invasive railroad circus gave its scattered consumers a shared knowledge base about the world —creating a palimpsest for subsequent American media representations of the globe. The circus's itinerancy and its three-ring scramble of time, space, and habitat imbued it with a kind of frenetic placelessness that prepared its audiences for the ascendancy of disembodied modern media technologies: Hollywood movies, radio, television, and the internet. In effect, the pervasive railroad circus—and its animal-bedecked trains, flipping cars, aerial bicycles, glowing electric generators, costly spectacles, and exotic performers—helped hasten the nation's move toward a mass consumer culture. Its constant action in three rings, two stages, and an outer hippodrome track made for a whoozy, "too big to see at once" sensory experience that prompted customers to spend their money to see it again and again. Despite the country's demographic shift from rural to urban, provincial residents participated fully in the making of this modern mass culture—as their wholesale and often raucous participation in Circus Day has shown throughout this book.

Circus Day was a major community celebration at the beginning of the twentieth century. But in the 1920s, its physical presence began to diminish. The morning street parade—a highly visible, ritualistic encounter between the circus and a community—disappeared at the big railroad outfits. The ethnological congress and up-to-date spectacles of recent foreign events also vanished.[4] Despite the continued popularity of individual stars like Lillian Leitzel, Alfredo Codona, and Bird Millman, the totalizing physical presence of the circus was fading. Towns no longer shut down in its midst. By the 1930s audience numbers were also in decline. Urban development and the rise of the suburbs pushed the show grounds away from the vicinity of the rail yards, making it difficult for the huge railroad shows to navigate efficiently.[5] But more significantly, the circus no longer had a monopoly on novelty or current events. Movies, radio, and (from the late 1940s) television provided audiences with compelling and immediate images that displaced the circus as an important source of information about the world.[6] In 1938 the nonagenarian circus trouper W. E. "Doc" Van Alstine remarked, "But the kids of today ain't so wide-eyed and amazed at what they see at a circus as they was a quarter of a century ago. So many marvelous things goes on all the time in this day and age that kids probably expect more from a circus now than it's humanly possible to give."[7] In 1956 just thirteen circuses existed in America—down from ninety-eight in 1903.[8]

The rise of industrial unionism during the New Deal also modified the circus. Harmonious scenes of burly, singing African American roustabouts setting up the big top in Disney's film *Dumbo* (1941) belied the heated labor disputes of the period. After workingmen and performers in 1937 joined the American Federation of Actors (an affiliate of the American Federation of Labor), the threat of strikes was common, especially after workers walked off the job in Scranton, Pennsylvania, in 1938 owing to John Ringling North's decision to cut wages by 25 percent. (In Janesville, Wisconsin, managers broke sympathy picket lines with elephants.)[9] No longer able to depend on an unorganized (and consequently cheap) roustabout labor pool, showmen further scaled back their lavish, labor-intensive operations. After experiencing a bruising (yet unsuccessful) organizational drive led by Jimmy Hoffa and the Teamsters Union in 1955 and 1956, John Ringling North cut his workforce drastically in July 1956 when he abandoned the canvas tent (a fixture of the traveling circus since 1825) in favor of indoor arenas and stadiums: 900 people lost their jobs in the process, and the outfit temporarily traveled by truck until 1960, when it returned to the railroad.[10]

After North's shutdown, *Life* magazine mourned the disappearance of what had now become an emblem of an older, simpler way of life: "But to Americans the circus means the Big Top—the predawn arrivals, the pounding elephants, the sweet and acrid smells, the tanbark, the jungle of nets, poles and moorings under the canvas. All this was no more. Amid the profound sadness and great memories the tent circus made its doleful journey to oblivion."[11] Ironically, just fifty years before, the gigantic railroad tent show had been a manifestation of modernization; its physical flowering into a transcontinental behemoth was dependent upon the nation's development into a mature industrial society. But by 1956 the rambling railroad tent show no longer had a home. Against a backdrop of television's facsimile verisimilitude and the sprawling Cold War suburban landscape, railroad tent showmen found it increasingly difficult to find performance sites or audiences willing to leave the predictable comforts of home for Circus Day's volatile community offerings. In 1959 the circus of old was memorialized as history with the establishment of the State of Wisconsin Historical Society's Circus World Museum in Baraboo—home to the Ringling Bros. circus from 1884 until 1918. Today, periodic newspaper articles still bemoan the circus's transformation into historical artifact. "Remember when the circus wasn't history?" asked Ron Seely in July 2001 while reporting the route of Circus World Museum's Great Circus Train as it rumbled to Milwaukee for the city's annual Great Circus Parade.[12]

Although the contemporary cultural landscape of the United States is nearly circus free, the fin-de-siècle railroad circus lives on in other ways. At the beginning of the twentieth century, the railroad circus beckoned its audiences to explore the exotic through an act of consumption. This impulse lives on at Sea World, Walt Disney World, shopping malls, and countless ethnic restaurants across America.[13] The circus's imperative to discover the world through consumerism has also been reincarnated at Las Vegas. Amid the city's constant slot-machine clatter (comparable to the shell games on the old circus midway) one can shop at Chinatown Plaza, sit next to an opulent Roman fountain at Caesars Palace, visit the Luxor, a sleek, black pyramid hotel outlined in neon (with a giant Sphinx included), watch a fake sun course through a lurid velvet-blue-and-crimson sky in forty-five minutes flat at the Forum Shops, or stroll on the streets of "New York—New York," a cluster of hotels replicating the New York skyline.

Las Vegas also possesses modern counterparts to the "human menagerie" of old. Its "show girls" echo the circus ballet girl: wearing brief, sparkling g-strings and bikini tops, towering feathery headdress, and thick makeup, these mostly white women play a sexually charged Other as an iconic part of the Las Vegas family entertainment scene. "Wild" white tigers perform with Siegfried and Roy at the Mirage Hotel, where one can also encounter an educational "Dolphin Habitat and the Secret Garden of Siegfried and Roy," which—just like the circus—compresses the natural world into a knowable commodity, consumed "at a glance": "33 white tigers. 7 white lions. A 3-ton elephant. And you . . . Face to face in an exotic jungle setting"—according to one advertisement.[14] The exhibition has a direct link to the circus business as well: the show is produced by Kenneth Feld, owner of the Ringling Bros. and Barnum & Bailey circus. The disembodied ethnological products at Las Vegas are complemented by the permanent presence of an actual circus (albeit one that is animal free), Cirque du Soleil, and twenty-four-hour circus acts at the loud, pink-and-white candy-striped, child-beckoning hotel, Circus Circus. In an age when disability rights activists have stopped most sideshows, Robert Wadlow, the "world's tallest man," and a pantheon of wax replica freaks from the past glisten silently at Ripley's Believe It Or Not! museum.[15] In Las Vegas, one can also purchase physical thrills reminiscent of dangerous big-top acts: bungee jumping, sky diving, or floating aloft at Flyaway Indoor Skydiving.

The contradictory circus image of the respectable "lady dainty" has also endured in American popular culture. Inside the ring, the circus has remained a free space for strong women. In 1947 a new curriculum on circus training helped ease Florida State University, previously a women's college, into its new position as a coeducational institution. According to *Life* magazine, "to speed transition . . . the administrators began searching for an activity in which both men and women could participate. The unusual, star-spangled circus course was the answer."[16] The circus, however, still offers multiple readings of women's position in public life. At most early-twenty-first-century shows, female performers dress in bright, sparkling thong leotards with loads of décolleté. Women commonly work as assistants for male big-top players—a job which calls for little beside wearing scant costuming, swaying suggestively to a synthesized bass beat, and tossing out balls or rings to the principal player. In 1995 Mattel captured the circus's normative impulses when it unveiled its new Circus Star Barbie. Wearing spangled micro garb, the doll was anorectic, big-haired, and buxom, bearing little resemblance to the muscular female circus athlete. The well-known toy store F.A.O. Schwarz created a huge display of Barbie-related circus scenes: Barbie walked the tightrope, "sawed Ken in half," and threw knives at Ken, who was strapped to a rotating wheel. Revealing the circus's subversive potential, the Barbie scenes drew immediate criticism. A Mattel executive rushed to justify the company's marketing strategy: "People have been shot out of cannons and survived—it's not as if she has a gun to Ken's head. It's part of a play pattern—the circus is a spectacular dream world beyond our normal lives. It shouldn't be taken as 'Barbie' throwing knives at 'Ken'—it's all just part of the circus act."[17]

This sort of bodily reconfiguration—the strong circus athlete transformed into the "lady dainty"—is also alive and well in the world of women's sport, in media images of the popular petite gymnasts Olga Korbut and Mary Lou Retton, and the figure skaters Nancy Kerrigan and Tara Lipinski. The titillating female athlete (Katarina Witt and Anna Kournikova come to mind) is also ubiquitous, garnering lucrative magazine spreads and product endorsements, while unabashedly strong and hefty sportswomen such as the Olympic weightlifting bronze medalist Cheryl Hayworth remain marginal to the profitable world of endorsements. The same has held true for openly bisexual or lesbian athletes like Billie Jean King and Martina Navratilova, and for women of color such as Jackie Joyner-Kersee.[18]

Turn-of-the-century circus constructions of the exotic Other also continue to shape American popular representations of the world, transforming tropical zone cultures into colorful, "natural" primordial sites of eroticism, danger, and modernistic self-discovery. In the 1990s Robert James Waller's best-selling romance novels capitalized on these themes. In *The Bridges of Madison County* (1992), the protagonist, Robert Kincaid, is a photographer for *National Geographic* who has been sexually seasoned through worldly encounters with the "silk merchant's daughters and their knowing ways." In *Slow Waltz in Cedar Bend* (1993), Michael Tillman remembers sexual relationships with "the women in Bangkok with their long hair and compliant ways." Tillman's lover, Jellie Braden, runs off to India, which reminds her of "jasmine on Bengali night winds, dark hands across her breasts and along the curve of her back..." Tillman follows her and meets a tiger in the Indian wilderness. "Michael began to take pleasure in just staring back at the tiger, in the simple purity of contemplating its existence, in knowing not everything wild and strong had been snuffed out by condos and shopping malls."[19] The couple's passion is anchored through their mutual contact with "primitive" India. Waller represents South Asia as an alluring, mysterious place for masculine renewal by using nostalgic racial and gendered discourses about dangerous, virile animals, alluring "oriental" women, sexually potent men of color, and "timeless" civilizations that circulated across the nation at the turn-of-the-century circus.

The fin-de-siècle circus helped shape the ways that the American news media represent the world, transforming developing nations into a bizarre ethnological congress of humans and animals. In 1990 the *Wisconsin State Journal* reported, "An Iranian hunter was shot to death Monday near Tehran by a snake that coiled itself around his shotgun as he pinned the reptile to the ground."[20] A year later the newspaper noted, "Armed with cats, traps and poisons, Bangladesh on Wednesday launched a two-week extermination campaign against rats, branding them public enemy No. 1."[21] In late May 2001, newspapers in the United States and media throughout the world were smitten by news that a mysterious, four-foot-tall, steel-clawed "Monkey Man" was terrorizing impoverished residents of New Delhi. Reportedly, the creature had caused three deaths from falls as people scrambled to escape. The *Los Angeles Times* noted, "And Monkey Man isn't even the latest of mythological creatures to crop up. In northeast India, Bear Man has appeared, with more than a dozen people claiming they've been attacked. Like they need to make up problems?"[22] On May 23, 2001, the Monkey Man was the subject of David Letterman's nightly "Top Ten List": "Top Ten Signs Your Neighbor Is the Monkey Man."[23] Few American media outlets thoroughly cover culture or politics in the developing world; instead, disjointed snippets consistently represent these nations as a sideshow. Modern depictions of the developing world are filled with Darwinian images of teeming populations controlled only through natural selection—typically a hurricane, mudslide, earthquake, or bus wreck. News blurbs detail thousands of anonymous deaths.

A handful of actual sideshows still perform, including the Bros. Grim troupe at the Great Circus Parade show grounds in July 2000 in Milwaukee, but apart from the circus.[24] The Bearded Lady, Fannie Bryson ("The Personality Fat Girl"), and two contemporary sideshow icons, Thenigma (a nearly all-blue tattooed sword swallower) and the fire-eater Tim Cridland (also know as "Zamora the Torture King"), travel nationwide. In contrast to the middle-class Victorian sideshow fan whose family photo album in the parlor likely contained postcards of his or her favorite freaks, the typical sideshow lover is now part of a youthful, hard-rock subculture.[25] But the sideshow is still deeply embedded in mainstream popular culture. Several scholars rightly implicate the daytime talk shows of Jerry Springer, Ricki Lake, Montel Williams, and Jenny Jones as critical sites of abnormality, sexual spectacle, and personal disclosure.[26] The spectacular exhibition of unusual bodies is omnipresent in professional sport, with seven-foot-plus basketball centers, hulking, too-huge football tackles, and oversized wrestling stars, like the late André the Giant. Medical spectacles of cutting-edge technologies, such as "The World's Most Shocking Medical Videos" on Fox TV, entertain, educate, bedazzle, and celebrate the superiority of technology to heal and normalize; its episodes have featured "Two-Headed Chang" ("conceived as a twin, he grew up a horror"), conjoined twins from Russia, parasitic twins from Yunnan Province in China, a human ear that grew attached to a genetically engineered mouse, and the story of Beck

Weathers, a survivor of a disastrous Mount Everest expedition in 1996, whose irreparably frostbitten nose was restored when surgeons attached it to his forehead and then back to its proper place. Weathers made sense of his temporary bodily displacement by turning to the circus sideshow: "I pretty much looked like Jo Jo the Dog-Faced Boy [a sideshow staple in the late nineteenth century]."[27] "Guinness World Records Prime Time," also on Fox, showcases dangerous interactions between people and animals and acts of extreme self-flagellation. The show recalls Poline the Hindoo Fakir's weight-lifting feats with his eyelids in 1906, and assorted glass-eaters and rubber men, as it features people like Lotan Baba "the rolling saint," a South Asian sadhu who stood chained for seven years and then rolled ten miles a day across the subcontinent for eight months, covering 2,485 miles in all.[28] In book form, *Guinness World Records* is a veritable sideshow of superlative human difference: fattest dog, longest nails, smartest, youngest professor, tallest, smallest twins, longest beard . . . and on and on. Unlike the sideshow of old, however, Guinness enables the audience to gaze at these amazing bodies from a distance, outside the ostensible realm of indignity.

In addition to these potent, symbolic linkages to the giant railroad show of old, circuses still tour the nation—albeit on a far smaller scale than in 1900. None of the circuses have a regular sideshow, and the presence of animal performers has dropped sharply in an age of vigorous animal-rights activism. In 1999 there were twelve animal-free circuses in the United States alone.[29] The small, intimate, one-ring tent show that dominated the antebellum era is booming today. Cirque du Soleil, the Big Apple Circus, and the recent Barnum's Kaleidoscape (a Kenneth Feld production, now defunct) have capitalized on the purchasing power of new upscale audiences—wealthy, middle-aged suburbanites who prospered in the bull market of the 1990s. Aside from the physical intimacy of these new productions, another feature connects them to the antebellum circus: *adults*, not children, are their primary audience. Individualized performances of bodily prowess (free from the distracting clutter of three rings and two stages of constant activity) have proven to be extremely popular with these well-to-do spectators, who also enjoy the plush appointments, cappuccino, fajitas, bottled water, and fancy salads. Kitschy circus staples—cotton candy, pink lemonade, and peanuts—are in short supply.[30] Cirque du Soleil, created in 1984 and based in Montreal, now charges over $50 per ticket (on average), sells out 93 percent of the time, and generates annual revenues of over $119 million and annual profits ranging from 15 to 20 percent.[31] Clearly, one-ring shows like this do not aspire to "instruct the minds of *all* classes" like the railroad circus of old.[32]

Today, people continue to "run away" with the circus for potential economic gain just as they did at the last turn of the century. In the mid-1990s the Career Resource Center at Paul Robeson High School in inner-city Chicago posted advertisements for Ringling Bros. and Barnum & Bailey's "Clown College" (now defunct), offering poor African American kids potential financial mobility: "For the career of a lifetime—look into it!!! There is a tuition free policy to remove any potential financial barriers that might prevent clown hopefuls from realizing their dreams."[33] In St. Paul, Minnesota, a youth circus, Circus of the Star, has provided a creative place for kids of diverse backgrounds since 1994. Basia Zaklika, a teenaged hoop spinner, observed: "You can see how much we pull together and how much want to do this."[34] In the 1980s and 1990s the circus owner Wayne Franzen, a former schoolteacher from Wisconsin, offered ex-convicts (one of whom had been arrested forty-eight times) a new life with employment on his show.[35] In 1997 the television show "48 Hours" broadcast the story of Kanuk Nadji, a talented in-line skater from the Bronx who made it big at the circus. Performing in one of his numbers as an Egyptian dancer in drag, Nadji had a lucrative first year with the show, earning $30,000—an enormous boon to his family, even though some of the roles he was hired to play bore a strong resemblance to the exoticized spectacles of racial difference found a century ago.[36]

People still join the circus for personal empowerment, just as marginalized members of society "ran away" to the circus a hundred years ago. Sedric Walker's UniverSoul Big Top, founded in 1993, is an exclusively African American circus designed as a traveling testament to black artistic achievement—a reclamation of agency in an entertainment that has historically limited African Americans from positions of power. The Ringling Bros. and Barnum & Bailey circus, for one, hired its first black ringmaster only in 1998—the lyric tenor Jonathan Lee Iverson.[37] In Melbourne, Australia, the Women's Circus, created in 1991, offers women of all ages and ranges of physical ability a chance "to address issues that affect women." One performer has written, "The Women's Circus has enabled me to work with my body in a safe environment. . . . Now when I perform, I feel joy more than I feel fear and I feel proud rather than shameful."[38] Pope John Paul II offers a similar assessment of the circus: "[Circus performers are] at peace with their own bodies and also with animals."[39] Even Princess Stephanie Grimaldi of Monaco, who hardly qualifies as disadvantaged, "ran away" from relentless media scrutiny, her estranged father, and siblings to live on the road with the Swiss Circus Knie in the spring of 2001. She and her three children stayed in a trailer alongside that of her lover, Franco Knie, the show's co-owner and elephant trainer. Her father and siblings were reportedly outraged, but circus people applauded the princess's arrival. Kenneth Feld remarked: "Look at what she's been through. The tragedy, the press. Circus people don't judge."[40]

The shared experience of living on the margins of society as a circus performer still elicits a kind of cosmopolitan solidarity among former show folk. Since 1936 Gibsonton, Florida, has functioned as a home for sideshow performers. Living mostly in mobile homes, retired players such as Melvin Burkhart, "the human blockhead" (who died in November 2001), and Monica Barris, a cooch stripper formerly known as "the flame of New Orleans," would gather each day to talk about their health and the old days with the sideshow. Jeannie Tomaini, who performed as "the world's only living half-girl" (who passed away in August 2000), succinctly articulated their enduring sense of community: "I guess you could say we were attracted to each other, because we a sort of stick together."[41] In nearby Sarasota, circus artists still gather at Showfolks Club, a cafeteria-style dining hall adjacent to a small, dark bar crammed with hundreds of framed, autographed glossies of mostly obscure entertainers. During a research trip to the John and Mable Ringling Museum of Art in 1995, I ate a dinner of pickled beef, schnitzel, and cheesecake at Showfolks, prepared by Jenny Wallenda, of "Flying Wallendas" fame. There I met dozens of performers, past and present, who represented an extraordinary scale of human diversity. People spoke frankly about their years with the circus, the numbing— but addictive—travel, and the tough working conditions (most impressively, one elderly acrobat described the aerial feats she performed while seven months pregnant and the corsetlike apparatus that hid her condition from the audience). And the performers clearly regarded each other with an affection that was virtually familial.[42] For several decades now, countless people have also landed at Circus World Museum in search of familial connections. As a researcher on the grounds, I listened to scores of polite genealogical queries from people looking to the circus to make sense of their own past.

What still attracts audiences to the circus, given that its educative power and its ability to capitalize on novelty have been eclipsed by other media? For one, the circus remains a live community experience. In June 1999 I watched, dumbfounded, along with several thousand others, while a member of the Wuhan Flyers, wearing tall stilts, sprang from a teeterboard, somersaulted several times, and then landed upright—stilts and all.[43] The internet and countless "reality" television programs offer consumers a privatized, in-home reality, divested of immediate consequences. Yet the circus forces its audiences to face risk. Even the Great Circus Parade in Milwaukee, cheerfully headed by Ernest Borgnine as its "grand clown marshal," unpredictably spills outside the realm of safe historical reenactment: one year an eight-pony team pulling the Cinderella float wagon became spooked and plowed into a media broadcast platform; another year, a team of draught horses broke free, barreling into the sidelines, where no one was seriously injured.[44] At the circus performers still occasionally fall, and elephants and tigers sometimes get loose.[45] On October 11, 1994, "Good Morning America" declared that the most dangerous occupation in the United States was not that of police officer or firefighter—instead it was that of elephant trainer.[46] When the circus proprietor and animal trainer Wayne Franzen was mauled to death by one of his tigers during a performance in front of two hundred school children at Broad Top City, Pennsylvania, on May 7, 1997, all the national television networks covered the story, and all marveled at the circus's dangers.[47] The circus makes us take pause: to acknowledge the powerful and occasionally perilous relationship between people and animals. And in the end, the circus endures because it beckons us to contend with our own fragility and potential.

NOTES

Abbreviations

The following abbreviations are used throughout the notes:

BRTC
  Billy Rose Theatre Collection, New York Public Library for the Performing Arts
BBPCB
  Barnum & Bailey Press Clippings Books, Circus Collection, John and
  Mable Ringling Museum of Art, Sarasota, Fla.
CAH
  Center for American History, University of Texas, Austin
HCCM
  Hertzberg Circus Collection and Museum, San Antonio
JFDC
  J. Frank Dobie Collection, Harry Ransom Humanities Research Center,
  University of Texas, Austin
JMRMA
  John and Mable Ringling Museum of Art, Sarasota, Fla.
JTMCC
  Joseph T. McCaddon Circus Collection, Manuscripts Division, Department of
  Rare Books & Special Collections, Princeton University Library Theatre Collection
LAMNH
  Library of the American Museum of Natural History, New York
LTLBBS
  L. T. Lee Barnum & Bailey Scrapbook, Todd-McLean Physical Culture Collection,
  University of Texas, Austin
RLPLRC
  Robert L. Parkinson Library and Research Center, Circus World Museum,
  Baraboo, Wis.
SHSW
  State Historical Society of Wisconsin, Madison
WFCC
  William F. Cody Collection, Harold McCracken Research Library, Buffalo Bill
  Historical Center, Cody, Wyo.

Preface

1. Said, *Orientalism*.

2. Pink lemonade is truly a circus original—so the story goes. According to the circus historians Fred Dahlinger and Stuart Thayer, a peanut and lemonade seller named Pete Conklin was working with a Wisconsin-based outfit, the Mabie Brothers circus, down in hot, dusty Texas in 1859. Conklin ran out of water one day, walked into one of the dressing tents, and snatched a bucket of water from one of the performers who was washing her red tights. He poured the pink water into his lemonade tub and quickly created a new circus sensation: strawberry lemonade. Dahlinger Jr. and Thayer, *Badger State Showmen*, 11.

3. Hemingway quoted in "Ring of Fire," Edward Hoagland, review of *The Circus Fire: A True Story*, by Stewart O'Nan, in *New York Times Book Review*, July 2, 2000, 13.

4. The circus ring itself was originally designed for the horse. In the mid-eighteenth century the Englishman Philip Astley built an arena forty-two feet in diameter, which he estimated was the minimum size necessary to accommodate a horse with a person dancing on its back. Diana Starr Cooper has noted that without the horse, for which the ring was originally designed, there is no circus. The carnival—which is often confused with the circus—has no ring. Evolving out of the midway at the late-nineteenth-century circus and world's fair, the carnival is itinerant and sometimes still has a freak show. But unlike the circus, the carnival is almost solely composed of mechanical and animal rides, games of chance, and concessions. The carnival, in essence, is a smaller, itinerant version of an amusement park. See Truzzi, "The Decline of the American Circus," 315; Cooper, *Night After Night*, 1–2; Blackstone, *Buckskins, Bullets, and Business*, 38–39.

5. Other secondary works on the history of the circus include Carlyon, "Dan Rice's Aspirational Project"; Chindahl, *A History of the Circus in America*; Dahlinger, *Trains of the Circus, 1872–1956, Show Trains of the Twentieth Century*; Dahlinger and Thayer, *Badger State Showmen*; Mischler, "The Greatest Show on Earth"; Saxon, *Enter Foot and Horse*; Speaight, *A History of the Circus*; Thayer, *Traveling Showmen, Annals of the American Circus*, 2 vols. For exacting coverage of circus topics see *Bandwagon*, a publication of the Circus Historical Society. For a structuralistic rather than historical treatment, see Bouissac, *Circus and Culture*. On P. T. Barnum see Adams, *E Pluribus Barnum*; Cook, *The Arts of Deception*; Harris, *Humbug*; Kunhardt, *P. T. Barnum: America's Greatest Showman*; Saxon, *P. T. Barnum: The Legend and the Man*; Slout, *A Royal Coupling*. Books on the freak show include Bogdan, *Freak Show*; Fiedler, *Freaks: Myths and Images of the Secret Self*; Thomson, *Freakery*.

6. Mannheim, *Ideology and Utopia*, 40, 73.

7. Ibid., 107.

Chapter 1

1. The show's unobtrusive entry into Austin conformed to rules set forth by Ringling Bros. management in the mid-1990s: the Ringling Bros. circus (like many others) no longer publicizes its arrival time or its "animal walk" from the rail depot to the cir-

cus site. Mike Allen, "Secret Circus Parades," *Wisconsin State Journal*, Feb. 21, 1995, 6A (citation courtesy of Jean Davis).

2. "Local News," *Austin* (Tex.) *American-Statesman*, June 12, 1999, sec. B, p. 2.

3. P. S. Maddox, ed., "Route Book, Adam Forepaugh Shows Season, 1892" (Buffalo: Courier, 1892), 106, Route Book Collection, RLPLRC.

4. "Schools to be Closed Circus Day," *Bridgeport* (Conn.) *Telegram*, June 12, 1907, BBPCB 1907; "School Boys' Union Won a Great Fight," *Paterson* (N.J.) *Evening News*, n.d., BBPCB 1904, JMRMA.

5. Hugh Coyle, "Circus Notes of the Past and Present," *Billboard*, Mar. 16, 1907, 35, Periodicals Collection, HCCM.

6. Alfonso Fernandez, "Reminiscences of Clifton, Arizona," typescript, n.d., Greenlee County Historical Society, Clifton-Morenci, Ariz. (citation courtesy of Linda Gordon).

7. "The Circus Tax," *Billboard*, Nov. 1, 1899, 9.

8. Anderson, *Tar*, 161.

9. *Ashland* (Wis.) *Daily Press*, Aug. 18, 1904, Newspaper Collection, RLPLRC.

10. Charles Andress, ed., "Route Book of Barnum & Bailey, 1905" (Buffalo: Courier, 1905), 43, 47, Route Book Collection, RLPLRC.

11. "Ringling Bros.' Circus," *New Orleans Daily Picayune*, Nov. 19, 1898, 12, CAH.

12. "Barnum & Bailey Parade Tonight," *New York Daily Tribune*, Mar. 25, 1891, Scrapbook 8, Barnum & Bailey, 1890–91, 46, JTMCC.

13. Film of circus parade from Barnum & Bailey's Greatest Show on Earth, Waterloo, Iowa, 1904, Film and Video Collection, RLPLRC.

14. The Robinson show introduced these wagons for the 1897 season. "Robinson Circus Opening," *Billboard*, May 5, 1900, 3.

15. "How Would You Like to Be a Circus Ticket Seller," *Billboard*, Dec. 2, 1899, 2.

16. Sandburg, *Always the Young Strangers*, 192.

17. "World's Greatest Circus in Town," *Racine* (Wis.) *Daily Times*, July 31, 1908, Archival Collection, Ringling Bros. Press Afterblasts Only, 1900–1909, Folder 1; "Ringling Bros. Official Program, 1908," n.p., Program Collection, all from RLPLRC.

18. Garland, *A Son of the Middle Border*, 111.

19. Twain, *The Adventures of Huckleberry Finn*, in *The Family Mark Twain*, 546.

20. "'Big Fred' Tells a Tale: A Baptism that Didn't Take," Fred Roys, interview by Wayne Walden, New York City, Nov. 1, 1938. Retrieved Aug. 16, 2001, from the World Wide Web at *American Life Histories: Manuscripts from the Federal Writers' Project, 1936–1940, American Memory*, <http://memory.loc.gov/ammem/wpaintro/wpahome.html>.

21. "Too Big to See at Once," *Republican-Register*, n.p., June 21, 1890, Newspaper Advertisement Collection, RLPLRC.

22. In 1901 a newspaper in Ashland, Wisconsin, observed that "Over Two Thousand People from Other Towns are Here" and published a list of trains which carried the strangers to the circus: the Wisconsin Central special train brought 390 people from Medford; the Wisconsin Central regular train brought 120 from other surrounding counties; the Northwestern regular brought 155 people and the Bayfield scoot 450 people to Ashland. Hundreds of people also arrived by boat from Washburn and Bayfield. "Over Two Thousand People from Other Towns Are Here," n.p., July 27, 1901, Archival Collection, Ringling Bros. Press Afterblasts Only, 1900–1901, Box 1, RLPLRC.

23. "It Was Circus Day," n.d., Mount Pleasant, Iowa, 1894, 2, Newspaper Advertisement Collection, RLPLRC.

24. F. S. Redmond, ed., "Route Book, Adam Forepaugh Shows, Season, 1893" (Buffalo: Courier, 1893), 81–82, Route Book Collection, RLPLRC.

25. "A Circus Calamity," n.p., June 21, 1893, Newspaper Collection, RLPLRC.

26. "A Big Parade," *Sherman* (Tex.) *Weekly Democrat*, Nov. 1, 1900, 4, CAH.

27. Truzzi, "The Decline of the American Circus," 315, 319.

28. Wiebe, *The Search for Order, 1877–1920.*

29. Ewen, *Immigrant Women in the Land of Dollars*, 21.

30. Foner, *The Story of American Freedom*, 116.

31. Henry Ford, quoted in Susman, *Culture as History*, 136.

32. Anderson, *A Story Teller's Story*, 220.

33. Adams, *The Education of Henry Adams*, 499.

34. Trachtenberg, *The Incorporation of America*, 80.

35. Green, *The World of the Worker*, 21.

36. Nasaw, *Going Out*, 6–9.

37. $3,696 in 2000. This calculation and all subsequent dollar conversions throughout the book are taken from the *Inflation Calculator.* Retrieved August 26–28, 2001, from the World Wide Web: <http://www.westegg.com/inflation/>.

38. Slout, *Olympians of the Sawdust Circle*, 197.

39. Peiss, *Cheap Amusements.*

40. Green, *The World of the Worker*, 43.

41. Rosenberg, *Spreading the American Dream*, chapters 1–5. See also McCormick, *The China Market.*

42. *Boston Evening Transcript*, quoted in Woodward, *The Strange Career of Jim Crow*, 55.

43. Trachtenberg, *The Incorporation of America*, preface, 123; Harris, *Humbug*, 291.

44. Bouissac, *Circus and Culture*, 9.

45. Joy Kasson, *Buffalo Bill's Wild West.*

46. There are several excellent studies of William F. Cody and the Wild West show: see Blackstone, *Buckskin, Bullets, and Business;* Kasson, *Buffalo Bill's Wild West;* Moses, *Wild West Shows and the Images of American Indians, 1883–1933;* Russell, *The Lives and Legends of Buffalo Bill.*

47. "Adeline Blakeley," interview by Mary D. Hudgins, Fayetteville, Ark. (date of interview not listed). Project no. 30325. *Arkansas Narratives*, vol. 2, pt. 1:191–92. Retrieved Aug. 16, 2001, from the World Wide Web: *Born in Slavery: Slave Narratives from the Federal Writers Project, 1936–1938, American Memory,* <http://memory.loc.gov/ammem/snhtml/snhome.html>.

48. Garland, *A Son of the Middle Border*, 111.

Chapter 2

1. Isaac, *The Transformation of Virginia, 1740–1790*, 247.

2. Thayer, *Annals of the American Circus* 1:1–4, 2.

3. For a lovely structural analysis of the circus ring itself see Cooper, *Night after Night.*

4. Pfening, "The Frontier and the Circus," 17.

5. Thayer, "Some Class Distinctions in the Early Circus Audience," 21.

6. Thayer, *Annals of the American Circus* 1:27, 66.

7. May, *The Circus from Rome to Ringling*, 28.

8. Thayer, *Traveling Showmen*, 1.

9. Ibid.

10. Thayer, "The Anti-Circus Laws in Connecticut 1773–1840," 18; "Legislating the Shows: Vermont, 1824–1933," 20.

11. Coup, *Sawdust and Spangles*, 139, Published Circus Memoir Collection, RLPLRC.

12. Chindahl, *A History of the Circus In America*, 1–2.

13. Thayer, "One Sheet," 23.

14. Thayer, *Traveling Showmen*, 2.

15. For a wonderful analysis of the Crowninshield Elephant and its relationship to then-prevailing ideologies about republicanism, respectability, and national identity, see Mizelle, "The Downfall of Taste and Genius"; see also Goetzmann, *West of the Imagination*, 131.

16. Thoreau, *A Year in Thoreau's Journal*, 143.

17. For a thorough account of the business machinations leading to Barnum and Bailey's merger in 1880, see Slout, *A Royal Coupling.*

18. Chindahl, *A History of the Circus in America*, 55–56.

19. "Personal Diary of Master James A. Bailey, 1863," Box 11, Item 1, JTMCC.

20. Dahlinger Jr., "The Development of the Railroad Circus," pt. 1:6–11.

21. Dahlinger Jr. and Thayer, *Badger State Showmen*, 82.

22. Dahlinger, "The Development of the Railroad Circus," pt. 1:7.

23. Jules Turnour, *The Autobiography of a Clown, as Told to Isaac F. Marcosson*, 59–60, Circus Memoir Collection, RLPLRC.

24. Author conversation with Fred Dahlinger Jr., fall 1999.

25. Dahlinger Jr., "The Development of the Railroad Circus," pt. 1:10; Chindahl, *A History of the Circus In America*, 89–92.

26. Dahlinger Jr., "The Development of the Railroad Circus," pt. 1:6.

27. Dahlinger Jr. and Thayer, *Badger State Showmen*, 82–83, 93.

28. Gollmar, *My Father Owned a Circus*, 49.

29. "Among the Wild Beasts," *New York Tribune*, Apr. 7, 1895, BBPCB 1895, JMRMA.

30. Speaight, *A History of the Circus*, 137.

31. Gossard, "A Reckless Era of Aerial Performance," 123–24.

32. Nasaw, *Going Out*, 4.

33. $20 and $10 in 2000.

34. "Special to Cuero," *Cuero* (Tex.) *Daily Record*, Nov. 9, 1898, 5, CAH.

35. Dahlinger Jr., "The Development of the Railroad Circus," pt. 2:18.

36. Stoughton, Wis., n.p., Friday, May 19, 1899, Gollmar Brothers Circus Reviews, 1890–1900, RLPLRC.

37. Howells, *Literature and Life*, 126, 128, 200.

38. Susman, *Culture as History*, 111; for other perspectives on the growth of modern

visual culture see Brumberg, *The Body Project;* Peiss, *Hope in a Jar;* Musser, *The Emergence of Cinema;* Boorstin, *The Image,* 13; Oriard, *Reading Football.*

39. Howells, *Literature and Life,* 200.

40. Harris, *Humbug,* 290–91.

41. Consequently, this study is indebted to a group of labor, social, and cultural historians who use ethno-cultural methods of analysis and also those studying how working-class culture (especially time away from the workplace) has shaped class consciousness. Thompson, *The Making of the English Working Class;* Peiss, *Cheap Amusements;* Cohen, *Making a New Deal;* Couvares, *The Remaking of Pittsburgh;* Roy Rosenzweig, *Eight Hours for What We Will;* see also Lipsitz, *Class and Culture in Cold War America;* Kelley, *Race Rebels.*

42. Writing in the United States during the 1940s and 1950s, Frankfurt School theorists (refugee scholars from the Institute of Social Research in Frankfurt who fled the Nazis in the 1930s) used mass culture to explain popular support for Hitler. By focusing on the cultural "superstructure" as an agent of history (rather than the traditional Marxian emphasis on the economic "base"), Theodor Adorno, for one, stressed the role of popular music, among other parts of the culture industry in creating passive, unquestioning "authoritarian" cultural consumers. In order to make the relationship between producers and cultural consumers more dialectical, Stuart Hall and other Marxian theorists at the Centre for Contemporary Cultural Studies at Birmingham (England) argue that the relationship between producers and consumers reflects and fuels class conflict. These theorists rightly define popular culture as a site of conflict, but they assume that classes (and by association, cultural producers and consumers) are discrete and homogeneous, a paradigm which ignores cross-class alliances and makes class conflict (to the exclusion of other factors) the primary locus of identity and impetus for social change. For a helpful overview of the Frankfurt School see Jay, *The Dialectical Imagination.* For an introduction to cultural studies literature see Hall, "Notes on Deconstructing 'the Popular'"; Hall and Jefferson, *Resistance through Rituals;* Mukerji and Schudson, *Rethinking Popular Culture.*

43. See Roediger, *The Wages of Whiteness; Towards the Abolition of Whiteness; Black on White;* Jacobson, *Whiteness of a Different Color;* Harris, "Whiteness as Property"; Saxton, *The Rise and Fall of the White Republic;* Allen *The Invention of the White Race;* Omi and Winant, *Racial Formation in the United States.*

44. William Shearer to Otto Ringling, Chicago, Jan. 27, 1903, Archival Collections, Correspondence, RLPLRC.

45. Bailey was a Confederate nurse and spy during the Civil War who became the sole owner of the "Mollie Bailey Shows" after her husband's death in 1896. Immensely popular in Texas, Bailey was known as the "Pioneer Southern Show Queen," who always admitted Confederate and Union veterans into her show for free. Bailey, *Mollie Bailey,* 63–69.

46. Mathias B. Harpin, "Hefty and Glad of It Is Rhode Island Circus Queen," *Providence* (R.I.) *Sunday Journal,* Feb. 23, 1936, 3, 3A82 Sideshow Pamphlet Collection, VS—Sideshow Fat People, HCCM.

47. Butler, *Bodies that Matter; Gender Trouble;* Grosz, *Volatile Bodies;* Foucault, *The His-*

tory of Sexuality; Douglas, *Purity and Danger;* see also Brumberg, *The Body Project;* Rooks, *Hair Raising;* Peiss, *Hope in a Jar;* Haiken, *Venus Envy.*

48. For an excellent overview of freak shows see Thomson, *Freakery;* Bogdan, *Freak Show.*

49. See Said, *Orientalism; Culture and Imperialism;* Bhabha, *The Location of Culture;* Gilman, "Black Bodies, White Bodies"; Poole, "A One-Eyed Gaze"; Lutz and Collins, *Reading National Geographic;* Pratt, *Imperial Eyes.*

50. The anthropologist James Clifford suggests that a "multivocal field of intercultural discourse" in which the "'objects' of observation . . . begin to write back" has radical epistemological potential to transform a centuries-old, strictly Eurocentric, "West-rest" *Weltanschauung.* I hope to contribute to this field by including the voices of various circus "Others": roustabouts, sideshow "freaks," acrobats, clowns, bit players. Clifford, *The Predicament of Culture,* 256.

51. See works by Coco Fusco and Jane Desmond on the dialectical relationship between self and Other. In 1992 Fusco and her partner drew upon the historical legacy of five centuries of ethnographic display to perform in a cage as "newly discovered" Amerindians from the fictitious country of "Guatinau." Fusco writes, "The performance was interactive, focusing less on what we did than on how people interacted with us and interpreted our actions." Fusco found that white audiences responded to the exhibition using deeply entrenched stereotypes about the racial Other (their authenticity was rarely questioned) as erotic and animalistic. Fusco, *English Is Broken Here,* 40; Desmond, "Dancing Out the Difference."

52. "Beneath White Tents: a Route Book of Ringling Bros., World's Greatest Shows, Season 1894" (Buffalo: Courier, 1894), 98–101, Route Book Collection, RLPLRC.

53. Dickinson, "Poems, Second Series, XVIII," *Collected Poems,* 107.

54. Bakhtin, *Rabelais and His World,* 34. See also Lipsitz, *Time Passages.*

55. Eagleton, *Walter Benjamin,* 148.

56. Advertisement in the *Sherman* (Tex.) *Weekly Democrat,* Nov. 1, 1900, State and Regional Newspapers Collection, CAH.

57. $5,985 and $19,950 in 2000.

58. The Georgia law stipulated that a circus must pay $1,000 per day ($19,950 in 2000) to show in a city with a population over 20,000; $400 a day ($7,980 in 2000) in a town with a population over 5,000; and $300 a day ($5,985 in 2000) in a smaller town. *Billboard* speculated that Georgians would simply travel to border states in order to see the circus, and spend their money in South Carolina, Alabama, Florida, and Tennessee. Although few taxed the circus to this degree, other communities shared the sentiments of the Georgia legislature. The *Evening Crescent,* in Appleton, Wisconsin, observed: "Practically all of [our money] went out of the city with the circus last night"; "The Circus Tax," *Billboard,* Nov. 1, 1899, 9; "The Big Circus Is a Disappointment," *Evening Crescent,* Aug. 3, 1907, Archival Collections, Press Afterblasts, RLPLRC.

59. Allen, *Horrible Prettiness,* 35.

60. "Waiting for Circus Train; Start Fire," n.p., Appleton, Wis., Aug. 19, 1910, Archival Collections, Ringling Bros. Press Afterblasts Only — 1910 and Beyond, RLPLRC.

61. $462 in 2000.

62. "Storm Was a Severe One," *Clinton* (Iowa) *Daily Herald*, n.d., 1905, Newspaper Collection, RLPLRC.

63. "Horses Stolen," *Sherman* (Tex.) *Weekly Democrat*, Nov. 1, 1900, 1, CAH.

64. "Be Careful To-Morrow," *Arkansas Democrat* (Little Rock), Sept. 30, 1898, 3.

65. "The Circus Here," n.p., n.d., Mount Pleasant, Iowa, 1894, Newspaper Advertisement Collection, RLPLRC.

66. Frediani, "The Fredianis, from the Early Years to America, 1866–1908," 34–37.

67. Ernest Cook (show manager) to M. W. J. Jackson, St. Louis, Oct. 28, 1908, VI:B—Personnel, Box 1, Folder 6, WFCC.

68. Gail O. Downing to Orilla Downing, May 19, 1913, VI:B—Personnel, Box 1, Folder 2, WFCC.

69. "Joined the Circus," *North Adams* (Mass.) *Evening Herald*, July 27, 1903, BBPCB 1902–3, JMRMA.

70. "Lynn Girls Run Away: Found with the Circus," *Boston American*, July 11, 1907, BBPCB 1907, JMRMA.

71. *Arkansas Democrat*, Oct. 19, 1898, 2.

72. "Official Route Book of Ringling Bros. World's Greatest Railroad Shows, Season 1892" (Buffalo: Courier, 1892), 77, Route Book Collection, RLPLRC.

73. Stallybrass and White, *The Politics and Poetics of Transgression*, 19, 53; Davis, *Parades and Power*.

74. According to a press release for Buffalo Bill, "there is no taking down of trapezes and the like to annoy the crowd as with circuses." "Wild West Here," n.p., La Crosse, Wis., Aug. 11, 1898, Newspaper Collection, RLPLRC; see also "The Rough Riders Return," *New York Daily Tribune*, Apr. 24, 1900, 7; Joy Kasson, *Buffalo Bill's Wild West*, 42.

75. In his influential work, *Highbrow/Lowbrow* (1988), Lawrence Levine traces the transformation of Shakespearean drama and the fine arts from shared popular forms in the nineteenth century to exclusively elite entertainment by the beginning of the twentieth. Robert Allen's study of nineteenth-century burlesque is also concerned with cultural hierarchy, but challenges Levine's assumption that all nineteenth-century Americans interpreted the same "shared" cultural forms in the same way. Allen argues that shared cultural experiences were actually contested among the different classes who attended the same entertainment. M. Alison Kibler suggests that in vaudeville, the performers themselves consciously crossed the boundaries between the "high" and "low," thus contradicting their managers' desire to attract reputable audiences. Fredric Jameson contends that cultural hierarchy is less a force in late capitalist America, where cultural workers like Philip Glass and Thomas Pynchon signify an "increasing interpenetration of high and mass cultures." Levine, *Highbrow/Lowbrow*; Allen, *Horrible Prettiness*; Kibler, *Rank Ladies*; Snyder, *The Voice of the City*; Erenberg, *Steppin' Out*; Butsch, *For Fun and Profit*; Jameson, "Reification and Utopia in Mass Culture," 14.

76. Invitation to President Theodore Roosevelt to inaugural performance of Barnum & Bailey's Greatest Show on Earth, Madison Square Garden, Mar. 18, 1903, Correspondence, Box 16, Folder 1, JTMCC. Most route books contain references to audiences from local asylums; see "Friday, July 14, 1893, St. Peter, Minnesota . . . several hundred patients from the insane asylum at this place visited the show in a body." "Route Book,

Ringling Bros., Season of 1893" (Buffalo: Courier, 1893), 61, Route Book Collection, RLPLRC.

77. "Review of Circus Day," *Galveston Daily News*, Nov. 13, 1898, 7, CAH.

78. $20 and $40 in 2000.

79. Speaight, *A History of the Circus*, 149.

80. Thayer, "The Birth of the Blues," 24–26.

81. Woodward, *The Strange Career of Jim Crow*, 81–87.

82. Dixie Willson, a ballet girl with the Ringling Bros. circus in 1921, commented on the color line in the South: "Never shall I forget my first sight of a straw house in the South. I hadn't thought about the dividing color line, and no one had spoken of it. I had traveled halfway around the hippodrome track on the elephant in the tournament, when suddenly I found myself facing a solid half circumference of faces! I can't describe the impression of it—so unexpected! So much of it all together! So terrifically shady!" Newspapers also mentioned the practice of segregation under the big top. The *New York News* reported in 1903 that Barnum & Bailey's date at Madison Square Garden had reserved seats in a segregated area for an African American boy's birthday party. Dixie Willson, *Where the World Folds Up at Night*, 61, Published Circus Memoir Collection, RLPLRC; "Pickaninnies at Circus," *New York News*, Apr. 10, 1903, BBPCB 1902–3, JMRMA.

83. C. Vann Woodward, *The Strange Career of Jim Crow*, 84; Alf T. Ringling, "With the Circus, A Route Book of Ringling Bros.' World's Greatest Shows, Seasons of 1895–1896" (St. Louis: Great Western, 1896), 78, 112–13, Route Book Collection, RLPLRC.

84. "Entrance for colored persons on the west side of building," *New Orleans Bee*, Jan. 13, 1835; "Persons of color admitted Wednesday and Saturday evenings (only), except those attending as servants," *Charleston* (S.C.) *Courier*, Jan. 19, 1835; "The pit is entirely enclosed for blacks," *Charlotte* (N.C.) *Journal*, Oct. 6, 1840 (all references courtesy of Stuart Thayer).

85. Virtually all route books mention the presence of Native American spectators. In the interest of brevity, here are two examples: "Official Route Book, Adam Forepaugh and Sells Bros. Combined Circuses, Season 1898" (Columbus, Ohio: Landon, 1898), 79, Route Book Collection, RLPLRC; "Official Route Book of Ringling Bros. World's Greatest Railroad Shows, Season 1892," 88.

86. $30 in 2000.

87. "Chinese at the Circus," *New York Mail and Express*, Mar. 26, 1903, BBPCB 1902–3, JMRMA.

88. A ten-cent, twenty-five-cent, and thirty-cent ticket in 1900 would cost $2, $5, and $6 in 2000.

89. Dahlinger Jr. and Thayer, *Badger State Showmen*, 89.

90. Kibler, *Rank Ladies*, 25–28; Nasaw, *Going Out*, 36–37.

91. John Kasson argues that the proliferation of large amusement parks such as Coney Island at the turn of the century heralded an emergent mass culture that rejected middle-class manners and taste in favor of interactive, and bawdy (albeit controlled) fun. As part of this burgeoning mass culture, the circus vigorously embraced "middle-class" values just as much as it flouted them. See John Kasson, *Amusing the Million*. For

an excellent discussion of Coney Island in the context of working-class women's leisure at the turn of the century, see Peiss, *Cheap Amusements*, chapter 5.

92. For a perceptive historical overview of cultural hierarchy in twentieth-century America, see Kammen, *American Culture, American Tastes*; see also Butsch, *The Making of American Audiences*.

93. Nasaw, *Going Out*, 3.

94. "Children Going to the Great Circus," *New York Daily Tribune*, Apr. 12, 1894, BBPCB 1894, vol. 2, JMRMA; see also "Big Crowds at the Circus," *New York Herald*, Apr. 2, 1895; "Orphans' Day at the Circus," *New York Daily Tribune*, Apr. 3, 1895; "Orphans at the Circus," *New York Morning Journal*, Apr. 3, 1895; "Children's Day at the Circus: Those from the Orphan Asylum Will Have a Treat Next Monday," *New York Times*, n.d., BBPCB 1895, JMRMA.

95. "With Secret Delight: A Court of Florida Boys Review the B&B Circus Parade: All the Makings of the Earth: Tody Hamilton Made the Boys Believe That Emperors, Kings, Queens, and Princes Were the Genuine Article: Dr. Diddier's Patients Are Now Cured," *New York Recorder*, Mar. 28, 1895; "Boys Bound Homeward: The Three Recorder Youngsters Rejoicing in Tears Start for Florida: Glorious Hours at the Circus: Willy Anderson Was Too Tired to Attend the Greatest Show on Earth but His Chums Went, and Bid Good-bye to Dr. Diddier and to Other Beloved Friends and Away They Go." *New York Recorder*, Mar. 29, 1895, BBPCB 1895, JMRMA.

96. C. Durand Chapman, "A Cut-Out Circus for the Children," *Ladies' Home Journal*, June 1913, p. 38; July 1913, p. 24; Aug. 1913, p. 22; Sept. 1913, p. 18, Oct. 1913, p. 18, Nov. 1913, p. 36; see also Dan Beard, "A Circus in the Attic, *Ladies' Home Journal*, Oct. 1899, p. 20; "A Wild West Show in the House," *Ladies' Home Journal*, Nov. 1899, p. 19.

97. The growing popularity of toys like the teddy bear and Kewpie doll underscored the nexus between mass-marketing and new child-centered parenting philosophies. Cross, *Kids' Stuff*, 28–34, 86–87.

98. Boyer, *Urban Masses and Moral Order in America, 1820–1920*, 242–49.

99. Cross, *Kids' Stuff*, 37–40.

100. See Dower, *War without Mercy*; Enloe, *Bananas, Beaches and Bases, The Morning After*; Gosse, *Where the Boys Are*; Hoganson, *Fighting for Manhood*; Iriye, "Culture and Power," "Culture (A Round Table: Explaining the History of American Foreign Relations)"; McEnaney, "He-Men and Christian Mothers"; Plummer, *Haiti and the U.S.: The Psychological Moment, Rising Wind*; Renda, *Taking Haiti*; Rosenberg, *Spreading the American Dream*; "Gender (A Round Table: Explaining the History of American Foreign Relations)"; "'Foreign Affairs' after World War II"; Von Eschen, *Race against Empire*; Wexler, *Tender Violence*.

101. Antonio Gramsci analyzes how "cultural hegemony" works at multiple levels, as people from different classes become complicit partners in solidifying popular support for oppressive regimes. Yet Gramsci also argues that "historical blocs" led by largely non-elite "organic intellectuals" can disrupt and overturn the existing order. Gramsci, *Selections from the Prison Notebooks*; see also a reinterpretation of this idea in Lears, "The Concept of Cultural Hegemony"; and a counterpoint in Denning, "The End of Mass Culture."

102. Rydell, *All the World's a Fair; World of Fairs*. See also LaFeber, *Michael Jordan*

and the New Global Capitalism; collected essays in Kaplan, *Cultures of United States Imperialism;* Slotkin, *Gunfighter Nation;* British historians have also examined the relationship between popular culture, imperialism, and public opinion. See the collection of essays in MacKenzie, *Imperialism and Popular Culture,* especially Springhall, "'Up Guards and at Them!'"; Summerfield, "Patriotism and Empire."

Chapter 3

1. Kelly, *Clown,* 42, Published Circus Memoir Collection, RLPLRC.

2. Sandburg, *Always the Young Strangers,* 189.

3. *Detroit Free Press,* Aug. 11, 1890, 4.

4. Harvey L. Watkins, "Barnum and Bailey Official Route Book, Season of 1893" (Buffalo: Courier, 1893), 126–27, Route Book Collection, RLPLRC.

5. "Ringlings' Great Shows Afternoon and To-night," *Eau Claire* (Wis.) *Daily Telegram,* 1908, Archival Collections, Ringling Bros. Press Afterblasts Only, 1900–1909, Folder 1, RLPLRC.

6. Williams, *Marxism and Literature.*

7. Lamoreaux, *The Great Merger Movement in American Business, 1895–1904,* introduction.

8. Gary B. Nash et al., *The American People,* 605–6.

9. In 1948 the U.S. Supreme Court reduced the power of the Big Five, by ruling that production company ownership of movie theaters and the practice of "block booking" violated the antitrust laws. Baughman, *The Republic of Mass Culture,* 22–24, 40–41.

10. "Monopoly Lodge Box Party at the Circus," *New York American,* Mar. 26, 1907, BBPCB 1906–7, JMRMA.

11. According to Nate Salsbury's and W. F. Cody's agreement with James Bailey, the parties split the show's profits equally, and each shouldered certain expenses. Cody and Salsbury were to pay for everything pertaining to the actual performance: performers' salaries, livestock, props, lighting, paper on which to print advertisements, and tickets, feed, and groceries for the whole company. Bailey provided the cars, baggage wagons, baggage stock, tents, seats, and the salaries of the workers concerned with advertising and moving the show; he also paid for the lot licenses and railroad transportation. The agreement stipulated that Cody was to perform "in the saddle" at every show. After Cody poured his capital into several bad financial schemes, James Bailey's financial stake in the Wild West show grew to one half, plus ownership of the Buffalo Bill title, after he bailed Cody out of bankruptcy during the show's disastrous European tour (1902–6), when valuable ring stock was lost during a glanders epidemic in France, and some Native American performers accused Cody of mistreatment. For more see Blackstone, *Buckskins, Bullets, and Business,* 22–30; Joy Kasson, *Buffalo Bill's Wild West,* 144–51.

12. $7,577,105 in 2000. For original figures see Chindahl, *A History of the Circus in America,* 148–49.

13. That figure is nearly $20 billion in 2000.

14. Nash et al., *The American People,* 731.

15. Weeks, *Ringling,* 232–56.

16. "Circus Combine a Gigantic Trust," *Toledo* (Ohio) *News*, May 30, 1907, BBPCB 1906–7, JMRMA.

17. Ibid.; see additional coverage in *Variety*, 1906–8.

18. "Winter Circus in Indian Territory," *Wausau* (Wis.) *Record*, June 26, 1907, Archival Collections, Ringling Bros. Press Afterblasts, 1900–1909-2, RLPLRC.

19. Dahlinger Jr., "The Development of the Railroad Circus," pt. 2:17.

20. Yet laborers were less sanguine about the purported benefits of scientific management. They felt that the "Taylor system" created a frenzied, fragmented workplace in which people raced the clock each day. Workers felt bound to specialized (and monotonous) tasks that destroyed their autonomy. They protested that the wage incentives for increased production were hardly worth the utter exhaustion and constant hounding for hours on end. After iron workers at the Watertown Arsenal in Massachusetts went on strike against the Taylor system in 1911, the House of Representatives held special hearings to investigate the integrity of scientific management in October 1911. In January 1912 Taylor testified before the special committee at length, and repeatedly stated that the task system equally benefited capital and labor. Although the House committee vehemently criticized time-motion studies in its report, it made no recommendations for legislation. Consequently, the results of the investigation were negligible. Taylor, *Scientific Management*, 5–287; Kanigel, *The One Best Way*, 443–84.

21. "Circus Life in America," *Music Hall* (London), Apr. 5, 1907, BBPCB 1906–7, JMRMA; Dahlinger Jr. and Thayer, *Badger State Showmen*, 77.

22. Untitled newspaper article, *Greenville* (Tex.) *Evening Banner*, Oct. 12, 1900, 2, CAH.

23. In 1892 the Ringling brothers had several nasty encounters with Barnum & Bailey workers at Kansas City and Topeka, culminating in a near-knifepoint showdown in Milwaukee, because Barnum & Bailey believed that the Ringling Bros.' show date (July 18) was far too close to their own on August 15. Dahlinger Jr. and Thayer, *Badger State Showmen*, 83.

24. "Beneath White Tents: A Route Book of Ringling Bros.' World's Greatest Shows, Season 1894" (Buffalo: Courier, 1894), 20–21, Route Book Collection, RLPLRC.

25. $18,482 in 2000.

26. Charles Theodore Murray, "In Advance of the Circus," *McClure's Magazine*, Aug. 1894, 252, Circus Scrapbook Collection, MWEZ+N.C.6312, BRTC.

27. Ibid., 253.

28. Ibid.

29. Dahlinger Jr., "The Development of the Railroad Circus," pt. 3:32.

30. Murray, "In Advance of the Circus," 256–58.

31. F. B. Hutchinson, "Official Route Book of the Adam Forepaugh Shows, Season 1894" (Buffalo: Courier, 1894), 45, Route Book Collection, RLPLRC.

32. Harvey L. Watkins, "The Barnum & Bailey Official Route Book," Season of 1893, 24–25.

33. Ibid., 28–29.

34. Murray, "In Advance of the Circus," 251.

35. "Billing Like a Circus," *Billboard Advertising*, Sept. 1, 1896.

36. Over $2.4 million in 2000.

37. *Billboard,* Dec. 1, 1897, 9.

38. Newspapers in Oshkosh and Monroe, Wisconsin, made identical comments about the Ringling Bros.' female snake charmer, on August 29, 1906, and September 1, 1906: "And the pretty snake charmer all dressed in black drew forth exclamations of pleasure and gratification on the part of the crowd after its long wait. Even the large snake did some interesting writhing and curling up and down the bars of his cage to show just how charmed he was to coil at the feet of the girl in black." N.p., Archival Collections, Ringling Bros. Press Afterblasts Only, 1900–1901, pt. 1, RLPLRC.

39. Murray, "In Advance of the Circus," 260; "Beneath White Tents: A Route Book of Ringling Bros.' World's Greatest Shows, Season 1894," 35.

40. Murray, "In Advance of the Circus," 260.

41. Cleveland Moffett, "How the Circus Is Put Up and Taken Down," *McClure's Magazine,* June 1895, 49–50 (citation courtesy of Fred Dahlinger Jr.).

42. Ibid., 49–54.

43. The circuses of Cooper and Bailey and of W. W. Cole were the first circuses in the United States to illuminate their big tops with electricity in 1879, using bulky steam-powered electric generators and open arc lights. However, the biggest railroad showmen soon found that this electrical system was bulky and difficult to transport. Consequently, they returned to open flame lighting within a couple of years. In 1896 Buffalo Bill's Wild West began using electricity for spotlighting, and then around 1905, Barnum & Bailey started illuminating its canvas big top with electricity generated by a gasoline engine and dynamo. Ringling Bros. followed suit in 1909. See "Electrified Circus," *Redding* (Conn.) *Independent,* July 28, 1882, 3, Michael Sporrer Collection, RLPLRC; "Circus to Have Electric Lights," *Billboard,* Apr. 3, 1909, 17, Ringling Bros. Clippings, RLPLRC; see also Speaight, *A History of the Circus,* 46. Thanks to Fred Dahlinger for filling in the gaps on electrification. Author conversation with Dahlinger, Aug. 28, 2001.

44. Moffett, "How the Circus Is Put Up and Taken Down," 56–57.

45. Ibid., 57–61.

46. "Al Rosboro, Ex-Slave 90 Years Old," Al Rosboro, interview by W. W. Dixon, Winnsboro, S.C., n.d. Project no. 1655, *South Carolina Narratives,* vol. 14, pt. 4:38–39. Retrieved Aug. 16, 2001, from the World Wide Web: *Born in Slavery: Slave Narratives from the Federal Writers' Project, 1936–1938, American Memory,* <http://memory.loc.gov/ammem/snhtml/snhome.html>.

47. Wilder, *By the Shores of Silver Lake,* 18–19.

48. Thompson, "Time, Work-Discipline, and Industrial Capitalism," 61, 75.

49. Schivelbusch, *Railway Journey;* Trachtenberg, *The Incorporation of America,* 60.

50. Trachtenberg, *The Incorporation of America,* 80.

51. Ibid., 91.

52. From 1841 to 1865 Barnum sold 38 million admission tickets; he sold an additional 4 million tickets at his second museum from 1865 to 1868. These sales were all the more remarkable, even when taking duplicate sales into account, given that the U.S. population was approximately 35 million in 1865! Saxon, *P. T. Barnum,* 107–8.

53. Although Barnum reaped a fortune from Lind's 1850–51 tour, his financial situation was often uncertain. Barnum's Great Asiatic Caravan, Museum, and Menagerie (1851–54) was a flop, and he was swindled by business partners in a failed clock company.

To make matters worse, several of Barnum's properties, including the American Museum, were completely destroyed by fire. After his second American Museum burned in 1868, Barnum left the museum business altogether. Harris, *Humbug*, 173.

54. Over $4.2 billion in 2000.

55. Slout, *A Royal Coupling*, 19.

56. Harris, *Humbug*, 207.

57. Adams, *E Pluribus Barnum*.

58. By 1884 every large American circus had several elephants, thus diminishing the beasts' novelty. In order to make pachyderms more marketable, P. T. Barnum attempted to present a more exotic elephant in 1884. In contrast to the seemingly humdrum grey African and Indian elephants that were usually exhibited, Barnum purchased a "rare," "sacred" white Burmese elephant named Toung Taloung at "great cost" from King Thibaw of Burma in 1883. Toung Taloung was shown in London for over a month before the opening of Barnum, Bailey, and Hutchinson at Madison Square Garden in March. Not to be outdone by his bitter rivals, Adam Forepaugh displayed a fraudulent white elephant of his own on March 22, 1884, in Philadelphia— just six days before Toung Taloung landed in new York City. Forepaugh and his animal keepers scrubbed a grey Indian elephant with plaster of paris and used peach-colored tint around the animal's ears, trunk, and feet. Barnum, Bailey, and Hutchinson immediately denounced the "Light of Asia" as a fake. In a letter to the editor of the *Chicago Inter-Ocean*, they protested: "In the interests of the public, of decency and of right, in the interests of all classes who may possibly pay their money to see what they are misled to believe a genuine white sacred elephant, Barnum, Bailey and Hutchinson now call upon the press of the country to exercise that duty and aid them in exposing Forepaugh's swindling deceptive, cheating, false and fraudulent elephant which he is now knowingly, willfully, and criminally imposing upon the community as a genuine white sacred one from Siam." However, the American public vastly preferred Forepaugh's dazzling white fake to the genuine, pale grey Burmese elephant of Barnum, Bailey, and Hutchinson. Ultimately, this controversy generated enormous free publicity for both circuses. Still, in November Forepaugh retreated from the white elephant "war" when he announced that the "Light of Asia" had suddenly died. In fact, his paint had been removed and he became an ordinary circus elephant again. Barnum, Bailey and Hutchinson, "Forepaugh's Whitewashed Hoax," letter to the editor, *Chicago Inter-Ocean*, June 12, 1884; "The Light of Asia," n.p., Aug. 12, 1884, George Chindahl Collection, Box 1, Folders 4–5, RLPLRC.

59. "Howard's Letter," n.p., Nov. 27, 1887, Barnum, Cole, Hutchinson, and Cooper Press Book, 1887, JMRMA.

60. Barnum, *Life of P. T. Barnum*, 261.

61. $369,829 in 2000.

62. Joseph T. McCaddon, quoted in Saxon, "New Light on the Life of James A. Bailey," 8. This fascinating article is based upon McCaddon's 600-page unpublished memoir, recently found in the basement of one of his descendants, of his professional and personal life with Bailey.

63. Personal Diary Master James A. Bailey, 1863, James A. Bailey Papers, Box 11, Item 1, JTMCC.

64. McCaddon, quoted in Saxon, "New Light on the Life of James A. Bailey," 7.

65. Ibid.

66. "Barnum's Right Bower: James A. Bailey, the Best All-Around Showman in the World," n.d., n.p., Scrapbook 8, Barnum & Bailey, 1890–91, 87, JTMCC; "Ringling Bros. and Barnum & Bailey Combined Shows, Magazine and Daily Review, Season 1920," Madison Square Garden (New York: Select, 1920), 4, Program Collection, RLPLRC.

67. "Story of Biggest Man in Circus Today," *New York World*, Mar. 30, 1903, BBPCB, 1902–3, JMRMA.

68. Saxon, "New Light on the Life of James A. Bailey," 8.

69. Ibid., 7.

70. Joseph T. McCaddon, "The New Circus," July 17, 1935, Correspondence, Box 10, Folder 7, JTMCC.

71. Kasson, *Buffalo Bill's Wild West*, 123.

72. Ibid., 13.

73. Blackstone, *Buckskins, Bullets, and Business*, 17.

74. W. F. Cody to Sam, Sept. 2, 1879, I:B—Correspondence, Box 1, Folder 6, WFCC.

75. W. F. Cody to Julia Cody Goodman, June 14, 1905, I:B—Correspondence, Box 1, Folder 21, WFCC.

76. Harry Webb, "Buffalo Bill, Saint or Devil?" original manuscript (to be published in *Real West Magazine*), n.d., VI:B—Personnel, Box 1, Folder 5, WFCC.

77. Upon his return to the United States from Europe in 1906, Cody's financial troubles multiplied. James Bailey's executors claimed that Cody owed over $150,000 to Bailey's estate. Although it was uncertain whether he had already paid his debt before Bailey's death, Cody still honored the terms of the estate. In 1908 Bailey's widow Ruth Louisa Bailey sold a third of Buffalo Bill's Wild West to a rival Wild West showman, Gordon W. Lillie (also known as Pawnee Bill). By 1909 Pawnee Bill had purchased a two-thirds interest in the show and a two-year guarantee that Cody would appear "in the saddle" during performances. Cody embarked on a lucrative "farewell tour" in 1910 in celebration of his impending retirement. The season was profitable and Cody was able to purchase a half interest in his show from Lillie. Nevertheless, Cody's debts were so large that he came out of "retirement" the very next season. Many audiences felt cheated and business was poor. Cody turned to an unscrupulous Denver newspaper owner, Harry Tammen, who also owned partial interests in the Sell-Floto circus. Tammen claimed to own the title to Buffalo Bill's Wild West, and his bad business decisions further eroded Cody's holdings. The bankrupt Cody was forced to tour from 1914 to 1916 as an old man. He became seriously ill in November 1916 and died on January 10, 1917. Conover, "The Affairs of James A. Bailey," 11–12; Blackstone, *Buckskins, Bullets, and Business*, 30–35.

78. W. F. Cody to C. L. Hinkle, Apr. 27, 1901, Box 1, Folder 16; unsigned letter, May 17, 1901, Box 1, Folder 16; also n.d. unsigned letter, Box 1, Folder 19, all from I:B—Correspondence, WFCC.

79. "Buffalo Bill's Wild West and Congress of Rough Riders of the World," Historical Sketches and Programme (New York: Fless and Ridge, 1899), 27, Program Collection, RLPLRC.

80. Blackstone, *Buckskins, Bullets, and Business*, 2; Joy Kasson, *Buffalo Bill's Wild West*, 164.

81. William F. Cody to M. R. Russell, Dec. 27, 1899, I:B—Correspondence, Box 1, Folder 17, WFCC.

82. "Buffalo Bill's Wild West Show Program, 1887" (London: Allen, Scott, 1887), VI:A—Programs, Etc., Box 1, Folder 6, WFCC.

83. Ringling, *Life Story of the Ringling Bros.*, 40, Circus Memoir Collection, RLPLRC.

84. Ibid., 233–34.

85. Thayer, "The Circus That Inspired the Ringlings," 23–25.

86. Ringling, *Life Story of the Ringling Bros.*, 40.

87. Dahlinger Jr. and Thayer, *Badger State Showmen*, 72.

88. Ringling, *Life Story of the Ringling Bros.*, 117.

89. Between $17,817 and $21,380 in 2000.

90. Dahlinger Jr. and Thayer, *Badger State Showmen*, 72.

91. $356 in 2000.

92. $1.4 million in 2000.

93. Bank of Baraboo (Wis.) Loan Records, 1883–1905, transcriptions courtesy of Fred Dahlinger Jr., RLPLRC.

94. Ringling, *Life Story of the Ringling Bros.*, 239.

95. Dahlinger and Thayer, *Badger State Showmen*, 73.

96. Al Ringling's wife Louise, or "Lou," also worked at the family's circus. She performed variously as an equestrienne, trapeze artist, snake charmer, mind-reader, and "half-living body," and in later years she worked as the show's wardrobe mistress. Bradna, *The Big Top*, 66–73.

97. By 1919, when the Ringling Bros. merged operations with their other biggest title, Barnum & Bailey's circus, John and Charles were the only surviving brothers. Dashing John was flamboyant, hired world-famous circus stars, and filled his Sarasota mansion with original Rubens, while Charles was a micro-manager who was closely connected to the show's daily operations. The two brothers became rivals, and split into separate camps. Although John had no children, his nephew, John Ringling North, the son of Ida Ringling and Henry Whitestone North, controlled the circus for several decades after Ringling died in 1936. Johnny North had been "Mr. John's" apprentice throughout the 1920s, working as a billposter while he learned the circus business. Bitter relatives ejected Johnny North from circus leadership in 1943 and installed Charles and Edith Ringling's son Robert, an opera singer, as the show's new owner. But after the devastating big top fire at Hartford, Connecticut, on July 6, 1944, which killed 168 audience members, the Ringling organization unraveled and five circus officials were charged with manslaughter. Within three years Johnny North acquired a 51 percent share of the Ringling Bros. and Barnum & Bailey stock, and he retained control of the circus until he sold it to the drugstore and record magnates Irvin and Israel Feld in 1968 (Irvin's son Kenneth Feld owns the show today). Upon John Ringling's death, North interred John and his deceased wife Mabel in a crypt in New Jersey, despite Ringling's wishes to be buried in Sarasota. Finally, after North's death in 1985, his brother Henry Ringling North—after much quarreling with other family members—buried John and Mabel Ringling at Sarasota in 1991, sixty-three years after Mabel's death and fifty-five years after John's! Hammarstrom, *Big Top Boss*, 20–22.

98. During this same period, the vaudeville proprietor B. F. Keith enacted his own

"Sunday School" circuit of vaudeville shows. Movie makers and theater owners also consciously moved away from producing kinetoscope "peep shows" and "cheap nickel dumps," originally made for a working-class audience, in favor of full-blown historical epics (e.g. *Birth of a Nation*) which were shown in fancy, "respectable" movie palaces designed for middle-class spectators.

99. Although the Ringling show was "cleaner" than most, the show had its own internal scam artists. "Honest John" Kelley, the Ringlings' attorney with the show from 1903 through the late 1930s, was convicted of filing $3.6 million (nearly $42 million in 2000) worth of fraudulent income tax returns for the show from 1918, when the federal government began taxing circus income, to 1932. "Honest John" claimed depreciation and abandonment on the show's assets in order to fudge the numbers. Although the Ringling brothers probably did not know that Kelley falsified the show's tax data, the U.S. Treasury Department billed the show for back taxes, and in 1938 Kelley was sentenced to two years in prison and fined $10,000 ($115,903 in 2000). The show's stockholders demanded that the federal Board of Tax Appeals review the Treasury Department's assessment, a tedious process; the Treasury Department reduced its bill to $800,000 (over $9.2 million in 2000), and the show paid up. The government's other option was to take the circus as a form of payment, but it readily declined the monumental task of running a show. Irey, *The Tax Dodgers*, 197–203, John M. Kelley Vertical File, RLPLRC.

100. Alexander Saxton emphasizes that the rise of the Pinkerton agent is tied to the rise of the railroad. Railroad companies hired detectives to combat interstate crime among employees and passengers. Other private industrial corporations (like the circus) quickly followed suit. Saxton, *The Rise and Fall of the White Republic*, 335.

101. MacGilvra was born in Baraboo in 1893. E. E. MacGilvra, Interview by Charles C. Patton, July 4, 1973, Sheridan, Mont., Archives Division, Manuscript Collection, SC 3009, SHSW.

102. Alger, *Ragged Dick*, 14, 123.

103. Al Mann and Irene Mann, interview by author, Oct. 22, 1994, tape recording, Bageley, Wis.

104. Gollmar, *My Father Owned a Circus*, 103.

105. Many African American musicians in the sideshow band later became successful jazz musicians: Jo Jones, for example, became a drummer for the Count Basie band. Korall, *Drummin' Men*, 117–63 (citation courtesy of Joel Dinerstein).

106. Tiny Kline, "Showground-Bound," unpublished memoir, 157, Manuscript Collection, RLPLRC.

107. Murray, "In Advance of the Circus," 254.

108. Most performers were promoted to live in a stateroom once they became big, center-ring attractions. Although they were already center-ring circus stars, Ella and Fred Bradna initially received their stateroom from the Ringling Bros.' equestrian director, Bud Gorman, who moved into the bachelor car in exchange for free nightly dinners cooked by Fred. Bradna, *The Big Top*, 82–83.

109. Ibid., 81.

110. Dahlinger Jr., "The Development of the Railroad Circus," pt. 4:30–31; Gollmar, *My Father Owned a Circus*, 102–3.

111. Kline, "Showground-Bound," 144.

112. Ibid., 145, 168.

113. Dahlinger Jr., "The Development of the Railroad Circus," pt. 4:31.

114. Robinson, *The Circus Lady*, 120, Published Circus Memoir Collection, RLPLRC.

115. "Official Program of Ringling Bros. and Barnum & Bailey Combined Shows, Season 1923," Madison Square Garden edition (New York: Select, 1923), Program Collection, RLPLRC.

116. Kline, "Showground-Bound," 153–57.

117. $307 in 2000.

118. Zunz, quoted in Nash et al., *The American People*, 622.

119. Derks, *The Value of a Dollar*, 2.

120. $185 and $924 in 2000.

121. $100 and $120 in 2000.

122. Records for billposters from later years were unavailable. Gollmar Brothers Circus Collection, Family Accounting Ledgers, 1902–12, RLPLRC.

123. Murray, "In Advance of the Circus," 254.

124. $3,053 in 2000.

125. $2,519 in 2000.

126. Contract for Lillian Leitzel, Season 1917, signed Oct. 11, 1916, to begin Apr. 7, 1917, Ringling Bros. Box 2A, Archival Collections, Work Contracts, RLPLRC.

127. $763 in 2000.

128. Contract for Krao Farini, Side Show, signed Oct. 9, 1916, to begin Apr. 10, 1917, Ringling Bros. Box 2A, Archival Collections, Work Contracts, RLPLRC.

129. $924 in 2000.

130. $277 in 2000.

131. Gollmar Brothers Family Accounting Ledger, 1906, RLPLRC.

132. $647 in 2000.

133. $462 in 2000.

134. Gollmar Brothers Family Accounting Ledger, 1904–5, RLPLRC.

135. $140 in 2000, based on 1902 rate.

136. $200 and $299 in 2000, based on 1902 rate.

137. This an average figure based on an overview of multiple circus contracts from 1896 to 1975 on file at RLPLRC.

138. $148 in 2000.

139. $139 in 2000.

140. Peiss, *Cheap Amusements*, 52.

141. Kline, "Showground-Bound," 167–68.

142. Gollmar, *My Father Owned a Circus*, 126.

143. "Circus in Town Today," *Baltimore News*, May 13, 1903, BBPCB 1902–3, JMRMA.

144. $55 in 2000.

145. Walter L. Main Circus Salary List, Geneva, Ohio, Mar. 10, 1892, Walter L. Main Circus Vertical File, RLPLRC.

146. $277 in 2000.

147. $100 and $200 in 2000.

148. Walter L. Main Circus Salary List, 1892.

149. $437 in 2000.

150. Zunz, quoted in Nash, ed., *The American People*, 622.

151. Harriet Quimby, "The Feminine Side of Sawdust and Spangles," *Leslie's Weekly*, Apr. 16, 1908, 374, Periodicals Collection, RLPLRC.

152. Hornaday, *The Minds and Manners of Wild Animals*, 52, JFDC; Roediger, "White Looks: Hairy Apes, True Stories and Limbaugh's Laughs," 40; Joseph T. McCaddon to editor of the *Spectator* (London), Feb. 2, 1931, Correspondence, Box 7, Folder 7, JTMCC.

153. "Official Route book of Ringling Bros.' World's Greatest Railroad Shows, Season 1892" (Buffalo: Courier, 1892), 61, Route Book Collection, RLPLRC.

154. Bill "Cap" Curtis to George Chindahl, Oct. 10, 1950, George Chindahl Papers, Box 2, Folder 6, RLPLRC.

155. "Too Much Prosperity Hurts the Big Circus," *Washington Times*, Nov. 8, 1903, BBPCB 1903–4, JMRMA.

156. $37 in 2000.

157. $55 in 2000. Walter L. Main Circus Salary List, 1892.

158. $767 in 2000.

159. Gollmar Brothers Family Accounting Ledger, 1903.

160. Hammarstrom, *Big Top Boss*, 271.

161. Dahlinger Jr. and Thayer, *Badger State Showmen*, 101–2.

162. Naturally, life on the road with a circus could be just as monotonous as life on the farm or in the factory, despite the circus's ever-changing environment. Several circus performers kept diaries, noting the dull moments as well as those more exciting ones. The star bareback rider May Wirth noted, "Long jumps for the show were always scheduled for Sunday so we traveled all day, loafing around our living room with the Sunday papers, my sister Stell at the piano and Mother fussing with dinner. . . . That was the *usual* routine." Balinda Spencer, an equestrienne for the John Robinson show, wrote several entries which typified her life on the road. On September 5, in Gallipolis, Ohio, she noted that "Sally got a chicken in her egg. No one to dinner or supper." "Outline for Script for Interview between May Wirth and Adelaine Hawley, 'Woman's Page of the Air,'" on Station WABC, 485 Madison Avenue, New York City, July 29, 1942. May Wirth Vertical File; Balinda Merriam Spencer, "Diary of a Circus Lady, Balinda (Estella) Merriam Spencer, Equestrienne (1849–1899)," Manuscript Collection; John Robinson's Ten Big Shows Combined, Season, 1899, Route Book Collection, all from RLPLRC.

163. Al Mann interview by author.

164. Irene Mann, who joined her husband as a circus and rodeo performer in the early 1930s, recalls that circus performers had much more solidarity than rodeo people: "Circus people [were] friendlier than rodeo people. They're not contesting. Rodeo people are competing against each other for [prize money]. . . . But circus people, they have their own act, and they do it, and they're not competing with somebody else. They're more friendly." Irene Mann interview by author.

165. Allan Harding, "The Joys and Sorrows of a Circus Fat Lady," *American Magazine*, 62, 63, 100, 103, 3A82 Sideshow Pamphlet Collection, VF—Sideshow Fat People, HCCM.

166. M. B. Bailey, ed., "Official Souvenir, Buffalo Bill's Wild West and Congress of Rough Riders of the World" (Buffalo: Courier, 1896), 267, Program Collection, RLPLRC.

167. "Cowboys in a Small Riot at Buffalo Bill Show," *Brooklyn Citizen*, May 6, 1892, 2, VI: G—Ephemera, Box 3, Folder 2, WFCC.

168. "Circus Days and Ways," W. E. "Doc" Van Alstine, interview by A. C. Sherbert, July 1938, Portland, Ore., Project no. W13864. Retrieved Aug. 16, 2001, from the World Wide Web: *American Life Histories: Manuscripts from the Federal Writers' Project, 1936–1940, American Memory,* <http://memory.loc.gov/ammem/wpaintro/wpahome.html>.

169. Zora, *Sawdust and Solitude,* 96.

170. "Official Route Book of Ringling Bros. Greatest Railroad Shows, Season 1892," 92.

171. Kelly, *Clown,* 82–94.

172. Mann, *My Home Was the Open Range of Wyoming,* 70.

173. Source materials regarding the wage scales of Native American Wild West performers are sketchy. One letter lists the back wages that Buffalo Bill's Wild West was forced to pay to six Native American performers who were stranded in Paris after protesting their wages and working conditions during the show's European tour of 1902–6. The Bureau of Indian Affairs ordered the show to pay each Native American female player four and a half months' salary of $12 per month ($239 in 2000), $13 in cash credit ($259 in 2000), and $41 for passage from Paris to Rushville, Nebraska ($818 in 2000). Four male performers each were to receive four and a half months' salary, ranging from $20 to $25 per month ($399–499 in 2000), $19 to $38 cash credit ($379–758 in 2000), and $67.75 to $71 for passage back to Nebraska ($1,352–$1,416 in 2000). The federal government also fined the show an additional $1,500 for breaking its contract with the Native Americans ($29,926 in 2000). From this scant information, one can see that the earnings of Native American female performers were generally about half those of their male counterparts, and less than half those of secondary white female big top players. (The above dollar conversions use 1902 as their base point.) C. F. Larrabee, Acting Commissioner of the Office of Indian Affairs, Department of the Interior, to the American Bonding Company of Baltimore, Feb. 26, 1906, Correspondence, Box 7, Folder 5, JTMCC.

174. Neihardt, *Black Elk Speaks,* 218.

175. Standing Bear, *My People the Sioux,* 245–47.

176. "Circus Days and Ways," interview with W. E. "Doc" Van Alstine, 1–3.

177. Ibid.

178. As a rule circus people ate well on the road. On August 13, 1892, the Walter L. Main circus menu included the following foods: (1) Meats—roast chicken and dressing, boiled leg of lamb and French peas, roast beef, lake trout with egg sauce (2) Vegetables—green corn, mashed potatoes, sweet potatoes, sliced tomatoes, stewed white onions (3) Desserts—fruit cake, Bartlett pears, cream layer cake, oranges, lemon meringue, grapes, ice cream, nuts. "Souvenir Dinner Tendered to the People of Walter L. Main's Monster Railroad Shows at Webb City, Mo.," August 13, 1892," Walter L. Main Circus Vertical File, RLPLRC.

179. "Huge Crowd Comes to See the Circus," n.p., Decatur, Ill., 1907, Newspaper Collection, RLPLRC.

180. But circus proprietors were quick to argue that it was the towns, not the shows,

that were riddled with crime. A Barnum & Bailey route book entry for Thursday, May 27, 1897, describes Nashville, Tennessee, as "[a] Town full of fakirs, who ply their swindling games unmolested." Harvey L. Watkins, "Barnum & Bailey Official Route Book, Season of 1897" (Buffalo: Courier, 1897), 77, Route Book Collection, RLPLRC.

181. "Battle of San Juan Hill," *Boston Traveler*, June 7, 1899, BBPCB 1899–1901, JMRMA.

182. Bradna, *The Big Top*, 88–89.

183. Suggestions and Rules, Employees, Ringling Bros., issued by Charles Ringling, early 1900s, Ringling Bros. Small Collection Box, "Rules" file, RLPLRC.

184. See assorted work contracts, Walter L. Main Small Collection Box, Ringling Bros. Small Collection Box, and George L. Chindahl Collection, RLPLRC.

185. Gollmar, *My Father Owned a Circus*, 103–4.

186. Cohen, *Making a New Deal*, 178–79.

187. Commentary in circus route books concerning employee drunkenness focuses almost entirely upon the workingmen.

188. Travel and Personal Diary and Expense Account of J. A. Bailey while agent for Lake's Circus, Box 11, Item 2, JTMCC.

189. Charles Theodore Murray, "On the Road with the Big Show," *Cosmopolitan*, June 1900, 118 (citation courtesy of Fred Dahlinger Jr.).

190. "Buffalo Bill's Wild West and Congress of Rough Riders of the World," Historical Sketches and Programme, 33.

191. At times, the circus actively assisted the military with its daily operations. When the Massachusetts militia had difficulty unloading six cars of cannon, they called on Byron Rose, Barnum & Bailey's transportation manager when the show was in Boston, who proceeded to unload the cannon easily with a gang of twenty-one circus workers. Moffett, "How the Circus Is Put Up and Taken Down," page number illegible, RLPLRC.

192. G. F. Humphrey, Quartermaster General, U.S. Army, to George Starr, Manager, Barnum & Bailey's Greatest Show on Earth, Washington, D.C., May 15, 1906, Correspondence, Box 7, Folder 5, JTMCC.

193. Henry G. Sharpe, Commissary General, to George Starr, Washington, D.C., May 15, 1906, Correspondence, Box 7, Folder 5, JTMCC.

194. "It Is a Great Circus," n.p., Ladysmith, Wis., July 25, 1919, Walter L. Main Vertical File, RLPLRC.

195. Ibid.

196. See Green, *The World of the Worker*, chaps. 1–3.

197. This discussion is indebted to Kelley, *Race Rebels*. Kelley rejects the notion that working-class struggle must be located solely in unions and organized civil rights groups to count as "legitimate" resistance. Thus he focuses on daily acts of resistance among working-class African Americans, arguing in favor of "try[ing] to make sense of people where they are rather than where we would like them to be . . . to reject formulaic interpretations in favor of the complexity of lived experience," 13.

198. Standing Bear, *My People the Sioux*, 256–59.

199. Allan Harding, "The Joys and Sorrows of a Circus Fat Lady," 62–63.

200. See, for example, employment contracts for the Carl Hagenbeck–Great Wallace

Circus Combined, early 1900s, and the Ringling Bros. circus, early 1900s, George Chindahl Collection, and Don Howland Scrapbook, 1945, pt. 1, RLPLRC.

201. When the small Hummel, Hamilton and Sells circus suddenly shut down at Morgan City, Louisiana, in 1897, a worker later reported that employees were redlighted. This notorious practice was occasionally violent. The most infamous example occurred in 1931 on the Robbins Brothers circus, a fifteen-car railroad show. On September 12, approximately fifty workingmen were thrown off the moving train outside of Mobile, Alabama, after they had protested not having been paid in three weeks. Several men were badly injured, one of whom died a few weeks later. The circus permanently disbanded that night, and one show owner, Fred Buchanan, ran off with a large sum of money. Although the Mobile Circuit Court charged the show owners with assault and attempted murder, the case was never brought to trial. One Robbins Brothers' worker later wrote that he and several other men were redlighted near Okalona, Mississippi, the day after the notorious Mobile incident. Bewildered and angry, the men walked to Okalona and got drunk. The town constable attempted to load them into an empty boxcar on the next freight train headed to St. Louis, but several workers traveled instead to the show's quarters in Lancaster, Missouri, where they nearly obliterated the circus train in revenge for the redlighting. Tom Duncan's popular novel *Gus the Great* (1947) was closely patterned after the Robbins Brothers circus season of 1931. Captain Curly Wilson to editor of *Billboard*, Apr. 11, 1936, 41, courtesy of Fred Dahlinger Jr.; Bradbury, "Robbins Bros. Circus, Season 1931," 10, 18–19; "The Fred Buchanan Railroad Circuses, 1923–1931," 12–26, especially 24–25.

202. $96 in 2000.

203. By 1906 the circus used the troublesome iron seats only on one side of the big top and collapsible bleacher seats elsewhere. "Circus Laborers Strike: Barnum & Bailey's Employees Delay Departure from City," *Washington Star*, May 13, 1903, BBPCB 1902–3; "Circus Men on Strike," *New York Sun*, May 13, 1903, BBPCB 1902–3; "Bleacher Seats Are Comfortable," *Huntington* (W.Va.) *Advertiser*, Sept. 5, 1906, BBPCB 1906, all from JMRMA.

204. "Street Parade Again," *Lowell* (Mass.) *Mail*, June 13, 1907, BBPCB 1907, JMRMA.

205. "No Labor Troubles with Barnum & Bailey's Circus," *Lancaster* (Pa.) *Labor Leader*, May 28, 1903; "Circus Employees Did Not Strike: No Truth in Story That Men Went Out Yesterday Morning," *Hartford Telegram*, June 26, 1903; "Why Canvasmen Do Not Strike," *Waterbury* (Conn.) *Republican*, July 1, 1903, BBPCB 1902–3, all from JMRMA.

206. "Circus Prodigies Protest," *New York Press*, n.d.; "Ultimatum of the Prodigies," *New York Sun*, Apr. 12, 1903; "Prodigies in Conference," *New York Times*, Apr. 12, 1903, all from BBPCB 1902–3; "Freaks to Form a Union," *New York Press*, Jan. 28, 1907, BBPCB 1906–7, all from JMRMA.

207. "Clowns Want In on Tariff Talk," *New York Telegraph*, Apr. 17, 1909, BBPCB 1909, JMRMA.

Chapter 4

1. "Official Program of Barnum & Bailey's Greatest Show on Earth," Madison Square Garden, Apr. 20–25, 1896, n.p., LTLBBS.

2. "Barnum Talks of the Shows of His Grandsons," n.p., 1896, LTLBBS.

3. "She Tosses Husband About Like Biscuit," n.p., 1911, GT-2 Barnum and Bailey Vertical File, RLPLRC.

4. Kelly, *Clown*, 102.

5. Stark, *Hold That Tiger*, 28–30, Published Circus Memoir Collection, RLPLRC.

6. Hughes and Meltzer, *Black Magic*, 69.

7. Michel Foucault's work helps illuminate how circus proprietors used respectability as a vehicle for erotic exhibition. Foucault analyzes how the so-called age of repression that accompanied the rise of the bourgeoisie in eighteenth-century Europe actually widened the parameters of sexual discourse. Here, sex became increasingly interconnected with power relations, made manifest through discourse and institutions: the insane asylum, hospital, and boys' secondary schools. Foucault, *The History of Sexuality*, vol. 1.

8. Allen, *Horrible Prettiness*, 47–48.

9. Thomas Skillman quoted in Thayer, *Traveling Showmen*, 84.

10. Thayer, "The Anti-Circus Laws in Connecticut 1773–1840," 18.

11. Johnson, *A Shopkeeper's Millennium*, 115.

12. Thayer, "Legislating the Shows: Vermont, 1824–1933," 20.

13. Barnum, *Life of P. T. Barnum*, 348.

14. Thayer, *Traveling Showmen*, 94.

15. Caroline Cowles Richards quoted in ibid., 85–86.

16. Dozens of prostitutes entered the theater through a separate entrance at least an hour before the main doors were opened. In the third tier, prostitutes drank at nearby bars, met new and regular patrons, and sometimes engaged in sexual relations during the show. Allen, *Horrible Prettiness*, 50, 52–53.

17. Advertisement for Raymond, Waring and Co., Circus, Chestnut Street Amphitheatre, Philadelphia, n.p., June 20, 1840, Newspaper Advertisement Collection, RLPLRC.

18. Allen, *Horrible Prettiness*, chapters 3–6.

19. Likewise, female sexual spectacle was acceptable to middle-class audiences in other amusements like Ziegfeld's Follies (1907) and Benjamin Franklin Keith's vaudeville shows. Ziegfeld's Follies divested the cabaret of its working-class origins with performances in spacious middle-class vaudeville theaters by silent, athletic, boyish female dancers. Though barely dressed, the dancers, like the female circus performers, were marketed in press releases as wholesome "girls next door." Ibid., 272.

20. For more on the gendered consequences of the "market revolution," see Douglas, *The Feminization of American Culture*; Ryan, *Cradle of the Middle Class*.

21. It is difficult to determine the actual proportion of women and children in the turn-of-the-century circus audience because shows did not keep attendance records or detailed notes about the gender composition of their audiences However, the visual evi-

dence—photographs, lithographs, and early films of live circus performances—as well as constant route book references, show that women and children attended the circus in large numbers.

22. By 1920 African American women accounted for the majority of domestic workers because racist hiring practices virtually barred them from the industrial workplace. With the rapid expansion of the industrial sector after the Civil War, the process of mechanization de-skilled many factory occupations previously performed by skilled (white male) craft unionists. Employers sought young, unorganized women (nearly all AFL-affiliated unions refused membership to women) who would accept low-paying, unskilled factory jobs in the garment industry among others. D'Emilio and Freedman, *Intimate Matters*, 189; Cott, *The Grounding of Modern Feminism*, 21–23.

23. Peiss, *Cheap Amusements*, introduction.

24. Addams, *The Spirit of Youth and the City Streets*, 5.

25. See Finnegan, *Selling Suffrage: Consumer Culture and Votes for Women*.

26. Robinson, *The Circus Lady*, 276–77.

27. "Ringling Bros. Help Wisconsin: Wives Are Members of Suffrage Society—Allow Campaigning on Circus Grounds," *Woman's Journal*, July 13, 1912 (citation courtesy of Susan Traverso).

28. D'Emilio and Freedman, *Intimate Matters*, 234.

29. Peiss, *Cheap Amusements*, 134–37.

30. Glenn, "'Give an Imitation of Me.'"

31. Rosen, *The Lost Sisterhood*, 33.

32. See Vertinsky, *The Eternally Wounded Woman*.

33. Evans, *Born for Liberty*, 147; Cahn, *Coming on Strong*, 23.

34. Dora Keen, "How I Climbed a 14,000-Foot Mountain," *Ladies' Home Journal*, Aug. 1913, 7, 41.

35. Green, *Fit for America*, 229–32.

36. Cahn, *Coming on Strong*, 15–16; Murray, "On the Road with the 'Big Show,'" 127–28.

37. Carol Smith-Rosenberg identifies the New Woman as an upper- or middle-class single professional white woman who chose higher education over motherhood. I define the New Woman more broadly here to include all women in the public sphere around the turn of the century, including activists and workers outside the home, like circus women, because many readily identified themselves as "New Women." Smith-Rosenberg, *Disorderly Conduct*, 265.

38. Ibid.

39. Roosevelt, *The Strenuous Life*, 4.

40. Bederman, *Manliness and Civilization*, especially chapters 3–5.

41. Green, *Fit for America*, 225.

42. See Todd, *Physical Culture and the Body Beautiful*.

43. O. S. Fowler, "Tight-Lacing," 11.

44. Green, *Fit for America*, 92, 242–45.

45. Ibid., 246.

46. Lutz and Collins, *Reading National Geographic*, 115–16.

47. Lutz and Collins observe that in the late twentieth century, *National Geographic*

editors used computer technology to darken a partly naked Polynesian woman, making her "look more native" so that her nudity would be more acceptable to American audiences. Ibid., 82, 115–16.

48. Frank S. Redmond, "Official Route Book of the Adam Forepaugh Shows, Season 1891" (Buffalo: Courier, 1891), 25–26, Route Book Collection, RLPLRC.

49. "Parade List Barnum & Bailey's Greatest Show on Earth, Season 1896," n.p., LTLBBS.

50. According to one female circus worker, "You never see a circus woman in a city after the season is over. She flees from them, I can tell you. She detests the noise and the hustle, and almost without exception, they live in the little country towns where they practice through the winter, go early to bed and are in fine condition when the season opens." "Life as a Woman of the Circus Is Not All Glitter: Hard Work and Discipline Her Lot," *New York Evening Telegram*, Aug. 27, 1902, BBPCB 1902–3, JMRMA.

51. "Is a Thing of the Past: Big Barnum & Bailey Exhibition Delighted Many Anderson People," *Anderson* (Ind.) *Herald*, Sept. 30, 1903, BBPCB 1902–3, JMRMA.

52. Robinson, *The Circus Lady*, 4–16.

53. Leitzel's father, a Hungarian army officer, became a dictatorial theatrical impresario who managed his family's career in Europe until they left him and worked on their own. Taylor, *Center Ring*, 222–23, Circus Memoir Collection, RLPLRC.

54. May's father, Johnny Zinga, trained her to tumble and perform as a contortionist when she was just three years old. Although May's early childhood was spent in poverty, she quickly became accustomed to a comfortable life with her new family. Her adoptive mother, Marizles Wirth, wrote in her diary that May was filthy when she came to live with the Wirths. Immediately, Marizles shaved May's head and made her wear a white cap; May cried when she saw her hair fall out but was comforted when her new mother told her that her hair would grow back curly. The Black Collection, Marizles Wirth's Diary, 56–57, JMRMA; Mark St. Leon, "An Unbelievable Lady: Bareback Rider May Wirth," 4–5.

55. "From Village Bell [*sic*] to Queen of the Arena: Famous Beauty Thrills Circus Spectators by Feats in Mid-Air," *New York Journal*, Mar. 25, 1903, BBPCB 1902–3, JMRMA.

56. "What Makes Auto in the L'Auto Bolide Twist Wrong Side Up?" *Rural Examiner*, n.p., Aug. 4, 1906, BBPCB 1906, JMRMA.

57. "The Carl Hagenbeck Greater Shows Embracing Grand Triple Circus East India Exposition," Ottumwa, Iowa, July 11, 1906 (Buffalo: Courier, 1906), Program Collection, RLPLRC.

58. "Quits Ballet for Fortune," *Chicago Tribune*, Sept. 9, 1903, BBPCB 1902–3; "The Women of the Circus," n.d., n.p., BBPCB 1906–7; "Big Circus Tents Cover a Very Pretty Circus Romance: Fair Italian Acrobat Wears Her Lover's Picture on Her Collar: He Holds a Trusted Position: Mama Objects but There May Be a Marriage before the Season is Over," *New York Times*, Apr. 28, 1903, BBPCB 1902–3, all from JMRMA.

59. Haraway, *Primate Visions*, 10–25.

60. $184,821 in 2000.

61. "Chiko Stuffed for Good," *New York World*, Jan. 16, 1895, BBPCB 1895, JMRMA.

62. "Chiko Wanted to Shake Hands But When Johanna Resented His Familiarity, He

Nearly Tore His Lady Chimpanzee's Ear Off," *New York World*, Dec. 19, 1893, BBPCB 1894, vol. 1, JMRMA.

63. "The Barnum & Bailey Show," n.p., n.d., Decatur, Ill., 1897, Newspaper Advertisement Collection, RLPLRC.

64. Charles Andress, "Day by Day with Barnum & Bailey, Season of 1904" (Buffalo: Courier, 1904), 76–78, Route Book Collection, RLPLRC.

65. "Johanna Back at the Park," *New York Recorder*, BBPCB 1895, JMRMA.

66. Harris, *Humbug*, 165–66.

67. "Circus Women Just Like All Human Beings: You Thought Them Mere Cigarette Smoking Chatter Boxes, Paint-Bedaubed, Etc.," *Des Moines News*, Sept. 3, 1907, BBPCB 1906–7, JMRMA.

68. Much of this quotation appeared verbatim in show programs over the next two decades. See, for example, "Circus Women and Children Healthy, Happy, 'Homey' Gypsies," Barnum & Bailey program, road edition, season 1914, 39–40, Program Collection, RLPLRC.

69. "Life as a Woman of the Circus Is Not All Glitter: Hard Work and Discipline Her Lot," *New York Evening Telegram*, Aug. 27, 1902, BBPCB 1902–3, JMRMA.

70. Harriet Quimby, "The Feminine Side of Sawdust and Spangles," *Leslie's Weekly*, Apr. 16, 1908, 374, Periodicals Collection, RLPLRC.

71. $183 in 2000.

72. Willson, *Where the World Folds Up at Night*, 80, Published Circus Memoir Collection, RLPLRC.

73. "Script for Interview between May Wirth and Adelaide Hawley, 'Woman's Page of the Air,'" Station WABC, 485 Madison Avenue, New York City, July 29, 1942, May Wirth Vertical File, RLPLRC.

74. Tiny Kline, "Showground-Bound" (unpublished memoir), 136–37, Manuscript Collection, RLPLRC.

75. Turnour, *The Autobiography of a Clown*, 77.

76. "From Home of Riches to the Bareback Ring," *St. Louis Post*, May 12, 1907, BBPCB 1906–7, JMRMA.

77. Robinson, *The Circus Lady*, 175–78.

78. Ibid., 216.

79. Caption on back of postcard of Annie Oakley, a "Shini Color" by "Colourpicture," Boston, Annie Oakley Vertical File, RLPLRC.

80. In a tribute to Oakley after her death, Paul Gould reminisced about Oakley's patriotism: "When [Oakley] toured Europe, she appeared before William Hohenzollern, Crown Prince of Germany, later to become Wilhelm II. At his imperial request she shot a cigarette from his mouth. Years later when the war broke out, Annie bitterly wrote the Kaiser saying she was sorry she had hit only the cigarette—and asked for another shot." From Paul Gould, "The Lady Could Shoot," *Buick Magazine*, Nov. 1946, 12, Annie Oakley Vertical File, RLPLRC.

81. "Darke County Plans a Statue," *Columbus* (Ohio) *Sunday Dispatch Magazine*, Apr. 24, 1949, 27, Annie Oakley Vertical File, RLPLRC.

82. Davis, "Shotgun Wedlock," 141–57.

83. Ibid.

84. These are two of many such examples: "Girl Tamer Faints at Mouse in Her Trunk: Kittie Florenz of Barnum & Bailey Circus Is Afraid of a Mouse, but Not of Her Lion Prince," *Boston Journal*, May 15, 1907, BBPCB 1907, JMRMA; also, see the newspaper reporter Courtney Ryley Cooper's description of Lucia Zora: "a woman whose appearance bespoke complexities; dominant, willing to face all and dare all, if by so doing an object might be attained; with the strength, the courage, the resourcefulness to pit one's self against adversaries before which men might quail, yet wholly a woman, to be frightened by a spider and cry for sympathy at the prick of an embroidery needle." Courtney Riley Cooper, preface to Zora, *Sawdust and Spangles*, viii, Circus Memoir Collection, RLPLRC.

85. Lucia Trevor Lee, "Women Who Conquer Beasts," *Twentieth Century Home*, Oct. 1904, 6–8, George Chindahl Collection, Box 1, Folder 21, RLPLRC.

86. Zora, *Sawdust and Spangles*, 39–40.

87. Stark, *Hold That Tiger*, 50–55.

88. Ibid., 124–25.

89. Ibid., 13. Stark also recounted her most serious accident, which occurred in 1928 with the John Robinson Company in Bangor, Maine. Because of weather delays, Stark's tigers had not been fed or watered in twenty-four hours, and had to perform immediately upon arrival. One tiger, Sheik, became surly during the show and initiated a group attack upon Stark, whose left leg was nearly severed. All of Stark's limbs were broken and she almost died, but miraculously she recovered after five weeks in the hospital, 230–45.

90. "Official Program of Ringling Bros. and Barnum & Bailey Combined Shows Season 1923," road edition (New York: Select, 1923), Program Collection, RLPLRC.

91. "Ringling Bros. and Barnum & Bailey Combined Shows, Magazine and Daily Review, Season 1920," Madison Square Garden (New York: Select, 1920), Program Collection, RLPLRC.

92. Flink, *The Automobile Age*, chapters 1–4.

93. See Tiny Kline, "Showground-Bound," for a description of the dangers of this stunt, 344.

94. "L'Auto Bolide," *Seattle Times*, Aug. 13, 1905, BBPCB 1905, JMRMA.

95. Charles Andress, ed., "The Official Route Book of Barnum & Bailey, Season of 1905" (Buffalo: Courier, 1905), 88, Route Book Collection, RLPLRC.

96. "Circus Act to Close Death's Plunge," *New York Journal*, Mar. 21, 1908, BBPCB 1908, JMRMA.

97. "How Circus Women Enjoy Life," *Sunday Leader*, n.p., June 17, 1906, Julia Lowande Vertical File, RLPLRC.

98. "Circus Women in the Winter Time," "Carl Hagenbeck Circus Official Program Magazine and Daily Review" (Chicago: W. F. Hall, 1916), 10, Program Collection, RLPLRC.

99. Robinson, *The Circus Lady*, 71, 175–78, 276–77.

100. Ibid., 173.

101. Al Mann interview by author.

102. Ringling Bros. circus work contract for Kathy Edwards, Oct. 1912, Archival Collections, Work Contracts, RLPLRC.

103. "Suggestions and Rules Employees Ringling Bros.," issued by Charles Ringling, early 1900s; Contract for Kathy Edwards; Contract for Miss Lulu Welsh, whose salary in 1914 was $9 a week ($151 in 2000), Archival Collections, Work Contracts, all from RLPLRC.

104. "Suggestions and Rules Employees Ringling Bros."

105. Kline, "Showground Bound," 169.

106. Ibid.

107. "Life as a Woman of the Circus Is not All Glitter: Hard Work and Discipline Her Lot," *New York Evening Telegram*, Aug. 27, 1902, BBPCB 1902–3, JMRMA.

108. "Official Program of Barnum & Bailey's Greatest Show on Earth," Madison Square Garden, April 20–25, 1896.

109. See, for example, "Circus Women Abandon Tights: Skirts of the Latest Style Are the Vogue among Performers of the Arena," *Pittsburgh Gazette*, May 15, 1904, BBPCB 1903–4; "The Passing of the Girl in Tights," *San Francisco Call*, Sept. 4, 1905, BBPCB 1905; "The Passing of Tights in Circus Life," *Palmyra* (Wis.) *Telegram*, Aug. 19, 1906, BBPCB 1906, all from JMRMA.

110. There is much evidence to support my contention that the "revolution in gowning" was fictional. Barnum & Bailey route books from 1904–6 do not mention any changes in costuming, nor do the memoirs of scores of performers, both female and male.

111. "The Passing of the Girl in Tights," *San Francisco Call*, Sept. 4, 1905, BBPCB 1905.

112. Gossard, "A Reckless Era of Aerial Performance," 96–97.

113. "Circus Women to Don Tights Again," *New York Telegraph*, Dec. 13, 1906, BBPCB 1906, JMRMA.

114. Ibid.

115. "A Circus in Undress," *New York Sun*, Mar. 26, 1895, BBPCB 1895, JMRMA.

116. "A Very New Woman," n.d., n.p., 1896, Newspaper Collection, RLPLRC.

117. $96 and $192 in 2000.

118. "News of the Theater: The Greatest Show on Earth Comes to Town," *New York Evening Sun*, Mar. 29, 1895, BBPCB 1895, JMRMA.

119. "A Very New Woman," 1896.

120. Kline, "Showground-Bound," 178–79.

121. Taylor, *Center Ring*, 121.

122. Harry V. Conlon to Ringling Bros., Feb. 14, 1903, Archival Collections, Correspondence, RLPLRC.

123. Bea Warfield, first hired as a ballet girl with the Ringling Bros. and Barnum & Bailey circus in 1942, remembers that at her audition, she was eager to demonstrate her skills as a tap and ballet dancer, but was disappointed when the audition consisted of simply walking around the stage in a swim suit. Author conversations with Fred Pfening Jr., July 17, 1996, Baraboo, Wis.; and Bea Warfield, May 19–20, 1995, Sarasota, Fla. See also the memoir of Connie Clausen, a former ballet girl, aptly titled *I Love You Honey, but the Season's Over.*

124. "One Thousand Women Storm Kiralfy for Ballet," n.p., New York, Mar. 7, 1903, BBPCB 1902–3, JMRMA.

125. "The Ballet at the Circus," *New York Sun*, Mar. 29, 1903, BBPCB 1902–3, JMRMA.

126. Al Ringling to Charles Ringling, Dec. 25, 1913, Baraboo, Wis., Archival Collection, Correspondence, RLPLRC.

127. See Dixon, *The Leopard's Spots: A Romance of the White Man's Burden, 1865–1900; The Clansman: An Historical Romance of the Ku Klux Klan;* Grant, *The Passing of the Great Race.*

128. Von Eschen, *Race against Empire*, 58.

129. Du Bois, *The Souls of Black Folk*, 167.

130. Said, *Orientalism*, 190.

131. "Imre Kiralfy's Sublime Nautical, Martial and Poetical Spectacle, 'Columbus and the Discovery of America', for the First Time Now Produced in Connection with the Barnum & Bailey Greatest Show on Earth" (Buffalo: Courier, 1892), 10, Program Collection, RLPLRC.

132. Ewen, *Immigrant Women in the Land of Dollars*, 218.

133. Dubuque, "The Original Miss Daisy," 26–28; see also Vertical File for Albert Hodgini Family, RLPLRC.

134. Using films like *Some Like It Hot, Victor, Victoria*, and *Tootsie*, Judith Butler demonstrates how heterosexual culture can appropriate drag to reaffirm normative ideologies about sexuality. "This is drag as high het [*sic*] entertainment . . . such films are functional in providing a ritualistic release for a heterosexual economy that must constantly police its own boundaries against the invasion of queerness." Butler, *Bodies That Matter*, 126.

135. See Ringling Bros. and Barnum & Bailey programs, 1920–30, including "Official Program of Ringling Bros. and Barnum & Bailey Combined Shows, Season 1920"; "Ringling Bros. and Barnum & Bailey Combined Shows, Magazine and Daily Review," Season 1922 (New York: Select, 1922); "Official Program of Ringling Bros. and Barnum & Bailey Combined Shows, season of 1926, Madison Square Garden, March 31–May 1, 1926 (New York: Select, 1926), all Program Collection, RLPLRC.

136. For the historical development of "personality," see Susman, "'Personality' and the Making of Twentieth-Century Culture," in *Culture as History.*

137. D'Emilio and Freedman, *Intimate Matters*, chapter 11, especially 239–74.

138. Leitzel's life, however, ended tragically. Several days after she fell during a performance at Copenhagen in 1931 (because one of her swivel rings snapped), Leitzel died. Thereafter, Codona became increasingly despondent and reckless in performances until an accident ended his career as the premier trapeze artist in the world. Eventually, he became an assistant to the owner of a gas station. By this time he had married a circus equestrienne, Vera Bruce, who sued him for divorce in 1937. Deeply depressed about the loss of Leitzel and his status, Codona murdered Bruce and then committed suicide in the office of his divorce lawyer. Before his death Codona immortalized himself and Leitzel in an opulent winged statue at Inglewood Park, California. Taylor, *The People of the Circus*, 243–46, 250.

139. Marie Lil, a fat lady for Barnum & Bailey, remembered that the first time she performed in the sideshow she was drenched in a cold sweat and was so scared that she could barely answer questions from the audience. "Stage Fright in the Sawdust Ring:

Everybody Has It, Including the Funny Man," *New York Evening Mail*, Mar. 30, 1907, BBPCB 1906–7, JMRMA.

140. Bogdan, *Freak Show*, 26–27, 64.

141. Thomson, "Introduction," 4.

142. There is an excellent body of scholarly work on ethnological exhibits, or "human zoos," and zoos more generally. See Baratay and Hardouin-Fugier, *Zoos: Histoire des Jardins Zoologiques en Occident*; Hoage, *New Worlds, New Animals*; Horowitz, "Seeing Ourselves through Bars"; Hyson, "Urban Jungles"; Lindfors, "Ethnological Show Business"; Rothfels, "Apes, Aborigines, and Ape-People."

143. Born in 1844, Carl Hagenbeck was the son of a fishmonger in Bremerhaven, Germany, who began to exhibit seals in 1848 to boost his fish business. The seals were so popular that Carl's father started his own small menagerie, which originally consisted of a raccoon, two American opossums, a few monkeys, and some parrots. The family soon moved to quieter quarters in Hamburg amid the explosive German revolution in 1848. At fifteen, Carl became involved in his father's menagerie business, and eventually he became the largest animal trader in the world, in addition to working as a circus owner and zoo owner. Carl built a forested, mountainous zoological park at Stellingham, Germany, in 1907 that served as the center for his wild animal trade, breeding experiments, and humane animal training methods. Carl Hagenbeck, *Beasts and Men*, 5, Published Circus Memoir Collection, RLPLRC.

144. Ibid., 16, 19.

145. H. L. Watkins, "The Barnum & Bailey Official Route Book, Season of 1894" (Buffalo: Courier, 1894), 27, Route Book Collection, RLPLRC.

146. Ibid.

147. Several Barnum & Bailey route books complain that fairs cut into the circus's business and that the Midway Plaisance at the World's Columbian Exposition of 1893 in Chicago was unfairly patterned after Barnum, Bailey and Hutchinson, which had introduced its first ethnological congress in 1886. Ibid.

148. "Marie Lil Tells Her Impressions," n.p., Mar. 15, 1903, BBPCB 1902–3, JMRMA.

149. "Inconveniences of Too Much Flesh," *Leslie's Weekly*, Apr. 9, 1903, BBPCB 1902–3, JMRMA.

150. The marriage of the midgets Tom Thumb (Charles Stratton) and Mercy Lavinia Warren Bump (Lavinia Warren) provides an earlier example of this sort of sexual display. On February 10, 1863, the midgets' wedding drew enormous publicity for P. T. Barnum. Photographs of the extravagant event were readily available, and the wedding ceremony was replicated throughout the country while the couple toured. Although the marriage was based upon genuine affection, the "birth" of the couple's daughter on December 5, 1863, was a hoax designed to maintain public interest in the couple. While on tour in Europe, Stratton and Warren exhibited with babies from various orphanages throughout England, France, and Germany. The charade eventually ended with the baby's "death" from inflammation of the brain. In 1901, eighteen years after Tom Thumb's death, Warren admitted that she had never given birth. Bogdan, *Freak Show*, 152–61.

151. Eventually Jonathan became so thin that he could no longer walk; thus, the couple retired to a farm in New Hampshire. "The Freaks That Made Barnum Famous,"

Birmingham (Ala.) *News*, July 9, 1916, 3A82 Sideshow Pamphlet Collection, VF—Sideshow Miscellaneous 1, HCCM.

152. When Clofillia and her husband wanted to marry, they had difficulty finding a minister to conduct the ceremony because no one believed that she was female. Therefore, according to press releases, Clofillia requested an official medical certificate which confirmed her sex. "The Bearded Lady of Geneva," *Gleason's Pictorial Drawing-Room Companion*, n.d., n.p.; "Women of Beards," *Portland* (Me.) *Transcript*, 1852, n.p., both from George Chindahl Collection, Box 5, Folder 39, RLPLRC.

153. "Esau," Isaac's son, sold his birthright to his brother, Jacob. Gen. 25–28, 32, 33, 36.

154. Bogdan, *Freak Show*, 228.

155. "Circus Press Agent to Wed the Bearded Lady," *Daily Times* (Brooklyn), Apr. 30, 1903, BBPCB 1902–3, JMRMA.

156. "He Yanked Off Her Beard: A Horseman with a Simple Twist of the Wrist Gave Bearded Lady a Cool Shave in a Lexington Restaurant," *Cincinnati Post*, Oct. 9, 1903, BBPCB 1902–3, JMRMA.

157. "The Bearded Lady of Geneva."

158. Russett, *Sexual Science*, 50, 74–75.

159. Bogdan, *Freak Show*, 224–26.

160. Obituary, *Billboard*, Jan. 18, 1913, 6, Ella Ewing Vertical File, RLPLRC.

161. Edward Husar, "'Miss Ella,' a Gentle Giantess Remembered," *Herald-Whig* (Quincy, Mass.), section E, Nov., 16, 1980, Ella Ewing Vertical File, RLPLRC.

162. $516,234 in 2000.

163. "The Tallest Woman in the World," *New York World*, Aug. 1901, George Chindahl Collection, Box 16, RLPLRC.

164. "The Saintly Giantess: A Maiden of Eight Feet Two Inches Famed among Bible Students," *Weekly Gate City*, n.p., July 11, 1895, Ella Ewing Vertical File, RLPLRC.

165. Ibid.

166. "How She Charms Snakes," *New York World Herald*, n.d., Barnum, Cole, Hutchinson and Cooper Press Book, 1887–88, JMRMA.

167. Ibid.

168. Bogdan, *Freak Show*, 256–59.

169. "The Snake Charmer at the Show: Her Life Sapped by the Fetid Breath of Her Serpents," *New York Daily News*, Apr. 19, 1891, Scrapbook 8, 142, Barnum & Bailey 1890–91, JTMCC.

170. One can still find snake charmers at any tourist site throughout India who will perform with their bedraggled cobras for a few rupees.

171. "She Likes Snakes," *New York Morning Journal*, Mar. 16, 1888, Barnum, Cole, Hutchinson and Cooper Press Book, 1887–88, JMRMA.

172. "Handling Serpents," n.d., n.p., George Chindahl Collection, Box 1, Folder 22, RLPLRC.

173. Parry, *Tattoo*, 41, 54–55.

174. Bogdan, *Freak Show*, 224–29, 241–56.

175. Ibid., 63–65.

176. Parry, *Tattoo*, 44–45, 66–69.

177. Gollmar, *My Father Owned a Circus*, 190.

178. Kline, "Showground-Bound," 103.

179. Ibid., 103–4.

180. "Big Crowd Sees Gollmar Circus," n.d., n.p., 1913, Gollmar Brothers Circus Collection, Circus Reviews, 1890–1900, RLPLRC.

181. McClintock borrows this term, "monarch of all I survey," from Pratt, *Imperial Eyes*, McClintock, *Imperial Leather*, 57–58, 121–22.

182. "Krao—A Missing Link," *Continent* (location unlisted), Feb. 1884, George Chindahl Collection, Box 5, Folder 40, RLPLRC; clipping, n.d., n.p., Scrapbook 6, Barnum, Bailey and Hutchinson, Winter 1882–83, JTMCC.

183. "Krao, 'The Missing Link,'" n.p., Portland, Me., Feb. 24, 1883, George Chindahl Collection, Box 5, Folder 40, RLPLRC.

184. Postcard of Krao Farini, in A. J. Barker, pamphlet, n.d., n.p., 36, Lionel the Lion-Faced Boy Vertical File, RLPLRC.

185. Shane Peacock, "Farini the Great," 13–20; George Arlington, "Barnum & Bailey Greatest Show on Earth and Book of Wonders Combined" (New York: Courier, 1903), 4th ed.; Charles Andress, "Day by Day with Barnum & Bailey, Season 1903–1904" (Buffalo: Courier, 1904); Charles Andress, "Barnum & Bailey Annual Route Book and Illustrated Tours, 1906" (Buffalo: Courier, 1906), 116, all route book citations from Route Book Collection, RLPLRC.

186. Harding, "The Joys and Sorrows of a Circus Fat Lady," *American Magazine*, 62–63, 3A82 Sideshow Pamphlet Collection, VF-Sideshow Fat People, HCCM.

187. Andress, "Day by Day with Barnum & Bailey, Season of 1904," 28.

188. "The Carl Hagenbeck Greater Shows Embracing Grand Triple Circus East India Exposition," 1906.

189. Melville, *Typee*, 123.

190. Yet Melville, like the later Progressive intellectual John Dewey, also presented preindustrial cultures in a positive manner, to critique the alienation and time clock discipline of his own industrial society.

191. "Barnum & Bailey's Greatest Show on Earth Show Program," Madison Square Garden edition (New York: J&H Mayer, 1895), Circus Programs, Box 13, Folder 6, JTMCC.

192. Memene's father, Kakala, a local chieftain, joined her at Barnum & Bailey's ethnological congress. "Around New York with a Cannibal Girl," *New York World*, Apr. 21, 1895, BBPCB 1895, JMRMA.

193. Morgan, *American Slavery American Freedom*, chapter 3.

194. Russett, *Sexual Science*, 144; Bederman, *Manliness and Civilization*, 134–50.

195. "Inside Asiatic Life," *New York Daily Tribune*, Feb. 25, 1895; "The Wild Men of New Guinea," *New York Press*, n.d., both from BBPCB 1895, JMRMA.

196. "Like Circling Birds of Death," *New York Herald*, Apr. 9, 1894, BBPCB 1894, vol. 2, JMRMA.

197. N.d., n.p., clipping, BBPCB 1895, JMRMA; Melville, *Typee*, 48–49.

198. "The Barnum & Bailey Greatest Show on Earth Show Program," Madison Square Garden edition, Mar. 26–Apr. 23, 1894 (New York: J&H Mayer, 1894), Circus Programs, Box 13, Folder 6, JTMCC.

199. David Roediger argues that the ideology of white supremacy had existed across Europe for centuries as an integral part of the European imperial project; consequently, many European immigrants arrived in America already believing in white supremacy, despite their own deeply divided ethnic and class identities. In the United States, popular cultural forms like the circus probably helped hasten the process of ethnic amalgamation and Americanization. Roediger recalls how his late grandmother-in-law, an immigrant from eastern Europe, saw an ethnological village of "dog-eating" Igorots from the Philippines at the Louisiana Purchase Exposition at St. Louis in 1904 and through the gaze recognized her own whiteness. Conversation with David Roediger, Madison, Wis., Nov. 10, 1995.

200. "Seven Sarakaba Beauties Arrive, Proud of Lips as Long as Duckbills," *New York Telegraph*, Mar. 31, 1930, George Chindahl Collection, Box 4, Folder 86, RLPLRC.

201. Taylor, *Center Ring*, 80–81; Robert Bogdan, *Freak Show*, 194–95.

202. Taylor, *Center Ring*, 80.

203. $14,968 in 2000.

204. "African 'Beauties' Here to Join Circus," *New York Times*, Apr. 1, 1930, George Chindahl Collection, Box 4, Folder 86, RLPLRC.

205. "Official Program of Ringling Bros. and Barnum & Bailey Combined Shows, Season of 1930," New York Coliseum, Mar. 27–Apr. 6, 1930, Program Collection, RLPLRC.

206. Sander Gilman, "Black Bodies, White Bodies," 223–61; Abrahams, "A Khoisan Contribution to Western Science," seminar paper (courtesy of Karen Dubinsky).

207. McClintock, *Imperial Leather*, 22.

208. Brown, "An Interview with Tom Barron, 'World's Tallest Clown,'" 24.

209. $37 in 2000.

210. Thus by the 1930s, medical technology was a partner in the sideshow's spectacular presentation of the Giraffe-Neck Women of Burma. In a section entitled "Primitive Superstition Meets Modern Science," the Padaung's booklet described how one of the women was x-rayed at two prestigious medical centers in the United States. As the result of x-ray "exploration," doctors concluded the following: "In the process of elongating the neck, the woman had added four of her back vertebrae to the neck, and the whole thorax had been pulled up, causing the upper part of the lungs to collapse and the lower part to flair. . . . [S]he had good lung power despite this, but that if the rings were removed she probably would not be able to support her head. It would wobble about, utterly out of control." Howard Y. Bary, "Interesting Facts and Illustrations of the Royal Padaung Giraffe-Neck Women from Burma," n.p., 1933 (citation courtesy of Richard J. Reynolds III).

211. "Hagenbeck-Wallace Circus, 1934 Season Program" (Louisville: Louisville Color Gravure, 1934), Program Collection, RLPLRC.

212. Howard Y. Bary, "Interesting Facts and Illustrations of the Royal Padaung Giraffe-Neck Women from Burma."

213. Al Mann interview by author.

214. Howard Y. Bary, "Interesting Facts and Illustrations of the Royal Padaung Giraffe-Neck Women from Burma."

215. Irene Mann interview by author.

216. Friedman, "Prurient Interests," 3, 7–13, 14–15, 21–22, 62.

217. "Prowling Prudes: Prurient Preachers and Spouting Spinsters Who Object to the Nude in Art," *Billboard*, May 1, 1898, 13.

218. "Girls for the Circus," *New York Herald*, n.d., Scrapbook 8, Barnum & Bailey 1890–91, 33, JTMCC.

219. The Board of Aldermen in Boston demanded that all theater managers submit their posters of women to the board's License Committee. However, the board could not reach a consensus regarding a standard of decency. One member wanted to cover depictions of pink fleshlings (tights) with black paper on posters. (Yet black paper was scarce, so this suggestion was rejected.) "Prowling Prudes: Prurient Preachers and Spouting Spinsters Who Object to the Nude in Art," *Billboard*, May 1, 1898, 13.

220. Ibid. For other references see "None So Lewd as a Prowling Prude," *Billboard*, Nov. 1, 1897, 5; "Federation of Women's Clubs Want Advertisements Interdicted," *Billboard*, Mar. 1, 1899, 5.

221. I have reached this conclusion after examining scores of state statutes, circus route books (which would have noted such restrictions), and newspaper clippings. As mentioned in chapter 2, states, counties, and municipalities often attempted to protect provincial economic interests because the itinerant circus "drained" local money. In such instances, authorities required circuses and traveling shows to obtain sometimes costly licenses which indirectly curtailed a circus's ability to perform. In Wisconsin, for example, a circus proprietor was required to procure a state license as a public showman by describing to the secretary of state, in writing, how he or she intended to travel and the content of the show. The state license cost a showman $100 ($1,995 in 2000), payable to the state treasury. Local authorities simply dictated where in town a circus was allowed to perform, and the cost of local licenses. Occasionally, local governments passed ordinances which forbade circuses to perform while a local or county fair was on, because the circus would draw local money away from the fair. In order to protect the economic interests of the Central Louisiana Fair (October 9–14, 1923), for instance, the city council in Alexandria, Louisiana, passed an ordinance which prohibited any circus from exhibiting from September 20 through October 20, 1923. Some towns made licenses so expensive that circuses located their show grounds outside the city limits, but sometimes a show's attempt to avoid paying a local license backfired. When Adam Forepaugh's circus played outside the city limits of Johnston, Pennsylvania, on October 9, 1883, the show received no police protection. The spectators, who knew they would not be arrested if they engaged in criminal behavior, became drunk and raucous, stoned the performers, and destroyed several circus wagons. Forepaugh vowed that he would never show in Johnston again. In short, communities occasionally banned the circus for its perceived economic threat—but not specifically for its erotic content. "Circus Licenses," *Wisconsin Statutes of 1898*, vol. 1 (Chicago: Callaghan, 1898), chap. 67, sec. 1584, p. 718, Library Division, Government Publications Section, State Laws and Statutes, SHSW; "To Protect Fair," *Billboard*, May 5, 1923, 74; "Trouble in Town," n.p., George Chindahl Collection, Box 1, Folder 4, RLPLRC.

222. The circus historian Orin Copple King observes that in thirteen years of examining newspaper articles from the turn of the century, he has discovered only a couple of instances of newspaper commentary concerning prurient circus women. In one ex-

ample from Kansas, leering boys "stumbled" into the ladies' dressing room at the Sells-Rentfrow circus grounds in 1893. The *Dodge City Democrat* defended the boys' actions, arguing that the ladies' indiscretion in wearing only corsets in and outside the arena flustered the boys, who got "lost" in the dressing room. Orin Copple King, "Only Big Show Coming: Prodigal Profusion of Princely Paraphernalia," 33–41, especially 36.

223. Beisel, *Imperiled Innocents*, chapter 7.

224. Garrow, *Liberty and Sexuality*, 16. The Barnum scholar A. H. Saxon disagrees vehemently with Garrow's claim that Barnum spearheaded the passage of Connecticut's "Little Comstock" law. Based on an examination of legislative records, Saxon concludes that Barnum successfully urged his fellow representatives to defeat the first version of the bill. When an amended bill was reintroduced, Barnum reluctantly supported its passage because the majority voted for it. Saxon, *Barnumiana*, 35–37 (citation courtesy of Bluford Adams).

225. Beisel, *Imperiled Innocents*, 4.

226. Hunter, *The Gospel of Gentility*, 32.

227. "Church Protest Stops a Circus," *Chicago Tribune Sun*, Apr. 14, 1901, no page number listed, Newspaper Collection, RLPLRC.

228. "A New Kind of Censorship," *Billboard*, May 2, 1914, 28.

229. In the latter instance, the two veterans had been recently hired to work at the Wild West in the opening pageant. Webb, "Buffalo Bill, Saint or Devil?" 13–14.

230. Mann recalls that when he performed with the Robbins Brothers circus in 1931 (the same show that redlighted several workingmen—see chapter 3) he confronted one voyeur who became violent: "He was better than six feet, and weighed about two hundred pounds, and he said, 'Well, what are you going to do little man?' I was wondering whether to holler, 'Hey Rube!' or whether I should go and get my gun out the trunk. I had a 45. He was just standing there, and it was night, and he was very cocky, and . . . Jack Dempsey's tent was right there. And he heard my voice and all he said was, 'You think he's too big for me, Al?' . . . And some kid . . . in the distance around there, he was listening, he yelled, 'That's Jack Dempsey!' And that's all it took and [the voyeur] was gone." Al Mann interview by author.

Chapter 5

1. Harkness, *Andy the Acrobat*, 39, Circus Fiction Collection, RLPLRC.

2. See also Alger, *The Young Acrobat of the Great North American Circus*; May, *Cuddy of the White Tops*; Otis, *Toby Tyler; or, Ten Weeks with a Circus, The Wreck of the Circus*; Peck, *Peck's Bad Boy with the Circus*; Root, *Tommy with the Big Tents*; Darlington, *The Circus Boys in Dixie Land*, 225–27, all from Circus Fiction Collection, RLPLRC.

3. "Small Boy Schemes," *New York News*, Mar. 18, 1903, BBPCB 1902–3; "Heaven for Small Boys," *New York World*, Mar. 31, 1895, BBPCB 1895; "Elixer of Youth Provided by Circus Coming," *Hartford Post*, June 15, 1907, BBPCB 1906–7, all from JMRMA.

4. Norwood, *The Circus Menagerie*, Circus Fiction Collection, RLPLRC.

5. Anderson, *Tar*, 153, 163.

6. Bederman argues that gender itself is a "historical, ideological process." She dem-

onstrates that manliness and masculinity have had different historical meanings: before 1890, "masculine" was a neutral adjective used to describe any overarching, universal male quality applicable to all men, regardless of race or class. "Manliness," by contrast, was a qualitative term which encapsulated nineteenth-century male ideals. The term also had racial and class dimensions because it was generally used in reference to native-born, middle- and upper-class white men. Gradually, by the 1930s, "masculinity" became synonymous with physical strength, male sexuality, and aggressive behavior. Bederman suggests that men used the discourse of "civilization" to reconcile the contradictions inherent in their desire to be both "manly" and "masculine." Bederman, *Manliness and Civilization*, 7, 18, 23.

7. Hofstadter, *The Age of Reform*.

8. Bederman, *Manliness and Civilization*, chapters 1, 3, 5, conclusion; Green, *Fit for America*.

9. Anderson, *A Story Teller's Story*, 195.

10. Hall, *Youth*, 326.

11. Roosevelt, *The Strenuous Life*, 160.

12. Mosse, *The Image of Man*, 29–32.

13. Green, *Fit for America*, 253.

14. Macfadden, *The Virile Powers of Superb Manhood*, 15.

15. Kraut, *Annals of American Sport*, 307, AMNH.

16. Saxton, *The Rise and Fall of the White Republic*, 323, 334.

17. Kraut, *Annals of American Sport*, 307–10; Deloria, *Playing Indian*.

18. John Muir quoted in Vickery, *Wilderness Visionaries*, 74.

19. Some of Muir's titles include *The Mountains of California* (1894), *Our National Parks* (1901), and *My First Summer in the Sierra* (1911).

20. Roosevelt, along with his fellow Boone & Crockett founder George Bird Grinnell, editor of *Forest and Stream*, published copious volumes of wilderness stories, which sparked the establishment of clubs modeled after Boone & Crockett in the British empire. Morris, *The Rise of Theodore Roosevelt*, 383–85.

21. Vickery, *Wilderness Visionaries*, 98.

22. Bederman, *Manliness and Civilization*, chapter 5; Morris, *The Rise of Theodore Roosevelt*, 161–63, chap. 8.

23. Roosevelt quoted in Morris, *The Rise of Theodore Roosevelt*, 224.

24. Ibid.

25. London, *The Call of the Wild*, 113–26.

26. Ibid., 129–30.

27. Bederman, *Manliness and Civilization*, 218–32.

28. Isenberg, *The Destruction of the Bison*, 167.

29. Hornaday, *Tales from Nature's Wonderland*, 49–63, especially 53.

30. Ibid., 50.

31. Carmeli, "The Sight of Cruelty—The Case of Circus Animal Acts," 10–12.

32. Frank S. Redmond, "Official Route Book of the Adam Forepaugh Shows" (Buffalo: Courier, 1891), 26, Route Book Collection, RLPLRC.

33. Indeed, a defecating elephant during a pachyderm promenade received the big-

gest laugh of the show when the Ringling Bros. and Barnum & Bailey circus came to Austin, Texas, on June 12, 1999.

34. P. T. Barnum and Sarah J. Burke, *P. T. Barnum's Circus, Museum and Menagerie* (London: Frederick Warne, 1888), no page number, VI:D—Peter H. Davidson, Box 1, Folder, 10, WFCC.

35. Ibid.

36. Stallybrass and White, *The Politics and Poetics of Transgression*, 44–59, especially 47.

37. *Webster's New Universal Unabridged Dictionary*, deluxe 2nd ed., s.v. "Long pig," p. 1066.

38. Strasser, *Waste and Want*, 30; see also Hoy, *Chasing Dirt*.

39. Bouissac quoted in Stallybrass and White, *The Politics and Poetics of Transgression*, 59.

40. The American Society for the Prevention of Cruelty to Animals (ASPCA) was chartered by the New York legislature in 1866, with the local shipping heir Henry Bergh as its first president. Significantly, the legislature granted the new organization the powers of arrest. Three years after its incorporation, the society reported that from May 1869 to January 1870 it had arrested 267 individuals for "various offences against the laws of the organization." See "American Society for the Prevention of Cruelty to Animals," *New York Times*, Apr. 24, 1866, 4; "Mr. Bergh's Work," *New York Times*, Feb. 19, 1870, 2.

41. "American Society for the Prevention of Cruelty to Animals—Letter from Mr. Bergh," *New York Times*, Mar. 22, 1868, 11; "Cruelty on the Railroads," *New York Times*, Apr. 16, 1868, 4; "Mr. Bergh and His Work," *New York Times*, Aug. 8, 1869, 4; "The Cider Mill Dog," *New York Times*, Oct. 9, 1874.

42. "Mr. Bergh's New Idea," *New York Times*, Apr. 11, 1874, 12; "Dogs and Cats," *New York Times*, Mar. 15, 1877, 4.

43. "The Swill Milk Trade," *New York Times*, Jan. 14, 1869, 8; "Mr. Bergh's Work," *New York Times*, Feb. 19, 1870, 2; "The Condition of Cattle on Their Arrival at the New York Stock-Yards," *New York Times*, Aug. 27, 1870.

44. Bellamy, *Looking Backward*, 132; "Repaving Fifth-Avenue," *New York Times*, Apr. 16, 1872, 5; "Mr. Bergh's Proposal," *New York Times*, Jan. 27, 1873.

45. Ritvo, *The Animal Estate*, 39–40.

46. Turner, *Reckoning with the Beast*, 4, see also chapter 5.

47. George Page, *Inside the Animal Mind*, 8–12.

48. Frank S. Redmond, ed., "Official Route Book of the Adam Forepaugh Shows, Season 1891," 31.

49. M. B. Bailey, ed., "Official Souvenir, Buffalo Bill's Wild West and Congress of Rough Riders of the World" (Buffalo: Courier, 1896), 38, Program Collection, RLPLRC.

50. According to Sarah Blackstone, the outcome of "Race of the Races" was probably not fixed. Blackstone, *Buckskins, Bullets, and Business*, 57; George H. Gootch, "Route Book for Buffalo Bill's Wild West" (Kansas City: Hudson-Kimberly, 1900), 24, Route Book Collection, RLPLRC.

51. "Beneath White Tents: a Route Book of Ringling Bros., World's Greatest Shows,

Season 1894" (Buffalo: Courier, 1894), 86, Route Book Collection, RLPLRC; "Circus Almost Looses Missing Link Again," *New York Press*, Mar. 18, 1909, BBPCB 1909, JMRMA; "Barnum & Bailey's Greatest Show on Earth, Show Programme," Apr. 1–25 [1894/98], LTLBBS.

52. Henry Bergh initially clashed with P. T. Barnum over the second American Museum's practice of feeding live rabbits to boa constrictors. Barnum retaliated by procuring a letter from the famed naturalist Louis Agassiz, which stated that constrictors needed live prey to survive. Barnum published this correspondence in order to frame Bergh as a "sentimental fool." In 1880 Bergh and the ASPCA protested again when Barnum's horse, Salamander, jumped through a ring of fire. Once Barnum proved that the act was harmless to the horse, he and Bergh became friends, and Barnum became an active member of the ASPCA, even dubbing himself the "Bergh of Bridgeport." Saxon, *P. T. Barnum*, 235–38; see also Frank S. Redmond, "Official Route Book of the Adam Forepaugh Shows, Season 1891," 74. For a comparative intellectual history of the animal rights movement in the United States and England see Turner, *Reckoning with the Beast*.

53. Joseph T. McCaddon to the editor of the *Spectator* (London), Feb. 2, 1931, Correspondence, Box 7, Folder 7, JTMCC.

54. "Crack! Bang! The Bill Show's Open," New York City, n.d., n.p., 1911, Series VI: G—Ephemera, Box 3, Folder 14, WFCC.

55. Hornaday, *The Minds and Manners of Wild Animals*, 53, JFDC.

56. "A Complete and Illustrated History of the Wrecking of Walter L. Main's Circus on May 30, 1893, Near Tyrone Pennsylvania" (Harrisburg, Pa.: W. F. Raysor, 1893), Walter L. Main Circus Vertical File, RLPLRC.

57. See Darlington, *The Circus Boys in Dixie Land*, 225–27.

58. Barnum, *Lion Jack*, opening page; see also *Jack in the Jungle; Dick Broadhead*, Circus Fiction Collection, RLPLRC.

59. Daniel Boone, the great-grandson of the Kentucky frontiersman, worked for Barnum, Bailey and Hutchinson as a hunter and trader of live animals. One press release noted that Boone's right arm had been nearly torn from the socket while he wrestled with a wounded lion. "Daniel Boone's Great Grandson Here," *New York Recorder*, Friday, Feb. 20, 1891, Scrapbook 8, Barnum & Bailey, 1890–91, JTMCC.

60. Mayer, *Trapping Wild Animals in Malay Jungles*, 57–75, Published Circus Memoir Collection, RLPLRC.

61. In 1895 James A. Bailey contracted Charles Mayer for $12,000 ($230,307 in 2000) on the condition that Mayer purchase "one female elephant (to match one of the show's males), sound, healthy and perfect and to be *not less than fourteen feet and six inches in height* from shoulder to ground" (emphasis mine). Alas, Mayer was unable to procure an elephant of such prodigious size. Contract between Charles Mayer and James A. Bailey, June 27, 1895, Box 8, Folder 16; Charles Mayer to James A. Bailey, Aug. 6, 1895, Box 8, Folder 16, both from Correspondence, JTMCC.

62. P. T. Barnum and Sarah J. Burke, *P. T. Barnum's Circus, Museum and Menagerie*, no page number.

63. "The Barnum-Bailey Illustrated Hand-Book of Natural History," 6, Circus Programs, Box 13, Loose Items; see also "The Purser I. Lloyd Log of the 'Assyrian Dromedary' and 'Jumbo,'" Correspondence, Box 7, Folder 8, both from JTMCC.

64. "The Barnum-Bailey Illustrated Hand-Book of Natural History," 6.

65. Norwood, *The Circus Menagerie*, 193–212.

66. "A Monkey's Thumbs," *New York Morning Journal*, Nov. 12, 1893, BBPCB 1893–94, vol. 1, JMRMA.

67. "Chiko Smells Garner: He Must Have Thought Professor Was Insolent," *New York Morning Journal*, Apr. 22, 1894, BBPCB 1894, vol. 2, JMRMA.

68. "Chiko a Politician," *New York Morning Journal*, Oct. 26, 1893; "Chiko Now a Society Favorite," *New York Herald*, n.d., BBPCB 1893–94, vol. 1, JMRMA.

69. R. L. Garner to James A. Bailey, Correspondence, Box 7, Folder 2, JTMCC; n.d., n.p., BBPCB 1893–94, vol. 1, JMRMA.

70. "Chiko Stuffed for Good," *New York World*, Jan. 16, 1895, BBPCB 1895, JMRMA.

71. Haraway, *Primate Visions*, 41.

72. Ibid., 38.

73. Ibid., 30.

74. Johnson, *I Married Adventure*, 260, Circus Memoir Collection, RLPLRC.

75. James Turner suggests that the idea of pain as evil itself is a historical construction. With the advent of morphine, ether, chloroform, aspirin, and other anesthetics and analgesics in the nineteenth century, people began to construe pain as something morally repugnant, as opposed to previous generations who simply tolerated pain because of its inevitability. After Darwinian theory posited an inextricable link between animals and humans, Victorians transferred their revulsion toward pain to the animal world, and consequently animal rights activists argued that human beings had a moral imperative to protect animals from pain. See Turner, *Reckoning with the Beast*, chapter 5.

76. See, for example, Peter Wumbel Taylor, "Training Wild Animals for Circus and Stage Not Cruel," *Billboard*, June 30, 1923, 63, Periodicals Collection, RLPLRC.

77. Hagenbeck, *Beasts and Men*, 30.

78. Katherine Grier, "'Why Can't You Talk?'"

79. Bostock, *The Training of Wild Animals*, xv–xvi, Published Circus Memoir Collection, RLPLRC.

80. Hagenbeck, *Beasts and Men*, 40–41.

81. See Bouissac, *Circus and Culture*, chapter 7, especially 121–22.

82. Bostock, *The Training of Wild Animals*, 204–8.

83. Ibid., 185.

84. Ibid., 192.

85. "Dangers Described!" *Detroit Free Press*, Aug. 11, 1890, 4.

86. Moffett, *Careers of Danger and Daring*, 318–22, Circus Memoir Collection, RLPLRC.

87. Mabel Stark, *Hold That Tiger*, 84.

88. "Official Program of Ringling Bros. and Barnum & Bailey Combined Shows, Season 1920," Madison Square Garden Season New York, Mar. 25–May 1, 1920 (New York: Select, 1920), Program Collection, RLPLRC.

89. "Official Program of Ringling Bros. and Barnum & Bailey Combined Shows, Season 1922, "Madison Square Garden Season New York, Mar. 25–Apr. 29, 1922 (New York: Select, 1922), Program Collection, RLPLRC.

90. Bostock, *The Training of Wild Animals*, 31–32.

91. Ibid., 217–20.

92. "Fierce Battle for Life of Boy Crushed in Assassin Tiger's Jaws," *New York World-Monthly*, Apr. 1901, 24, Newspaper Collection, RLPLRC.

93. "Big Elephant Executed: Chief Forepaugh, the Slayer of Seven Men, Strangled," n.d., n.p., Philadelphia, 1888, George Chindahl Collection, Box 1, Folder 4, RLPLRC.

94. Barnum, *Life of P. T. Barnum*, 344.

95. "Big Elephant Tip Dead," *New York Times*, May 12, 1894, 1.

96. Ownby, *Subduing Satan*, 62.

97. Martin Johnson, "Taming Elephants," *Saturday Evening Post*, Jan. 5, 1929, George Chindahl Collection, Box 3, Folder 21, RLPLRC.

98. Bederman asserts that since most Northerners never witnessed an actual lynching, this method of murder became "an imaginary scenario," constructed and fueled by newspapers and literary accounts which enabled them to imagine themselves as participants to a lynching or at least witnesses—and thus enhance their own sense of racial and masculine privilege. *Manliness and Civilization*, 46.

99. Ibid., 47–49.

100. Barnum, *Life of P. T. Barnum*, 348–49.

101. "Official Program of Ringling Bros. and Barnum & Bailey Combined Shows' Season 1920," Madison Square Garden edition, 14.

102. Gossard, "A Reckless Era of Aerial Performance," 162.

103. "Risked Lives and Fearless Falls to Amuse the Public," *New York World*, Mar. 22, 1903; "Most Daring Cycle Feat," *Chicago Post*, Sept. 14, 1903, BBPCB 1902–3, vols. 1–2, JMRMA.

104. "Fall from Trapeze," *New York Advertiser*, n.d., BBPCB 1895; "Women Scream as Rider Falls," *New York World*, Mar. 25, 1904, BBPCB 1903–4, JMRMA.

105. Tiny Kline, "Showground-Bound" (unpublished memoir), 87–94, Manuscript Collection, RLPLRC.

106. "Beneath White Tents: A Route Book of Ringling Bros. World's Greatest Shows, Season 1894," 126.

107. "Circus on Friday," *Syracuse* (N.Y.) *Herald*, May 13, 1907, BBPCB 1907, JMRMA.

108. Japanese acrobats had been employed by American circuses since the 1860s, and male Asian acrobats were an integral part despite severe restrictions imposed by federal Asian exclusion acts (from 1882 onward). Employment opportunities were further limited by federal restrictions. In 1916 Barnum & Bailey had to put up a bond of $1,000 ($16,425 in 2000) with federal immigration authorities to guarantee that the Chinese performers would cause no trouble and remain with the show for the season. Kline, "Showground-Bound," 160; "Official Program of Ringling Bros. and Barnum & Bailey Combined Shows, Season 1920," Madison Square Garden edition; "Ringling Bros. and Barnum & Bailey Combined Shows Magazine and Daily Review," Washington, D.C., Newark, N.J., and Easton, Pa. (New York: Select, 1927), programs from Program Collection, RLPLRC.

109. Kline, "Showground-Bound," 162–64.

110. Speaight, *A History of the Circus*, 63.

111. McClintock, *Imperial Leather*, 103.

112. Speaight, *A History of the Circus*, 58.

113. Peacock, "Farini the Great," 15.

114. "Story Book and Program, Sells-Floto Circus, Home of 1001 Wonders" (U.S. Lithograph, 1913), VI: A—Programs, Etc., Box 2, Folder 9, WFCC.

115. "Official Route Book of Ringling Bros.' World's Greatest Railroad Shows, Season 1892" (Buffalo: Courier, 1892), 64, Route Book Collection, RLPLRC.

116. "Madamoiselle Cleveland, Circus Heroine, Is Really a Boy," *Chicago Daily Journal*, Oct. 15, 1904, BBPCB 1903–4, JMRMA.

117. Brown, "An Interview with Tom Barron, 'World's Tallest Clown,'" 20.

118. Chauncey, *Gay New York*, 9, 57.

119. Ibid., 13.

120. Laurence Senelick, "Lady and the Tramp," 27.

121. Bouissac, *Circus and Culture*, 164–66.

122. Bouissac characterizes American clowns as "somewhat less sophisticated than in European circuses," because they have to work in such enormous big tops: "[American clowns] have to rely on large visual devices rather than on a subtle combination of speech, facial expressions, miming and physical objects." Ibid., 154.

123. Turnour, *The Autobiography of a Clown*, 45–46.

124. $18,498 in 2000.

125. Turnour, *The Autobiography of a Clown*, 49–50; see also Carlyon, "Dan Rice's Aspirational Project."

126. Chindahl, *A History of the Circus in America*, 62–63.

127. Thayer, "The Circus Roots of Negro Minstrelsy," 43–45.

128. Lott, *Love and Theft*, 22–25.

129. Turnour, *The Autobiography of a Clown*, 57–59.

130. The term "rosin back" refers to the sticky substance applied to a horse's back to facilitate bareback stunts.

131. Mark Twain, *The Adventures of Huckleberry Finn*, 547.

132. "Clown Policemen Fooled Real Cop," *Trenton* (N.J.) *Evening Times*, May 12, 1907, BBPCB 1907, JMRMA.

133. Bakhtin, *Rabelais and His World*, introduction.

134. "Cowboys in a Small Riot at Buffalo Bill Show," *Brooklyn Citizen*, May 6, 1892, 2, VI: G—Ephemera, Box 3, Folder 2, WFCC.

135. On May 16, 1896, police in Evansville, Indiana, arrested the press agent Dexter Fellows (and two others) under suspicion of being confidence men. M. B. Bailey, ed., "Buffalo Bill's Wild West Route Book Season 1896" (Buffalo: Courier, 1896), 46, VI:A— Programs, Etc., Box 1, Folder 13, WFCC.

136. Williams, *Marxism and Literature*, 131; my thinking here is also indebted to Eric Lott's conceptualization of minstrelsy and racial formation.

137. Alf T. Ringling, "With the Circus: A Route Book of the Ringling Bros. World's Greatest Shows, Seasons of 1895 and 1896" (St. Louis: Great Western, 1896), 109, Route Book Collection, RLPLRC.

138. Ibid., 109–10, 112.

139. Slout, *Olympians of the Sawdust Circle*, 301.

140. Russett, *Sexual Science*, 50–55.

141. Jacobson, *Whiteness of a Different Color*, 56.

142. F. S. Redmond, "Route Book, Adam Forepaugh Shows, Season, 1893," (Buffalo: Courier, 1893), 23, Route Book Collection, RLPLRC.

143. Charles Andress, "The Barnum & Bailey Annual Route Book and Illustrated Tours" (Buffalo: Courier, 1906), 27, Route Book Collection, RLPLRC.

144. Kibler, *Rank Ladies*, 56.

145. Speaight, *A History of the Circus*, 89.

146. In a recent "Bizarro" cartoon, two women stand over an open casket and one says to the other: "It wasn't until after his death that we discovered the real reason Uncle Ed became a professional clown." The scary clown stereotype prompted one clown, Eric Coffman, to write a letter to the editor of the *Wisconsin State Journal* in defense of clowns and their good character. See Dan Piraro, "Bizarro," United Press Syndicate, November 23, 2000; "Clowns Possess Gentle Souls," *Wisconsin State Journal*, Aug. 24, 1996, 10A; Michael Goldman, "It's No Laughing Matter," *Wisconsin State Journal*, Apr. 13, 2000, 1F (citations courtesy of Jean B. Davis).

147. Kline, "Showground-Bound," 65–73; see also Senelick, "'Boys and Girls Together.'"

148. Charles Sherwood Stratton ("General" Tom Thumb) was a model for the construction of the turn-of-the-century freak. After Stratton began working for P. T. Barnum at the age of four, his mother was shocked to find that handbills advertising Charles to the public described the boy as an eleven-year-old general named "Tom Thumb" (after a mythical dwarf knight in King Arthur's court) who was born in London. Barnum explained that few people would be impressed with a four-year-old American-born midget, so he stretched the truth to attract customers. Throughout his long career, Stratton was marketed as a charming man (he captivated Queen Victoria, who met him three times during his European tours), a proper patriarch who married his fellow midget Mercy Lavinia Warren Bump in 1863, and a (faux) "father." Robert Bogdan, *Freak Show*, 149.

149. *Strange Human Beings*, clipping, n.d., n.p., 3A82 Sideshow Pamphlet Collection VF—Sideshow, Charles Tripp, HCCM.

150. Bogdan, *Freak Show*, 213–14.

151. James D. DeWolfe and H. P. Matlack, "Route Book Forepaugh and Sells Bros. Combined Shows" (New Orleans: L. Graham and Son, 1896), 16, Route Book Collection, RLPLRC.

152. Bogdan, *Freak Show*, 263.

153. Ibid., 241.

154. Tyng, *Before the Wind*, 49.

155. Bogdan, *Freak Show*, 241–56.

156. Parry, *Tattoo*, 60–63; Bogdan, *Freak Show*, 243–47.

157. Bogdan, *Freak Show*, 249.

158. Parry, *Tattoo*, 59.

159. The "human ostrich" was named after the giant bird because of its habit of swallowing large indigestible objects—usually stones—to pulverize its food for digestion.

160. Thanks to Fred Dahlinger for giving me more details about the art of eating glass, September 1999.

161. Kline, "Showground-Bound," 107.

162. Ibid.

163. See J. C. Oman, *The Mystics, Ascetics and Saints of India;* "Monthly Paper of the American Board of Commissioners for Foreign Missions" (citation courtesy of Kirin Narayan).

164. Narayan, *Storytellers, Saints, and Scoundrels,* 120; "From Self-Torturing Bodies to the Guru of Good Times: Representing American Representations of Hindu Ascetics, 1833–1990."

165. Gregor, with Cridland, *Circus of the Scars,* 10, 66–67.

166. James Cook offers a fascinating analysis of Johnson's popular late antebellum performances as a "nondescript." Cook suggests that by hiring Johnson to play a liminal character—neither animal nor human—Barnum deftly made him attractive to abolitionist and proslavery white artisans alike. Cook, "Of Men, Missing Links, and Nondescripts."

167. Scrapbook for "Zip," "Zip" Vertical File; Andrew J. Bakner, "Side Show Attractions," 34–38, "Lionel the Lion-Faced Man" Vertical File, RLPLRC; see also Cook, "Of Men, Missing Links, and Nondescripts," 144.

168. Scrapbook for "Zip."

169. See Bogdan, *Freak Show,* 134.

170. Kline, "Showground-Bound," 13–14.

171. "Congress of Nations," *New York Advertiser,* Mar. 22, 1894, BBPCB 1894, vol. 1, JMRMA.

172. Lindfors, "Circus Africans," 10; Bradford and Blume, *Ota, the Pygmy in the Zoo.*

173. Hornaday, *The Minds and Manners of Wild Animals,* 67; for a fascinating treatment of birds, human society, and the animal welfare movement see Doughty, *Feather Fashions and Bird Preservation.*

174. Hall, *Adolescence* 2:651.

175. "Long-Haired Mohaves," *Commercial New York Advertiser,* n.d., BBPCB 1895, JMRMA.

176. "Buffalo Bill's Wild West Combined with Pawnee Bill's Great Far East," courier (Buffalo: Courier, 1909), VI:A—Programs, Etc., Box 2, Folder 9, WFCC.

177. Looking back in 1955 to the turn of the century, the circus proprietor Lorenz Hagenbeck lamented the "demise" of an "authentic" folk over the preceding fifty years: "Progress cannot be halted, the wheel of history allows no reversal. Radio, the press and above all the cinema have touched the remotest corners of the world, including those from which our exhibitions came." Lorenz Hagenbeck, *Animals Are My Life,* 134, Published Circus Memoir Collection, RLPLRC.

178. "Too Much Hash Annoys Indian," n.d., n.p., VI: G—Ephemera, Box 3, Folder 14, WFCC.

179. M. B. Bailey, ed,. "Buffalo Bill's Wild West Route Book Season of 1896," 198–200.

180. Moses, *Wild West Shows and the Images of American Indians, 1883–1933.*

181. "The Barnum & Bailey Greatest Show on Earth Show Program," Madison Square Garden edition, Mar. 26–Apr. 23, 1894 (New York: J&H Mayer, 1894), Circus Programs, Box 13, Folder 6, JTMCC.

182. "Delegates from the East," n.p., Mar. 13, 1894, BBPCB 1894, vol. 1, JMRMA.

183. "Wild Men of New Guinea," n.d., n.p., BBPCB 1895, JMRMA.

184. In contrast to Buffalo Bill's cast of stoic, plainly dressed Euroamerican cowboys, Wild West couriers depicted the Mexican vaqueros (North America's original cowboys and part of the show's "ethnological equestrian congress") as "extravagant dandies" who spent lavishly on "high-pommeled saddles . . . silver spurs, richly-embroidered and bespangled velvet jackets, gorgeous silk sashes, and enormous sugar-loaf hats, weightily and brilliantly trimmed with gold and silver braid." "Buffalo Bill's Home Again from a Foreign Shore," courier (Buffalo: Courier, 1907), 27, VI:A—Programs, Etc., Box 2, Folder 7, WFCC.

185. Adams, *E Pluribus Barnum*, 191.

186. "Congress of Nations," *New York Advertiser*, Mar. 22, 1894, BBPCB 1894, vol. 1; "South Pacific Savages," *New York Sun*, n.d., but filed with clippings from Nov.–Dec. 1894, BBPCB 1895, JMRMA.

187. "A Symphony in Coffee," *New York Morning Journal*, Mar. 13, 1894, BBPCB vol. 1, 1894, JMRMA; "The Barnum & Bailey Greatest Show on Earth Show Program, Madison Square Garden, 1894," Circus Programs, Box 13, Folder 6, JTMCC.

188. "Wild Men of New Guinea."

189. Newspaper account cited in Lindfors, "Hottentot, Bushman, Kaffir," 24.

190. "Long-Haired Mohaves: Many Serious Specimens of Aborigines Arrive in This City," *Commercial New York Advertiser*, n.d.; "Long-Haired Mohaves: Additions to the Ethnological Congress of the Greatest Show on Earth," *New York Mail and Express*, Mar. 16, 1895; "Long-Haired Mohaves: Recent Additions to BB Ethnological Congress," *New York Daily News*, n.d.; "Indians for the Circus," *New York Evening Sun*, Friday, Mar. 15, 1895; "Indians by the Car Load: A Great and Glorious Addition to Barnum & Bailey's Mammoth Ethnological Congress," *New York Reporter*, Mar. 16, 1895; "'A Car Load of Indians': One Brave is 6'7″ and There's a Chief's Daughter," *New York World*, Mar. 16, 1895; "A Car Load of Indians: Numbers of Different Western Tribes for Barnum & Bailey," *New York Mercury*, n.d.; "All Sorts of People Will Be There," *New York Press*, Mar. 16, 1895; "A Group of Strange Americans: Numbers of Little-Known Indian Tribes to Be Exhibited This Year," *New York Sun*, Mar. 16, 1895; "Look Out for Scalps! Queer Red Men and Women Arrive from Far-Off Teepees," *New York County Journal*, Mar. 16, 1895; "'Smokestack' Sees the Town," *New York World*, Mar. 25, 1895; all from BBPCB 1895, JMRMA.

191. "The Romance of the Big Horn Basin Related by Buffalo Bill to Allen Kelly," *New York Daily News*, Dec. 20, 1903, BBPCB 1903–4, JMRMA.

192. Editorial, *New York Democrat*, June 5, 1899, reprinted in "Buffalo Bill's Wild West Show and Congress of Rough Riders of the World, Historical Sketches and Programme" (New York: Fless and Ridge, 1899), 11, Program Collection, RLPLRC.

193. "Buffalo Bill's Home Again from a Foreign Shore," 1907, 9.

194. Mark Twain to William F. Cody, quoted in "Buffalo Bill's Wild West, Show Programme, 1887" (London: Allen, Scott, 1887), VI:A—Programs, Etc., Box 1, Folder 6, WFCC.

195. Captain C. W. Briggs, letter to the editor, *Billboard*, June 2, 1900, 5.

196. Brown, "An Interview with Tom Barron, 'World's Tallest Clown,'" 15–16.

197. Edward Arlington, ed., "Official Route Book of the Walter L. Main's Grandest and Best Shows on Earth" (Buffalo: Courier, 1894), 69, Route Book Collection, RLPLRC.

198. "Official Route Book of Ringling Bros. World's Greatest Railroad Shows, Season 1892," 75.

199. "Official Route Book, Adam Forepaugh and Sells Bros. Combined Circuses, Season 1898" (Columbus, Ohio: Landon, 1898), 75, Route Book Collection, RLPLRC.

200. "Is a Thing of the Past," *Anderson* (Ind.) *Herald*, Sept. 30, 1903, BBPCB 1903–4, JMRMA; see "Route Book, Season 1890, Barnum & Bailey's Greatest Show on Earth" (Buffalo: Courier, 1890), 42, Route Book Collection, RLPLRC.

201. Charles Andress, "Day by Day with Barnum & Bailey, Seasons 1903–1904" (Buffalo: Courier, 1904), 89, Route Book Collection, RLPLRC.

202. "Snowballed the Circus," *New York World*, Mar. 20, 1892, BBPCB 1891–92, JMRMA.

203. "Route Book, Season 1890, Barnum & Bailey's Greatest Show on Earth," 42.

204. "Official Route Book of Ringling Bros. World's Greatest Railroad Shows, Season 1892," 50.

205. Kelly, *Clown*, 95.

206. "Route Book, Ringling Bros., Season of 1893" (Buffalo: Courier, 1893), 73–74, RLPLRC.

207. See Kaplan, "Is the Gaze Male?" 309–27; Mulvey, "Visual Pleasure and Narrative Cinema," 6–18. More recently, several feminist media scholars, including Mulvey, have examined the "omni-erotic" dimensions of the gaze. While fully acknowledging the ways that the gaze frames normative power relations, these works examine its subversive possibilities as well. See Mulvey, *Fetishism and Curiosity*; Rabinovitz, *For the Love of Pleasure*; Staiger, *Perverse Spectators, Interpreting Films.*

208. See Stallybrass and White, *The Politics and Poetics of Transgression.*

209. "Tom Throttle, The Boy Engineer of the Midnight Express; or, Railroading in Central America," *Pluck and Luck: Complete Stories of Adventure*, no. 336 (New York: Frank Tousey, 1904); "The Bradys in the Chinese Quarter; or, The Queen of the Opium Fiend," *Secret Service Old and Young King Brady, Detectives*, no. 137 (New York: Frank Tousey, 1902), all from VIII: Vincent Mercaldo, Box 2, folder unmarked, WFCC.

210. Hall, *Adolescence* 1:202.

Chapter 6

1. In the absence of uniformly conclusive primary source materials, it is difficult to assess how each railroad showman felt privately about the new empire: the Ringling brothers and James Bailey, for example, were intensely guarded about their political views, while Peter Sells and William F. Cody publicly supported overseas expansion. Despite the inability to make explicit connections between each showman's personal views toward the new empire and the content of his shows, it is safe to say that the public presentations were uniformly supportive of America's new role on the world stage.

2. "Official Route Book, Adam Forepaugh & Sells Brothers Combined Circuses, Season 1898" (Columbus, Ohio: Landon, 1898), 108–9, Route Book Collection, RLPLRC.

3. For example, in 1861 Dan Rice's Great Show performed "Ward's Mission to China," "introducing the games and royal festivities of celestial people, the Mandarin's

Court, and the royal reception together with the far famed *procession of lanterns*" (emphasis in original), clipping, n.p., July 16, 1861, Newspaper Advertisement Collection, RLPLRC.

4. "The Barnum and Bailey Greatest Show on Earth Show Program," Madison Square Garden edition, Mar. 26–Apr. 23, 1894 (New York: J&H Mayer, 1894), Circus Programs, Box 13, Folder 6, JTMCC.

5. "The Rough Riders Return," *New York Daily Tribune*, Apr. 24, 1900, 7.

6. "Buffalo Bill's Wild West and Congress of Rough Riders of the World, Historical Sketches and Programme" (New York: Fless and Ridge, 1899), 63, Program Collection, RLPLRC.

7. The circus cowgirl Irene Mann remembers the earthy sensations of riding a camel in an Indo-Arabic procession during the early 1930s; as she rode dressed in scant, glittery attire playing the "Queen of Sheba," Mann recalls how the camel following her constantly spat on her back, while the camel in front of her broke wind with every step: according to Mann, "It was just stinking." Audience members sitting close by, no doubt, could smell the camels. Irene Mann interview by author, tape recording, Oct. 22, 1994, Bagley, Wis.

8. "Barnum & Bailey Circus Next Week," n.p., Mar. 14, 1903, BBPCB 1902–3, JMRMA.

9. William Appleman Williams, *The Tragedy of American Diplomacy*, 13.

10. Consequently, circus spectacles of empire attest to the ways that popular forms give sustenance to Antonio Gramsci's formulation of "cultural hegemony," a term stressing the interconnectedness of popular culture, politics, and the dialectical process of social change. Gramsci, *Selections from the Prison Notebooks*.

11. "Freak Hunting in India," *New York Herald*, Apr. 1, 1894, BBPCB 1894, vol. 2, JMRMA.

12. Lorenz Hagenbeck, *Animals Are My Life*, 70–71.

13. Carl Hagenbeck, *Beasts and Men*, 53.

14. Coup refers to the hunter as "Paul Tuhe," which is probably a misprint. The Ruhe brothers were German animal agents, and in all likelihood Coup means one of them. Coup, *Sawdust and Spangles*, 27.

15. Advertisement clipping for P. T. Barnum's Asiatic Caravan, Museum and Menagerie, Aug. 18, 1851, Newspaper Advertisement Collection, RLPLRC.

16. See MacKenzie, *Imperialism and the Natural World*.

17. Pal and Dehejia, *From Merchants to Emperors*, 73.

18. Haraway, *Primate Visions*, 53–54.

19. Carl Hagenbeck, *Beasts and Men*, 22.

20. Lorenz Hagenbeck, *Animals Are My Life*, 65, 68.

21. "Affairs in Japan," *New York Times*, Nov. 20, 1858, 2 (citation courtesy of Stuart Thayer).

22. "Academy of Music," *New York Times*, Sunday, May 5, 1867, 7 (citation courtesy of Stuart Thayer).

23. The Japanese troupe was immensely successful, drawing 3,000 people to the Academy of Music in New York City on May 6, 1867. A boy slack rope dancer named "All

Right" charmed audiences, especially after he responded to a fall with pluck and grace. According to a show review, "We have had jugglers and conjurers and necromancers [sorcerers of the dead] in Yankeeland before—but we cannot just now recall any who so completely set all the laws of gravitation at defiance as these curious and ingenious strangers." Once they had completed a stint at the Paris Exposition in 1868, the troupe returned to the United States for another tour. Eight years later, All Right and other members of his family performed with Cooper and Bailey's circus during its international tour of Hawaii, India, Indonesia, Australia, and New Zealand until All Right suddenly died in October 1877. Soon after the Imperial Troupe's departure from the United States in 1868, another group of twelve Japanese acrobats and jugglers toured the United States and Great Britain for a year (their managers, Thomas F. Smith and Gustavus W. Burgess, having obtained permission from the Tycoon in Yokohama). After a successful season in New York City, the troupe prepared to return to Japan, but the entertainment manager Thomas Maguire then offered Smith and Burgess $15,000 ($178,252 in 2000) to extend the tour for a few weeks of travel from Buffalo to Chicago to New York City on the way home to Japan. When the troupe returned to New York City, Maguire, working through an interpreter, obtained the signatures of several of its members to extend their contract through February 1, 1868. By these terms, the troupe member Foo-Choo-Matz was to be given $2,500 in gold ($29,709 in 2000). However, once the terms of the contract were initiated, Foo-Choo-Matz refused to perform and demanded to go home to Japan. He claimed that he did not understand the terms of the contract, and that the troupe had been deceived. Maguire, in turn, charged Foo-Choo-Matz with breach of contract and had him arrested for embezzling the $2,500. Smith was sent to jail for ten days, and Maguire agreed to settle his larger claim for $10,000 ($118,835 in 2000), along with the rights to the performers for an extended period. Legally rendered helpless, the troupe was forced to tour across the entire United States. "The Japanese at the White House," *New York Times*, Apr. 19, 1867, 5; "Academy of Music—The Japanese Jugglers," *New York Times*, May 7, 1867, 4; "The Japanese in This Country—The Troubles of a Manager," *New York Times*, Nov. 1, 1867, 2 (citations courtesy of Stuart Thayer); New Zealand telegram from George Middleton, Oct. 26, 1877, Scrapbook 4: Cooper and Bailey Australia Tour, 1876–78, JTMCC; for a fuller secondary treatment, see Mihara and Thayer, "Richard Risley Carlisle: Man in Motion," 12–14.

24. "The Carl Hagenbeck Greater Shows Embracing Grand Triple Circus East India Exposition," Ottumwa, Iowa, July 11, 1906 (Buffalo: Courier, 1906), Program Collection, RLPLRC.

25. *Spirit of the Times*, Jan. 20, 1883, Scrapbook: Barnum, Bailey and Hutchinson, Winter 1882–83, JTMCC.

26. An official at the Australian Embassy in Washington stated that Tambo traveled home at the expense of the Australian government. On the 110th anniversary of his death, February 23, 1994, Tambo was finally buried during a funeral ceremony performed by the Palm Island community. In 1997 and 1998, Jimmy Tambo and his fellow captives were memorialized in another traveling show, an exhibit "performed" by the National Library of Australia and entitled "Captive Lives: Looking for Tambo and His Companions." Jay Maeder, "P. T. Barnum 'cannibal' to be buried with dignity," *Wisconsin*

State Journal, Nov. 27, 1993, 1; "Captive Lives: Looking for Tambo and His Companions," handbill for traveling exhibit, Australian National Library, n.d. (citation courtesy of Alison Kibler).

27. Over $2.6 million in 2000.

28. Harris, *Humbug,* 103–4.

29. "An Illustrated Catalogue and Guide Book to Barnum's American Museum" (New York: Wynkoop, Hallenbeck and Thomas), n.p., 110, Circus Programs, Box XIII, Folder 5, JTMCC.

30. As a reminder, all circus specs shared a defining characteristic: they told a story— either fictional (fairy tales, myths), biblical, or based on foreign affairs, past or present. By contrast, the pageant, or grand entry, simply contained costumed riders and animals parading around the arena. American circuses and theaters had presented spectacles since the late eighteenth century; these grew dramatically over time. In 1858 North's National Theatre presented a historical spec, "Fall of Delhi, or The Revolt in India," based on the Sepoy Mutiny in India in 1857. Seventeen principal actors took part, along with a small chorus of Europeans, Sepoys, Brahmins, "Hindoos," soldiers, slaves, and attendants. Barnum & Bailey's production "Columbus and the Discovery of America" (1892), by contrast, contained a cast of 1,200. A. Morton Smith, "Spec-ology of the Circus," *Billboard,* July 31, 1943, 51–55, Periodicals Collection; newspaper advertisement for North's National Theater, n.p., Feb. 19, 1858; "An Immense Spectacle, Columbus and the Discovery of America with 1,200 People," *Keokuk* (Iowa) *Constitution-Democrat,* 1892; preceding two references, Newspaper Collection, RLPLRC; Barker, *Bolossy Kiralfy.*

31. See also the British historian John Springhall's excellent essay, "'Up Guards and at Them!'" Springhall examines how British "little wars of empire" during the 1870s and onward provided exciting fodder for British artwork and entertainment, depicting small, colonial wars with relatively few British casualties.

32. Johannsen, *To the Halls of the Montezumas,* 221–22; for more on moving panoramas and westward expansion, see Goetzmann and Goetzmann, *The West of the Imagination,* 114.

33. The dime novel, a national cultural form like the circus and Wild West, also ignored the Civil War. Capitalizing upon national consensus, the dime novel depended for its salability upon popular topics like victorious little wars of empire on the trans-Mississippi West against Indians and other people of color. From 1860 to 1890 the house of Beadle and Adams in New York City published fifty-five novels about the Civil War and slavery (1 percent of all novels published by the house). By contrast, the firm published 2,710 frontier and western dime novels (48 percent of all novels), 1,238 of which focused on American Indian–related themes. Alexander Saxton observes that President Abraham Lincoln was virtually absent in the dime novel—although he actually was born in a log cabin in the Kentucky wilderness! Unlike dime novel staples such as Daniel Boone, William F. Cody, and Theodore Roosevelt, Lincoln hardly had jingoistic politics: he avoided killing Indians in the Blackhawk War, opposed the Mexican War, and was president during the fractious Civil War. Saxton speculates that the reasons for Lincoln's exclusion as a dime novel subject were political: although Lincoln supported western expansion by signing into law the Homestead and Transcontinental Railway Acts, he remained more closely linked to the whiggish eastern part of the Republican

party. Saxton, *The Rise and Fall of the White Republic*, 329, 334; Pfening, "The Big Show of the World," 5.

34. "Adam Forepaugh & Sells Brothers Book of Wonders and Official Program" (Buffalo: Courier, 1905), Program Collection, RLPLRC.

35. Over $9.2 million in 2000.

36. Almost $1.4 million, $4.6 million, and $924,105 in 2000.

37. "An Immense Spectacle, Columbus and the Discovery of America with 1,200 People," *Keokuk* (Iowa) *Constitution-Democrat*, 1892, Newspaper Collection, RLPLRC.

38. "Imre Kiralfy's Sublime Nautical, Martial and Poetical Spectacle, 'Columbus and the Discovery of America,' for the First Time Now Produced in Connection with the Barnum and Bailey Greatest Show on Earth" (Buffalo: Courier, 1892), 21, Program Collection, RLPLRC.

39. Hobsbawm, "Introduction."

40. *Philadelphia Claypoole's American Daily Advertiser*, May 12, 1798, Newspaper Collection, RLPLRC.

41. "The Adam Forepaugh Shows Program" (Buffalo: Courier, 1893), Circus Programs, Box 12, Folder 7, JTMCC.

42. F. S. Redmond, "Official Route Book, Adam Forepaugh Shows, Season 1893" (Buffalo: Courier, 1893), 35–37, Route Book Collection, RLPLRC.

43. "Buffalo Bill's 'Wild West' and Congress of Rough Riders of the World," England road edition (London: Partington Advertising, 1903), 46–47, Program Collection, RLPLRC.

44. Sarah J. Blackstone also notes that the name "Yellow Hand" was a poor translation of the Cheyenne chief's real name, Yellow Hair. During his stage dramas in the late 1870s, Cody displayed the actual fetish objects of war—the dried scalp, scabbard, warbonnet, and gun. Blackstone, *Buckskins, Bullets, and Business*, 69–70; Slotkin, *Gunfighter Nation*, 72–73.

45. Slotkin, *Gunfighter Nation*, 71–73.

46. "Buffalo Bill's Wild West," road edition (London: Calhoun, 1887), Program Collection, RLPLRC.

47. Richard Slotkin contends that Custer's Last Stand serves as a mythic model which has rationalized how people of color (like the Japanese during World War II), could defeat white American troops. *Gunfighter Nation*, 318–19.

48. Elizabeth B. Custer to William F. Cody, May 9, circa 1893, I:B—Correspondence, Box 2, Folder 29, WFCC.

49. "Buffalo Bill's Wild West and Congress of Rough Riders of the World" (New York: Fless Ridge, 1895), 11, Route Book Collection, RLPLRC.

50. Joy Kasson, *Buffalo Bill's Wild West*, 171–83.

51. Frank S. Redmond, Official Route Book of the Adam Forepaugh Shows, Season 1891" (Buffalo: Courier, 1894), 31, Route Book Collection, RLPLRC.

52. See Slotkin, *Gunfighter Nation;* Walter L. Williams, "United States Indian Policy and the Debate over Philippine Annexation;" for an extended treatment of the relationship between American Indians and empire consolidation, see Drinnon, *Facing West*, especially chapters 20–22.

53. Specifically, in the Supreme Court case, *Cherokee Nation v. Georgia* (1831), Chief

Justice John Marshall ruled that current treaties recognized the Cherokees as a "state," but not a foreign state; and thus, the Constitution allowed the federal government to govern an alien people without granting them U.S. citizenship (and all of its attendant constitutional rights). *Cherokee Nation v. Georgia* served as a precedent for the liminal status conferred upon the native residents of Alaska, Guam, Hawaii, Puerto Rico, and the Philippines.

54. Drinnon, *Facing West*, chapters 4–5.

55. Walter L. Williams, "United States Indian Policy," 825–28.

56. Roosevelt, quoted in Drinnon, *Facing West*, 299.

57. Blackstone, *Buckskins, Bullets, and Business*, 2–8.

58. See the collection of European dime novels about the Wild West at VII: Vincent Mercaldo, Box 2, Folders unmarked, in addition to other dime novel collections at WFCC.

59. Moses, *Wild West Shows and the Images of American Indians, 1883–1933*, 118.

60. "Buffalo Bill's Wild West and Congress of Rough Riders of the World" (Chicago: Blakely, 1893), 2, Program Collection, RLPLRC.

61. "Buffalo Bill's Wild West and Congress of Rough Riders of the World, Historical Sketches and Programme" (1899), iv.

62. "Buffalo Bill's Wild West Combined with Pawnee Bill's Great Far East," courier (Buffalo: Courier, 1909), VI:A—Programs, Etc., Box 2, Folder 9, WFCC.

63. Joy Kasson, *Buffalo Bill's Wild West*, 249.

64. "Buffalo Bill's Wild West and Congress of Rough Riders of the World" (New York: Fless and Ridge, 1898), iv–v, Program Collection, RLPLRC.

65. "Official Route Book, Adam Forepaugh & Sells Brothers Combined Circuses, Season 1898," 84.

66. LaFeber, *The American Age*, 195.

67. Ibid., 199–200.

68. Plummer, *Haiti and the U.S.*, 89–102; LaFeber, *The American Age*, 267.

69. William Appleman Williams, *The Tragedy of American Diplomacy*, 55.

70. Theodore Roosevelt to William F. Cody, June 22, 1903, Theodore Roosevelt Papers, microfilm series 2, reel 331, 220, Library of Congress, Microfilms Collection.

71. *New York World*, Apr. 3, 1901, 4.

72. "The Barnum & Bailey Greatest Show on Earth at Olympia," London edition (London: Walter Hill, 1899), 19, Program Collection, RLPLRC.

73. John D. Long, secretary of the Navy, to Whiting Allen, Feb. 6, 1902, Correspondence, Box 8, Folder 25, JTMCC.

74. Barnum & Bailey's circus paid $6,000 ($115,098 in 2000), for the construction of eleven models. The U.S. Navy Department stipulated that all models were to be made stationary, without movable parts or machinery; Barnum & Bailey's circus management (unspecified) to Charles E. Dressler and Brother, New York, Jan. 11, 1902, Correspondence, Box 8, Folder 25, JTMCC.

75. "Barnum & Bailey Greatest Show on Earth Official Program and Book of Wonders Combined, Season of 1904," road edition (New York: Self-published, 1904), Program Collection, RLPLRC.

76. "Barnum & Bailey Circus Next Week," n.p., Mar. 14, 1903, BBPCB 1902–3, JMRMA.

77. Blackstone, *Buckskins, Bullets, and Business*, 70–71.

78. "Buffalo Bill's Wild West and Congress of Rough Riders of the World, Historical Sketches and Programme" (1899), 34.

79. Although TR's Rough Riders eventually captured San Juan Heights, the actual Battle of San Juan Hill was disorganized, lethargic, and costly in terms of human lives. Morris, *The Rise of Theodore Roosevelt*, 632–57.

80. "Exciting Spectacle at the Wild West Opening," n.p., 1900, VI:G—Ephemera, Box 3, Folder 4, WFCC.

81. Over $191 million and nearly $4.8 million in 2000.

82. LaFeber, *The American Age*, 228.

83. *Public Opinion*, quoted in ibid., 230.

84. Ibid.

85. "Adam Forepaugh and Sell Brothers Book of Wonders and Official Program" (1905).

86. "The Great Adam Forepaugh & Sells Brothers' Enormous Shows United," courier (Buffalo: Courier, 1905), Program Collection, RLPLRC.

87. "The Barnum and Bailey Greatest Show on Earth Magazine of Wonders and Daily Review," road edition (Buffalo: Courier, 1906), 27–29, Route Book Collection, RLPLRC.

88. LaFeber, *The American Age*, 237–38; see also Cumings, *Korea's Place in the Sun*, chapter 3.

89. Lewis, *W. E. B. Du Bois*, 370; see also Plummer, *Haiti and the U.S.*, 75.

90. "Official Route Book, Adam Forepaugh & Sells Brothers Combined Circuses, Season 1898," 110.

91. Mahan, *The Influence of Sea Power upon History*, 26.

92. See Rosenberg, *Spreading the American Dream*, 28–34.

93. Hunter, *The Gospel of Gentility*, 5, 7–9.

94. LaFeber, *The American Age*, 208–9.

95. Blackstone, *Buckskins, Bullets, and Business*, 70–71.

96. "Buffalo Bill's Wild West and Congress of Rough Riders of the World, Historical Sketches and Programme" (Buffalo: Courier, 1901), 60, Program Collection, RLPLRC.

97. The spec producer John Rettig made earlier Christian productions for the Order of Cincinnatus, such as "Montezuma; or, The Conquest of Mexico" in 1889 (celebrating the arrival of Hernando Cortez in Mexico), and "Moses; or, The Bondage in Egypt" in 1890.

98. "The Circus Annual, Season 1903: A Route Book of Ringling Bros. World's Greatest Shows" (Chicago: Press of Central Printing and Engraving, 1903), 9, Route Book Collection, RLPLRC.

99. Ibid.

100. Green, *Fit for America*, 270–77.

101. "Official Route Book, Adam Forepaugh & Sells Brothers Combined Circuses, Season 1898," 108–9.

102. Anderson, *A Story Teller's Story*, 277–78.

103. At the end of the eighteenth century, Lord Richard Colley Wellesley, governor general of Bengal, deemed British missionaries in the region subversive because they worked diligently to translate, print, and teach the Bible in many Indian languages. Wolpert, *A New History of India*, 207–8.

104. See *Monthly Paper of the American Board of Commissioners for Foreign Missions*, no. 12, June 1833 (citation courtesy of Kirin Narayan).

105. Thayer, "One Sheet," 23.

106. Horseman, *Race and Manifest Destiny*.

107. Anderson points to three international crises in which British and American policymakers worked together: the Venezuela boundary crisis of 1894–95; the Spanish-American War (1898); and the Boer War (1899–1902), in which the British enjoyed American executive (and legislative) support. Anderson, *Race and Rapprochement*.

108. Those four shows were North's National Theater; Nixon's Great American Circus and Kemp's Mammoth English Circus; Sands, Nathans & Co.'s American Circus; and Rivers & Derious's Grand Dramatic Equestrian Company.

109. "Rivers & Derious's Grand Gymnastic, Acrobatic, Ballet, and Dramatic Establishment!" *Wilkes-Barre Record of the Times*, Wednesday, Sept. 8, 1858. This clipping and other information concerning American circus productions of the 1857 Sepoy Mutiny are courtesy of Stuart Thayer.

110. "Barnum & Bailey's Greatest Show on Earth, Show Program, 1896," Potsdam, N.Y., Aug. 31, 1896, n.p., LTLBBS.

111. Ibid.

112. "Barnum & Bailey's Greatest Show on Earth Show Program," Apr. 20–25, 1896, Madison Square Garden, n.p., LTLBBS.

113. "Barnum & Bailey Greatest Show on Earth, Show Program, 1894," Madison Square Garden edition.

114. See "Barnum & Bailey Greatest Show on Earth Official Program and Book of Wonders Combined, Season of 1904." The Great Floto Shows immediately copied the "Durbar" in its 1905 spectacle, "Mysterious India; or, the Durbah [*sic*] of Delhi," and Ringling Bros. and Barnum & Bailey revived it again in 1933 for several years thereafter. "The Great Floto Shows (The Circus Beautiful) Official Song Book" (Chicago and New York: Will Rossiter, 1905); "Ringling Bros.' Golden Jubilee Program," New York City edition (Louisville: Dearing, 1933); additional Ringling Bros. and Barnum & Bailey programs from 1934–38, all from Program Collection, RLPLRC.

115. "Barnum & Bailey Greatest Show on Earth, Book of Wonders and Official Program, Season of 1905," Madison Square Garden, 2d ed. (Buffalo: Courier, 1905), 27, Program Collection, RLPLRC.

116. The Indian National Congress was composed of well-educated, high-caste Hindus, Parsis, a few Englishmen, and Muslims from all provinces of British India. At its inception, the Congress announced its loyalty to the British Crown, demanding that Indians be granted a larger role in governing the nation. After Lord Curzon carelessly partitioned Bengal into two separate provinces in 1905 (which sparked a national outcry), the Congress began to champion Indian independence.

117. The orientalist image of the Other as static also held true elsewhere. For ex-

ample, John Dewey, who stayed in China from 1919 to 1921, described China as a "wonderland," "old" and "sleeping," yet also a site for potential reform. Israel, *Progressivism and the Open Door*, 182.

118. See Lears, *No Place of Grace*.

119. Green, *Fit for America*, 266–77; Lears, *No Place of Grace*.

120. While in the United States, Vivekananda founded the Vedanta Society in 1895, a religious organization which cultivated a holistic, nondualistic faith, based on the Vedas.

121. As part of this movement, the new Theosophical Society was popular among some American intellectuals. Founded in 1875 by a Russian countess, Madame Blavatsky, and an American, Colonel Henry Olcott, the Theosophists sought to combat the materialism of western industrial capitalism with the spiritualism of "Eastern Wisdom," based on the teachings of Gautama Buddha, Zoroaster, Confucius, and the Hindu scriptures, collectively known as the Vedas. The Theosophists were critical of missionaries, who they felt had degraded Asian religions; instead, the Theosophists called for a monistic, universal faith, comprising a "brotherhood of Humanity, wherein all good and pure men, of every race, shall recognize each other as the equal effects (upon this planet), of one Uncreated, Universal, Infinite and Everlasting Cause." Quoted in Milton Singer, "Passage to More Than India," 29.

122. LaFeber, *The American Age*, 225; Noer, *Briton, Boer and Yankee*; for a fascinating account of U.S. cultural exports to South Africa during the nineteenth and twentieth centuries, see Campbell, "The Americanization of South Africa."

123. "Buffalo Bill's Wild West and Congress of Rough Riders of the World, Historical Sketches and Programme" (1901), 62–63.

124. "The Circus Annual: A Route Book of Ringling Bros. World's Greatest Shows, Season 1901" (Chicago: Central, 1901), Route Book Collection, RLPLRC.

125. "List of Twelve Staged Productions," Circus Programs, Box 13, Folder 6, JTMCC; "The Barnum & Bailey Greatest Show on Earth, "The Mahdi,' or for the Victorian Cross' Program'" (London: Walter Hill, 1898), 3–15, Program Collection, RLPLRC; "Music for Entry and Mahdi," London, 1897–98, Merle Evans Papers, RLPLRC.

126. William Appleman Williams, *The Tragedy of American Diplomacy*, especially 8.

127. Charles Towne, quoted in Miller, *"Benevolent Assimilation,"* 109.

128. "An Address," the *Bonham* (Tex.) *News*, Oct. 26, 1900, 2, CAH.

129. Miller, *"Benevolent Assimilation,"* 105.

130. Drinnon, *Facing West*, 308–9.

131. Mark Twain's famous essay "To the Person Sitting in Darkness" (1901) turned the tables on the supposed distinction between "civilization" and "savagery" by arguing that U.S. missionaries in China and the U.S. military in the Philippines engaged in particularly "savage acts." Mark Twain, "To the Person Sitting in Darkness"; see also Bain, *Sitting in Darkness*; Miller, *"Benevolent Assimilation,"* especially chapter 7; Kantrowitz, *Ben Tillman and the Reconstruction of White Supremacy*, 262–64.

132. "The President's Negro Guest," *Atlanta Constitution*, Oct. 20, 1901, 16; "Sensation at Louisville," *Jackson Clarion-Ledger*, Oct. 24, 1901, 1.

133. This characterization of Washington's Tuskegee model comes from Du Bois, *The Souls of Black Folk*, 94.

134. "The South's Attitude," *Jackson Clarion-Ledger*, Oct. 31, 1901, 6.

135. "Colonies, and Not States," *Daily Picayune*, Nov. 20, 1898, 4.

136. "A Check on Imperialism," *Daily Picayune*, Nov. 17, 1898, 1.

137. See Bain, *Sitting in Darkness;* Miller, *"Benevolent Assimilation,"* especially chapter 7; Lasch, "The Anti-Imperialists, the Philippines, and the Inequality of Man"; Amy Kaplan, "Black and Blue on San Juan Hill"; Michaels, "Anti-Imperial Americanism."

138. "Work of a Negro Soldier," *Bonham* (Tex.) *News,* Oct. 28, 1898, 1, CAH; "Negro Soldiers So-Called," *Daily Picayune,* Nov. 16, 1898, 4; "Negro Regiment from Virginia Has Proved a Failure," *Daily Picayune,* Nov. 17, 1898, 9.

139. Drinnon, *Facing West,* 313.

140. See, for example, "De-Citizenization: Georgia Depriving Black Men of Prerogative of Bearing Arms for Native Land," *New York Age,* Mar. 30, 1905, 1.

141. After examining several African American newspapers from the turn of the century, I was surprised to find little critical coverage of the railroad circus. While some articles acknowledged a show's presence, or used the term "circus" as a synonym for chaos or mismanagement (see "A Traveling Circus," *New York Age,* Apr. 23, 1908, 3), none examined the circus's imperial or racial politics. Although route books, newspapers, and performers' memoirs documented the constant presence of black circus audiences, African American newspapers typically reviewed only those amusements produced by and starring African Americans—of which the largest railroad shows had neither. The railroad circus—despite its multiracial content and diverse audience base—was primarily advertised in Euroamerican newspapers targeting middle-class Euroamerican families.

142. H. L. Watkins, "The Barnum and Bailey Official Route Book, Season of 1894" (Buffalo: Courier, 1894), 27, Route Book Collection, RLPLRC.

143. Kaplan, "Black and Blue on San Juan Hill," 219–36.

144. "Red Wagon Annual: A Route Book of Ringling Bros. World's Greatest Shows, Season 1898" (Chicago: Central, 1898), 3, Route Book Collection, RLPLRC.

145. "Corean [*sic*] Twins May Be Split for War," *New York Press,* Mar. 26, 1904, BBPCB 1903–4, JMRMA.

146. "The Barnum and Bailey Greatest Show on Earth, Show Program, 1894," Madison Square Garden edition.

147. "Route Book Sells Brothers' Enormous United Shows, Five Continent Menagerie and Three-Ring Circus" (Waco, Tex.: Brooks and Wallace, 1890), 28; Route Book Collection, RLPLRC.

148. "Official Route Book, Adam Forepaugh & Sells Brothers Combined Circuses, Season 1898," 87.

149. Many spectacles had previously celebrated the fitness of the U.S. military in features like Buffalo Bill's 1901 demonstration, "United States Life-Saving Service Beach-Apparatus Drill."

150. "Buffalo Bill (Himself), and 101 Ranch Wild West Combined, with the Military Pageant 'Preparedness,' Magazine and Daily Review" (Philadelphia: Harrison, 1916), Program Collection, RLPLRC.

151. Tiny Kline, "Showground-Bound" (unpublished memoir), especially 287–88, 428, Manuscript Collection, RLPLRC.

152. William Appleman Williams, *The Tragedy of American Diplomacy,* 15.

153. Notes from McCormick Steele diaries, Sept. 22–23, 1954; and Sept. 15, 1954, to Nov. 18, 1955 (citation courtesy of Richard J. Reynolds III).

154. Mary Dudziak analyzes how U.S. policymakers touted the civil rights movement abroad in order to woo newly liberated Third World countries into America's political and economic orbit, and to offset Soviet propaganda concerning the United States' long and violent history of racial discrimination. Yet, as Penny Von Eschen demonstrates, African Americans, not U.S. policymakers, led the push for civil rights legislation in the United States. While U.S. leaders used civil rights rhetoric abroad to gain foreign support for U.S. Cold War objectives, they virtually ignored civil rights demands at home and indeed harassed and prosecuted black American activists. Dudziak, "Desegregation as a Cold War Imperative"; Von Eschen, *Race against Empire: Black Americans and Anticolonialism, 1937–1957.*

Chapter 7

1. For a concise overview of modernism, especially vis-à-vis modernization, see Daniel Joseph Singal, "Toward a Definition of American Modernism."

2. In a recent dissertation, Doug Mischler forcefully contends that the American circus represented a popular (i.e. non-elite) form of modernism that preceded the explosion of modernist thought in "high" culture during the years surrounding World War I. This book, by point of comparison, analyzes the circus as a product of modernization; moreover, Mischler does not fully account for the continued (and contradictory) presence of Victorian rhetoric and ideology at the turn-of-the-century circus. See Mischler, "The Greatest Show on Earth."

3. On "a nation of loosely connected islands" see Wiebe, *The Search for Order, 1877–1920.*

4. Audiences continued to flock to the big railroad circus in the 1920s, even though its breadth had diminished. On September 13, 1924, at Concordia, Kansas, 16,702 people—the largest tented audience in American history—converged for the Ringling Bros. and Barnum & Bailey circus. Al Mann witnessed this gathering: "[I]n the rolling, grassy hills of Kansas . . . as far as a person could see, the hills were black because there were Model T Fords there. All Model T Fords were black." Al Mann interview by author.

5. Joy Schaleben Lewis, "It's the Greatest Parade on Earth," *St. Petersburg* (Fla.) *Times,* June 23, 1991, 1E.

6. Movies often looked to the circus for subject matter, as in Charlie Chaplin's *Circus* (1928), Tod Browning's *Freaks* (1932) and *Dumbo* (1941), and Cecil B. DeMille's *The Greatest Show on Earth* (1952). Built in the first three decades of the twentieth century, movie palaces contained a panoply of orientalist motifs—blending Hindu, Mughal, Persian, Arabic, Chinese, and Greco-Roman styles under one roof. Like the canvas city, these palaces created the illusion of compressing the world into one sparkling amusement where the world could be easily consumed "at a glance." The theater designer John Eberson noted that his typical movie hall held "a magnificent amphitheater under

a glorious moonlit sky . . . an Italian garden, a Persian court, a Spanish patio, or a mystic Egyptian temple-yard . . . where friendly stars twinkled and wisps of cloud drifted." Eberson quoted in Hall, *The Best Remaining Seats*, 102.

7. "Circus Days and Ways," interview with W. E. "Doc" Van Alstine, 4.

8. Many critics contend that the American circus declined (both aesthetically and financially) because of its size. In 1947 Ringling Bros. and Barnum & Bailey traveled on 108 railroad cars—the highest number ever. Although losing money in 1953, the circus still carried 74,000 yards of canvas and 72 miles of rope and employed 1,289 workers from 35 countries. The circus scholar Ernest Albrecht writes, "Instead of spotlighting individual achievement and creativity, the circus, at least in the United States, with its emphasis on spectacle, came to be seen as promoting conspicuous consumption." John Ringling North tried to combat the circus's creative torpor in 1939–40 by hiring the industrial designer Norman Bel Geddes to update the midway, to design sideshow banners in the style of contemporary poster art, and to construct a sleek cage housing Gargantua the Gorilla and his "wife." In November 1941 North hired George Balanchine to choreograph a ballet for elephants (clad in tutus) for the following show season. North also commissioner Igor Stravinsky to write the *Circus Polka* (1942) for the "Ballet of the Elephants." According to the bandleader Merle Evans, the *Circus Polka* was choppy and difficult to master: "Let's just say it was Harvard music and let it go at that." In the late twentieth century, critics of the three-ring circus argue that the American circus has been resuscitated only with the advent of new one-ring shows. Inspired by stationary, one-ring circuses in Europe and Russia, street entertainment and ancient *saltimbanci* (itinerant performers), these one-ring circuses are rooted in communities across the country: the Pickle Family circus in San Francisco, Circus Flora in St. Louis, the Big Apple Circus in New York City, and Cirque du Soleil in Montreal (and now also in Las Vegas). Albrecht writes that many of the owners were part of the counterculture of the 1960s and view the tightly knit circus community as an ideal society. See Truzzi, "The Decline of the American Circus: The Shrinkage of an Institution," 315–16; Hammarstrom, *Big Top Boss*, 76–77, 94, 193; Albrecht, *The New American Circus*, 4–7; Plowden, *Merle Evans*, 136–37; Taper, *Balanchine*, 177–78.

9. "'The Men Scrambled Away': Charging Elephants Scatter Pickets at Janesville Circus," *Wisconsin State Journal*, July 20, 1938, Newspaper Collection, RLPLRC.

10. "Ringling Circus Exec Dies at 83," *Capital Times*, Oct. 5, 1993, 6A.

11. "Big Top Bows Out Forever," *Life*, July 30, 1956, 13.

12. In 1963 the circus executive Chappie Fox and Ben Barkin, owner of the Schlitz brewery, founded the annual Great Circus Parade in Milwaukee. The parade ran each July until Schlitz dropped its sponsorship in 1973. In 1984 Barkin, armed with funding from the city of Milwaukee, private donors, and Circus World Museum, spearheaded the parade's return. It has reenacted this critical historical feature ever since. Ron Seely, "Remember When the Circus Wasn't History?" *Wisconsin State Journal*, July 8, 2001, 1E; Natasha Kassulke, "Borgnine under the Big Top," *Wisconsin State Journal*, July 21, 1996, 1G.

13. See Price, *Flight Maps*, "Looking for Nature at the Mall: A Field Guide to the Nature Company"; Davis, "Touch the Magic."

14. Advertisement for the Secret Garden of Siegfried and Roy, "Today in Las Vegas," visitors' guide, Las Vegas, Nev., Mar. 5–11, 1998.

15. Lisa Gubernick, "The Last of the Politically Incorrect Country Fairs," *Wall Street Journal*, July 24, 1998, W1 (citation courtesy of Shelley Fisher Fishkin); Bogdan, *Freak Show*, chapter 10.

16. "*Life* Goes to a Course in Circus," *Life*, Mar. 31, 1952, 110.

17. Joanne Gruber, "Barbie, Hear Her Roar," *New York*, Apr. 17, 1995, 34 (citation courtesy of Jeremy Solomon).

18. See Lopiano, "Women's Sports."

19. Waller, *The Bridges of Madison County*, 8; *Slow Waltz in Cedar Bend*, 178.

20. "Snake Shoots Man," *Wisconsin State Journal*, Apr. 24, 1990, 5A; see also "Drunken Elephants are Killers in India," *Wisconsin State Journal*, Dec. 25, 1991, 7A; "Elephant Alert!," *Time*, Apr. 17, 1995, 19; "Indian Animals Want Work," *Wisconsin State Journal*, Sept. 17, 1992, 11A.

21. "14-Cent Bounty Put on Rat Tails," *Wisconsin State Journal*, Oct. 17, 1991, 7A.

22. Robert Burns, "Monkey Man and Other Sightings," *Los Angeles Times*, June 7, 2001, Tech Times, pt. T, 3.

23. "The Late Show with David Letterman," on CBS, May 23, 2001.

24. Jackie Loohauis, "Freak Sideshow Has Its Dreary Side," *Milwaukee Journal Sentinel*, July 13, 2000, 4E.

25. For a colorful account of the Jim Rose Sideshow, see Gregor, *Circus of the Scars*, and for more on middle-class Victorian sideshow lovers, see Bogdan, *Freak Show*.

26. See Gamson, *Freaks Talk Back*.

27. "The World's Most Shocking Medical Videos," on FOX, Feb. 18, 1999.

28. "Guinness World Records Prime Time," on FOX, July 19, 2001.

29. David Haldane, "All-Human, No-Animal Big Tops Flying with Greatest of Ease," *Austin* (Tex.) *American-Statesman*, June 13, 1999, K1.

30. Kenneth Feld's late circus, Barnum's Kaleidoscape, featured cushy red velvet seats and collapsible hardwood floors, in addition to its intimate one-ring performances. Glenn Collins, "Under the Little Top, a Designer Circus," *New York Times*, Feb. 9, 1999, B1.

31. Ibid.

32. The Ringling Bros. and Barnum & Bailey three-ring circus, by contrast, charges from $10 to $45 and the smallest family circuses charge even less.

33. Paul Robeson High School Career Resource Center employment notice, 1993 (citation courtesy of Monica Drane).

34. Peg Meier, "A Show of Generosity," *Star Tribune*, Aug. 9, 1999, B1 (citation courtesy of Kathy Messerich).

35. According to Franzen's sister, Kathy Franzen: "He gave them respect and they gave him respect back." Kathy Franzen and Kim Franzen, interview by author, transcript, July 20, 1998, Merrill, Wis.

36. "48 Hours," on CBS, May 22, 1997.

37. Glenn Collins, "From the Boys Choir to the Big Top: Ringmaster Is Breaking New Ground," *New York Times*, Dec. 18, 1998, A27.

38. Susan Hawthorne, "Circus with a Purpose," *Ms.*, Jan.–Feb. 1998, 11.

39. Peggy Polk, "Pope Extols 'Human Virtues' of Circus," *Chicago Tribune*, Dec. 17, 1993, 8 (citation courtesy of Monica Drane).

40. Michael Gross, "The Princess and the Showman," *Talk*, Aug. 2001, 70.

41. "Visit to Gibsonton—Home for Retired Sideshow Freaks," Segment 5, "All Things Considered," National Public Radio, transcript, July 2, 1994.

42. Author visit to Showfolks Club, Sarasota, Fla., May 20, 1995.

43. Ringling Bros. and Barnum & Bailey Circus (Red Unit) performance, June 12, 1999, Erwin Center, Austin, Tex.

44. Sheila Hotchkin, "Historic Circus Parade Creates Its Own Memories," *Associated Press Wire*, July 9, 1999.

45. See, for example, Alison Green, "No Wonder Tigers Fight Back: Life under the Big Top Is No Circus," *Milwaukee Journal Sentinel*, June 1, 1997, 3J; "Killer Elephant," AP Photo of rampaging circus elephant being shot in Honolulu, *Wausau* (Wis.) *Daily Herald*, Aug. 22, 1994, 10A.

46. "Good Morning America," on ABC, Oct. 11, 1994.

47. "Ex-Wisconsin Teacher Gored to Death by Tiger in His Circus," *Wisconsin State Journal*, May 9, 1997, 4C; "Wisconsin Trainer Mauled to Death," *Saint Paul Pioneer Press*, May 9, 1997, 6C.

BIBLIOGRAPHY

Primary Sources

ARCHIVAL COLLECTIONS

Austin, Texas
University of Texas, Center for American History
    State and Regional Newspaper Collections
University of Texas, Harry Ransom Humanities Research Center
    J. Frank Dobie Collection
University of Texas, Todd-McLean Physical Culture Collection
    L. T. Lee Barnum & Bailey Scrapbook

Baraboo, Wisconsin
Circus World Museum, Robert L. Parkinson Library and Research Center
    Archival Collections
        Correspondence
        Press Afterblasts
        Work Contracts
    George L. Chindahl Collection
    Merle Evans Papers
    Circus Fiction Collection
    Film and Video Collection
    Gollmar Brothers Circus Collection
        Circus Reviews, 1890–1900
        Family Accounting Ledgers, 1902–12
    Don Howland Scrapbook Collection
    Manuscript Collection
    Newspaper Advertisement Collection
    Newspaper Collection
    Periodicals Collection
    Poster Collection
    Program Collection
    Published Circus Memoir Collection
    Frank Robie Collection
    Route Book Collection
    Small Collections Box
        Walter L. Main
        Ringling Brothers
    Michael Sporrer Collection
    Vertical Files

Cody, Wyoming
Buffalo Bill Historical Center, Harold McCracken Research Library
William F. Cody Collection
Correspondence
Peter H. Davidson Collection
Ephemera
Vincent Mercaldo Collection
Personnel
Programs

Madison, Wisconsin
State Historical Society of Wisconsin
Archives Division
Manuscript Collection
Library Division
Government Publications Section
State Laws and Statutes
Wisconsin Center for Film and Theatre Research
Circus Films

New York, New York
Library of the American Museum of Natural History
New York Public Library for the Performing Arts
Billy Rose Theatre Collection
Circus Scrapbook Collection
Circus Programs
Circus Route Books
Magazine Clippings

Princeton, New Jersey
Princeton University Library, Department of Rare Books & Special Collections,
Manuscripts Division
Joseph T. McCaddon Circus Collection
James A. Bailey Personal Diary, 1863
Circus Programs
Circus Scrapbooks
Correspondence
Route Books

San Antonio, Texas
Hertzberg Circus Collection and Museum
3A82 Sideshow Pamphlet Collection

Sarasota, Florida
John and Mable Ringling Museum of Art
Circus Collection
Circus Business Records, Barnum & Bailey
Cash Books, 1890–96, 1900–1902

Payroll, 1900–1902
Receipts, 1899–1902
Glasier Photograph Collection
Manuscript Collections
Black Collection
Marizles Wirth Diary
May Wirth Diary
Press Clippings Books
Barnum & Bailey, 1892, 1894–1916
Barnum, Cole, Hutchinson and Barnum and London, 1886
Barnum, Cole, Hutchinson and Cooper, 1887

AUTHOR INTERVIEWS

Al Mann and Irene Mann, Bagley, Wis., October 22, 1994
Kathy Franzen and Kim Franzen, Merrill, Wis., July 20, 1998

INTERNET SOURCES

American Life Histories: Manuscripts from the Federal Writers' Project, 1936–40, American Memory, <http://memory.loc.gov/ammem/wpaintro/wpahome.html>.
Born in Slavery: Slave Narratives from the Federal Writers' Project, 1936–38, American Memory, <http://memory.loc.gov/ammem/snhtml/snhome.html>.
Inflation Calculator, <http://www.westegg.com/inflation>.

PUBLISHED BOOKS, ARTICLES, AND PAMPHLETS

Adams, Henry. *The Education of Henry Adams.* Boston: Houghton Mifflin, 1961 [1918].
Addams, Jane. *The Spirit of Youth and the City Streets.* New York: Macmillan, 1912.
Alger, Horatio. *The Young Acrobat of the Great North American Circus.* New York: American, 1889.
———. *Ragged Dick; or, Street Life in New York.* New York: Penguin, 1986 [1868].
Anderson, Sherwood. *A Story Teller's Story.* Garden City, N.Y.: Garden City, 1924.
———. *Tar: A Midwest Childhood.* Cleveland: Press of Case Western Reserve University, 1969 [1926].
Bailey, Olga. *Mollie Bailey: The Circus Queen of the Southwest.* Dallas: Harben-Spotts, 1943.
Barker, Barbara M., ed. *Bolossy Kiralfy, Creator of Great Musical Spectacles: An Autobiography.* Ann Arbor: U.M.I. Research Press, 1988.
Barnum, P. T. *Dick Broadhead: A Tale of Perilous Adventure.* New York: Platt & Nourse, 1888.
———. *Life of P. T. Barnum: Written by Himself.* Buffalo: Courier, 1888.
———. *Jack in the Jungle.* New York: Platt & Nourse, 1880.
———. *Lion Jack.* New York: G. W. Carleton, 1876.
Beal, George Brinton. *Through the Backdoor of the Circus.* Springfield, Mass.: McLoughlin Bros., 1938.

Bostock, Frank C. *The Training of Wild Animals.* New York: Century, 1903.

Bradna, Fred, as told to Hartzell Spence. *The Big Top: My Forty Years with the Greatest Show on Earth.* New York: Simon & Schuster, 1952.

Brown, Gordon. "An Interview with Tom Barron, 'World's Tallest Clown.'" *Bandwagon: The Journal of the Circus Historical Society* 40 (Nov.–Dec. 1996): 12–25.

Channell, Herbert Walter, and Velma E. Lowry. *Fifty Years under Canvas.* Hugo, Okla.: Acme, 1962.

Clausen, Connie. *I Love You Honey, but the Season's Over.* N.p. 1961.

Conklin, George. *The Ways of the Circus, Being the Memories and Adventures of George Conklin, Tamer of Lions.* New York: Harper & Bros., 1921.

Cooper, Courtney Riley. *Lions 'N' Tigers 'N' Everything.* Boston: Little, Brown, 1930 [1922].

Coup, W. C. *Sawdust and Spangles: Stories and Secrets of the Circus.* Chicago: Herbert S. Stone, 1901.

Court, Alfred. *Wild Circus Animals.* London: Burke, 1954.

Darlington, Edgar B. P. *The Circus Boys in Dixie Land.* Akron: Saalfield, 1911.

Day, Charles H. *Ink from a Circus Press Agent.* Ed. William L. Slout. San Bernardino, Calif.: Borgo, 1995.

Dickinson, Emily. "Poems, Second Series, XVIII." *Collected Poems.* Philadelphia: Courage, 1991.

Dixon, Thomas. *The Leopard's Spots: A Romance of the White Man's Burden, 1865–1900.* Ridgewood, N.J., 1967 [1902].

———. *The Clansman: An Historical Romance of the Ku Klux Klan.* New York: Grosset & Dunlap, 1905.

Du Bois, W. E. B. *The Souls of Black Folk.* New York: Signet, 1995 [1903].

Federal Writers' Project. *Wisconsin Circus Lore.* 2 vols. Madison: Works Progress Administration, 1937.

Fellows, Dexter W., and Andrew A. Freeman. *This Way to the Big Show: The Life of Dexter Fellows.* New York: Viking, 1936.

Forster, Nicholas. *Greatest Wonder in the World: History and Life of Jo-Jo the Dog-Faced Man.* New York: New York Popular, 1885.

Fowler, O. S. "Tight-Lacing, or the Evils of Compressing the Organs of Animal Life." In *Who was the Commander at Bunker Hill?,* ed. S. S. Swett, 1–16. Boston: John Wilson, 1850 [1844].

Garland, Hamlin. *A Son of the Middle Border.* New York: Penguin, 1995 [1923].

Gollmar, Robert H. *My Father Owned a Circus.* Caldwell, Idaho: Caxton, 1965.

Grant, Madison. *The Passing of the Great Race; or, the Racial Basis of European History.* New York: Charles Scribner's Sons, 1916.

Greenwood, Isaac J. *The Circus, Its Origin and Growth prior to 1835.* New York: Dunlap Society, 1898.

Hagenbeck, Carl. *Beasts and Men.* London: Longmans, Green, 1909.

Hagenbeck, Lorenz. *Animals Are My Life.* Trans. Alec Brown. London: Bodley Head, 1956.

Hall, G. Stanley. *Youth: Its Education, Regime and Hygiene.* New York: D. Appleton, 1908.

———. *Adolescence.* 2 vols. New York: D. Appleton, 1905.

Harkness, Peter T. *Andy the Acrobat; or, Out with the Greatest Show on Earth.* Cleveland: World Syndicate, 1907.

"Hindoo Devotees." *Monthly Paper of the American Board of Commissioners for Foreign Missions* 12 (June 1833): 45–49.

Hoagland, Edward. *Balancing Acts.* New York: Simon and Schuster, 1992.

Hornaday, William T. *The Minds and Manners of Wild Animals.* New York: Charles Scribner's Sons, 1922.

———. *Tales from Nature's Wonderland.* New York: Charles Scribner's Sons, 1924.

Howells, William Dean. *Literature and Life.* New York: Harper & Bros., 1902.

Jennings, John J. *Theatrical and Circus Life.* Philadelphia: Scammell, 1882.

Johnson, Osa. *I Married Adventure: The Lives and Adventures of Martin and Osa Johnson.* Garden City, N.Y.: Halcyon House, 1940.

Joint Committee on Ceremonies of the World's Columbian Commission and the World's Columbian Exposition. *Memorial Volume, Dedicatory and Opening Ceremonies of the World's Columbian Exposition.* Chicago: Stone, Kastler and Painter, 1893.

Kelly, Emmett, with F. Beverly Kelly. *Clown.* New York: Prentice-Hall, 1954.

Knight, Clifford. *The Affair of the Circus Queen.* New York: Dodd, Mead, 1940.

Kraut, John Allen. *Annals of American Sport.* Pageant of America Series, no. 15. New Haven: Yale University Press, 1929.

London, Jack. *The Call of the Wild.* New York: Macmillan, 1966 [1903].

Macfadden, Bernarr A. *The Virile Powers of Superb Manhood: How Developed, How Lost, How Gained.* New York: Physical Culture, 1900.

Mahan, Alfred Thayer. *The Influence of Sea Power upon History.* New York: Henry Holt, 1889.

Mann, Al. *My Home Was the Open Range of Wyoming.* Dodgeville, Wis.: Dodgeville Chronicle, 1993.

May, Earl Chapin. *Cuddy of the White Tops.* New York: D. Appleton, 1924.

Mayer, Charles. *Trapping Wild Animals in Malay Jungles.* Garden City, N.Y.: Garden City, 1920.

Mayo, Katherine. *Mother India.* New York: Blue Ribbon, 1927.

Melville, Herman. *Typee: A Peep at Polynesian Life.* New York: Penguin, 1986 [1846].

Moffett, Cleveland. *Careers of Danger and Daring.* New York: Century, 1901.

Morse, George. *Circus Dan.* Chicago: Goldsmith, 1933.

Neihardt, John G. *Black Elk Speaks: Being the Life Story of a Holy Man of the Oglala Sioux.* Lincoln: University of Nebraska Press, 1961 [1932].

Norwood, Edwin P. *The Other Side of the Circus.* Garden City, N.Y.: Doubleday, Doran, 1938.

———. *The Circus Menagerie: True Stories of Interesting Animals Told to a Boy.* Garden City, N.Y.: Doubleday, Doran, 1929.

Oman, J. C. *The Mystics, Ascetics and Saints of India.* London: T. Fisher Unwin, 1905.

Otis, James. *Toby Tyler; or, Ten Weeks with a Circus.* New York: Harper & Bros., 1903.

———. *The Wreck of the Circus.* New York: Thomas Y. Crowell, 1897.

Parry, Albert. *Tattoo: Secrets of a Strange Art as Practised among the Natives of the United States.* New York: Simon & Schuster, 1933.

Peck, George W. *Peck's Bad Boy with the Circus.* Chicago: Stanton & Van Vleet, 1905.

Ringling, Alfred. *Life Story of the Ringling Brothers*. Chicago: R. R. Donnelley, 1900.

"Ringling Bros. Help Wisconsin: Wives Are Members of Suffrage Society—Allow Campaigning on Circus Grounds." *Women's Journal*, July 13, 1912, 219.

Robinson, Josephine DeMott. *The Circus Lady*. New York: Thomas Y. Crowell, 1925.

Roosevelt, Theodore. *The Strenuous Life: Essays and Addresses*. New York: Century, 1902.

Root, Harvey W. *Tommy with the Big Tents*. New York: Harper & Bros., 1924.

Sandburg, Carl. *Always the Young Strangers*. New York: Harcourt, Brace, 1952.

———. *Good Morning, America*. New York: Harcourt, Brace, 1928.

Slout, William L., ed. *An Annotated Narrative of Joe Blackburn's A Clown's Log (as Compiled by Charles H. Day)*. San Bernardino, Calif.: Borgo, 1993.

Smith, Eleanor. *Satan's Circus*. Indianapolis: Bobbs-Merrill, 1931.

Snider, Denton. *World's Fair Studies*. Chicago: Sigma, 1895.

Standing Bear, Luther. *My People the Sioux*. Boston: Houghton Mifflin, 1928.

Stark, Mabel, as told to Gertrude Orr. *Hold That Tiger*. Caldwell, Idaho: Caxton, 1938.

Taylor, Frederick Winslow. *Scientific Management*. New York: Harper & Bros., 1947 [1939].

Thoreau, Henry David. *A Year in Thoreau's Journal, 1851*. New York: Penguin, 1993 [1906].

Turnour, Jules. *The Autobiography of a Clown as Told to Isaac F. Marcosson*. New York: Moffat, Yard, 1910.

Twain, Mark. *The Adventures of Huckleberry Finn* (1884). Repr. in *The Family Mark Twain*, 441–650. New York: Harper & Bros., 1935.

———. "To the Person Sitting in Darkness." *North American Review* (1901). Repr. in *The Family Mark Twain*. 1387–1403.

Tyng, Charles. *Before the Wind: The Memoir of an American Sea Captain*. New York: Penguin, 1999.

Waller, Robert James. *The Bridges of Madison County*. New York: Warner, 1992.

———. *Slow Waltz in Cedar Bend*. New York: Warner, 1993.

Webber, Malcolm. *Medicine Show*. Caldwell, Idaho: Caxton, 1941.

Wetmore, Helen Cody, and Zane Grey. *Last of the Great Scouts ("Buffalo Bill")*. New York: Grosset & Dunlap, 1899.

Wilder, Laura Ingalls. *By the Shores of Silver Lake*. New York: HarperCollins, 1971 [1939].

Willson, Dixie. *Where the World Folds Up at Night*. New York: D. Appleton, 1932.

Zora, Lucia. *Sawdust and Spangles*. Boston: Little, Brown, 1928.

Secondary Sources

PUBLISHED BOOKS AND ARTICLES

Adams, Bluford. *E Pluribus Barnum: The Great Showman and the Making of U.S. Popular Culture*. Minneapolis: University of Minnesota Press, 1997.

Albrecht, Ernest. *The New American Circus*. Gainesville: University Press of Florida, 1995.

Allen, Robert C. *Horrible Prettiness: Burlesque and American Culture.* Chapel Hill: University of North Carolina Press, 1991.

Alloula, Malek. *The Colonial Harem.* Trans. Myrna and Wlad Godzich. Minneapolis: University of Minnesota Press, 1986.

Anderson, Stuart. *Race and Rapprochement: Anglo-Saxonism and Anglo-American Relations, 1895–1904.* London: Associated University Presses, 1981.

Bain, David Haward. *Empire Express: Building the First Transcontinental Railroad.* New York: Viking, 1999.

———. *Sitting in Darkness: Americans in the Philippines.* New York: Penguin, 1984.

Bakhtin, Mikhail. *Rabelais and His World.* Trans. Hélène Iswolsky. Bloomington: Indiana University Press, 1984 [1968].

Baratay, Eric, and Elisabeth Hardouin-Fugier. *Zoos: Histoire des Jardins Zoologiques en Occident.* Paris: Découverte, 1998.

Bederman, Gail. *Manliness and Civilization: A Cultural History of Gender and Race in the United States, 1880–1917.* Chicago: University of Chicago Press, 1995.

Beisel, Nicola. *Imperiled Innocents: Anthony Comstock and Family Reproduction in Victorian America.* Princeton, N.J.: Princeton University Press, 1997.

Benedict, Burton, et al. *The Anthropology of World's Fairs: San Francisco's Panama Pacific International Exposition of 1915.* Berkeley: Scolar, 1983.

Berger, Peter L., and Thomas Luckmann. *The Social Construction of Reality: A Treatise in the Sociology of Knowledge.* New York: Anchor, 1967.

Blackstone, Sarah J. *Buckskins, Bullets, and Business: A History of Buffalo Bill's Wild West.* New York: Greenwood, 1986.

Bogdan, Robert. *Freak Show: Presenting Human Oddities for Amusement and Profit.* Chicago: University of Chicago Press, 1988.

Boorstin, Daniel. *The Image: A Guide to Pseudo-Events in America.* New York: Atheneum, 1987.

Bouissac, Paul. *Circus and Culture: A Semiotic Approach.* Bloomington: Indiana University Press, 1976.

Boyer, Paul. *Urban Masses and Moral Order in America, 1820–1920.* Cambridge: Harvard University Press, 1978.

Bradbury, Joseph T. "The Fred Buchanan Railroad Circuses, 1923–1931, Robbins Bros. Circus, Part Nine— 1931 Season." *Bandwagon* 27 (Sept.–Oct. 1983): 12–26.

———. "Robbins Bros. Circus, Season 1931." *White Tops* 31 (Nov.–Dec. 1958): 3–19.

Bradford, Phillips Verner, and Harvey Blume. *Ota, the Pygmy in the Zoo.* New York: St. Martin's, 1992.

Brumberg, Joan Jacobs. *The Body Project: A Social History of American Girls.* New York: Vintage, 1997.

Butler, Judith. *Bodies that Matter: On the Discursive Limits of "Sex."* New York: Routledge, 1993.

———. *Gender Trouble: Feminism and the Subversion of Identity.* New York: Routledge, 1990.

Butsch, Richard. *The Making of American Audiences: From Stage to Television, 1750–1990.* New York: Cambridge University Press, 2000.

————, ed. *For Fun and Profit: The Transformation of Leisure into Consumption.* Philadelphia: Temple University Press, 1990.

Cahn, Susan K. *Coming on Strong: Gender and Sexuality in Twentieth-Century Women's Sport.* New York: Free Press, 1994.

Campbell, James T. "The Americanization of South Africa." In *"Here, There and Everywhere": The Foreign Politics of American Popular Culture,* ed. Elaine Tyler May and Reinhold Wagnleitner, 34–63. Hanover, N.H.: University Press of New England, 2000.

Carmeli, Yoram C. "The Invention of Bourgeois Hegemony: A Glance at British Circus Books." *Journal of Popular Culture* 29, no. 1 (Summer 1995): 213–21.

Chandler, Alfred Dupont, ed. *The Railroads, the Nation's First Big Business: Sources and Readings.* New York: Harcourt, 1965.

Chauncey, George. *Gay New York: Gender, Urban Culture, and the Making of the Gay Male World, 1890–1940.* New York: Basic, 1994.

Chindahl, George. *A History of the Circus in America.* Caldwell, Idaho: Caxton, 1959.

Clifford, James. *The Predicament of Culture: Twentieth-Century Ethnography, Literature, and Art.* Cambridge: Harvard University Press, 1988.

Cohen, Lizbeth. *Making a New Deal: Industrial Workers in Chicago, 1919–1939.* New York: Cambridge University Press, 1990.

Conover, Richard E. "The Affairs of James A. Bailey: New Revelations on the Career of the World's Most Successful Showman." Xenia, Ohio: Self-published, 1957.

Cook, James W. *The Arts of Deception: Playing with Fraud in the Age of Barnum.* Cambridge: Harvard University Press, 2001.

————. "Mass Marketing and Cultural History: The Case of P. T. Barnum." *American Quarterly* 51 (Mar. 1999): 175–86.

————. "Of Men, Missing Links, and Nondescripts: The Strange Career of P. T. Barnum's 'What Is It'? Exhibition." In *Freakery: Cultural Spectacles of the Extraordinary Body,* ed. Rosemarie Garland Thomson, 139–57. New York: New York University Press, 1996.

Cooper, Diana Starr. *Night After Night.* Washington: Island, 1994.

Cott, Nancy, F. *The Grounding of Modern Feminism.* New Haven: Yale University Press, 1987.

Couvares, Francis. *The Remaking of Pittsburgh: Class and Culture in an Industrializing City, 1877–1919.* Albany: State University of New York Press, 1984.

Cross, Gary. *Kids' Stuff: Toys and the Changing World of American Childhood.* Cambridge: Harvard University Press, 1997.

Crunden, Robert M. *American Salons: Encounters with European Modernism, 1885–1917.* New York: Oxford University Press, 1993.

Cumings, Bruce. *Korea's Place in the Sun: A Modern History.* New York: W. W. Norton, 1997.

Dahlinger, Fred, Jr. *Show Trains of the Twentieth Century.* Baraboo, Wis.: Circus World Museum, 2000.

————. *Trains of the Circus 1872–1956.* Baraboo, Wis.: Circus World Museum, 2000.

————. "The Development of the Railroad Circus." Four parts, *Bandwagon* 27, no. 6

(Nov.–Dec. 1983): 6–11; 28, no. 1 (Jan.–Feb. 1984): 16–27; 28, no. 2 (Mar.–Apr. 1984): 28–36; 28, no. 3 (May–June 1984): 29–36.

Dahlinger, Fred, Jr., and Stuart Thayer. *Badger State Showmen: A History of Wisconsin's Circus Heritage.* Baraboo, Wis.: Circus World Museum, 1998.

Davis, Janet. "Spectacles of South Asia at the American Circus, 1890–1940." *Visual Anthropology* 6 (1993): 121–38.

Davis, Susan G. "Touch the Magic." In *Uncommon Ground: Rethinking the Human Place in Nature,* ed. William Cronon, 204–17. New York: W. W. Norton, 1995.

———. *Parades and Power, Street Theatre in Nineteenth-Century Philadelphia.* Berkeley: University of California Press, 1986.

Davis, Tracy C. "Shotgun Wedlock: Annie Oakley's Power Politics in the Wild West." In *Gender in Performance: The Presentation of Difference in the Performing Arts,* ed. Laurence Senelick, 141–57. Hanover, N.H.: Tufts University Press, 1992.

Deloria, Philip J. *Playing Indian.* New Haven: Yale University Press, 1998.

D'Emilio, John, and Estelle Freedman. *Intimate Matters: A History of Sexuality in America.* New York: Harper & Row, 1988.

Denning, Michael. "The End of Mass Culture." *International Labor and Working Class History* 37 (Spring 1990): 4–18.

———. *Mechanic Accents: Dime Novels and Working Class Culture in America.* New York: Verso, 1987.

Derks, Scott, ed. *The Value of a Dollar: Prices and Incomes in the United States, 1860–1989.* Detroit: Gale Research, 1994.

Desmond, Jane. "Dancing Out the Difference: Cultural Imperialism and Ruth St. Denis's 'Radha' of 1906." *Signs: Journal of Women in Culture and Society* 17 (Autumn 1991): 28–49.

Doughty, Robin W. *Feather Fashions and Bird Preservation: A Study in Nature Protection.* Berkeley: University of California Press, 1975.

Douglas, Ann. *The Feminization of American Culture.* New York: Alfred A. Knopf, 1977.

Dower, John W. *War without Mercy: Race and Power in the Pacific War.* New York: Pantheon, 1986.

Drinnon, Richard. *Facing West: The Metaphysics of Indian-Hating and Empire-Building.* Rev. ed. Norman: University of Oklahoma Press, 1997.

Dubuque, Barry L. "The Original Miss Daisy." *Bandwagon* 23 (Nov.–Dec. 1979): 26–28.

Dudziak, Mary. "Desegregation as a Cold War Imperative." *Stanford Law Review* 41 (Nov. 1988): 61–120.

Eagleton, Terry. *Walter Benjamin: Towards a Revolutionary Criticism.* London: Verso, 1981.

Enloe, Cynthia. *Bananas, Beaches and Bases: Making Feminist Sense of International Politics.* Berkeley: University of California Press, 1990.

Erenberg, Lewis. *Steppin' Out: New York Nightlife and the Transformation of American Culture, 1890–1930.* Westport, Conn.: Greenwood, 1981.

Evans, Sara, M. *Born for Liberty: A History of Women in America.* New York: Free Press, 1989.

Ewen, Elizabeth. *Immigrant Women in the Land of Dollars: Life and Culture on the Lower East Side, 1890–1925.* New York: Monthly Review, 1985.

Ewen, Stuart. *Captains of Consciousness: Advertising and the Social Roots of the Consumer Culture.* New York: McGraw-Hill, 1976.

Fiedler, Leslie. *Freaks: Myths and Images of the Secret Self.* New York: Simon & Schuster, 1978.

Fields, Barbara J. "Ideology and Race in American History." In *Region, Race, and Reconstruction*, ed. J. Morgan Kousser and James M. McPherson, 143–77. New York: Oxford University Press, 1982.

Foner, Eric. *The Story of American Freedom.* New York: W. W. Norton, 1998.

Flink, James J. *The Automobile Age.* Cambridge.: MIT Press, 1988.

Fonseca, Isabel. *Bury Me Standing: The Gypsies and Their Journey.* New York: Alfred A. Knopf, 1995.

Foucault, Michel. *The History of Sexuality.* Vol. 1, *An Introduction.* Trans. Robert Hurley. New York: Vintage, 1990 [1978].

———. *Discipline and Punish: The Birth of the Prison.* Trans. Alan Sheridan. New York: Vintage, 1995 [1977].

———. *The Archeology of Knowledge.* Trans. A. M. Sheridan Smith. New York: Pantheon, 1972.

Fox, Charles Philip. *A Ticket to the Circus: A Pictorial History of the Incredible Ringlings.* Seattle: Superior, 1959.

Fox, Richard, and T. J. Jackson Lears, eds. *The Culture of Consumption: Critical Essays in American History, 1880–1980.* New York: Pantheon, 1983.

Frediani, Lorenzo. "The Fredianis, for the Early Years to America, 1866–1908." *Bandwagon* 41 (Mar.–Apr. 1997): 34–37.

Fusco, Coco. *English Is Broken Here: Notes on Cultural Fusion in the Americas.* New York: New Press, 1995.

Gamson, Joshua. *Freaks Talk Back: Tabloid Talk Shows and Sexual Nonconformity.* Chicago: University of Chicago Press, 1998.

Garrow, David J. *Liberty and Sexuality: The Right to Privacy and the Making of Roe v Wade.* New York: Macmillan, 1994.

Gilman, Sander L. "Black Bodies, White Bodies: Toward an Iconography of Female Sexuality in Late Nineteenth-Century Art, Medicine, and Literature." In *"Race," Writing, and Difference*, ed. Henry Louis Gates Jr., 223–61. Chicago: University of Chicago Press, 1986.

Glenn, Susan A. "'Give an Imitation of Me': Vaudeville Mimics and the Play of the Self." *American Quarterly* 50 (Mar. 1998): 47–76.

Goetzmann, William H., and William N. Goetzmann. *The West of the Imagination.* New York: W. W. Norton, 1986.

Gossard, Steve. "A Reckless Era of Aerial Performance; the Evolution of Trapeze." 2d ed. Normal, Ill.: Self-published, 1994.

Gosse, Van. *Where the Boys Are: Cuba, Cold War America and the Making of the New Left.* New York: Verso, 1993.

Gramsci, Antonio. *Selections from the Prison Notebooks.* Ed. and trans. Quintin Hoare and Geoffrey Nowell-Smith. New York: International, 1971.

Green, Harvey. *Fit for America: Health, Fitness, Sport and American Society.* Baltimore: Johns Hopkins University Press, 1986.

Green, James. *The World of the Worker: Labor in Twentieth-Century America*. New York: Hill & Wang, 1980.

Gregor, Jan T., with Tim Cridland. *Circus of the Scars: The True Inside Odyssey of a Modern Circus Sideshow*. Seattle: Brennan Dalsgard, 1998.

Grosz, Elizabeth. *Volatile Bodies: Toward a Corporeal Feminism*. Bloomington: Indiana University Press, 1994.

Guttmann, Allen. *Women's Sports: A History*. New York: Columbia University Press, 1991.

Haiken, Elizabeth. *Venus Envy: A History of Cosmetic Surgery*. Baltimore: Johns Hopkins University Press, 1997.

Hall, Ben M. *The Best Remaining Seats: The Story of the Golden Age of the Movie Palace*. New York: Clarkson N. Potter, 1961.

Hall, Stuart. "Notes on Deconstructing 'the Popular.'" In *People's History and Socialist Theory*, ed. Raphael Samuel, 227–40. London: Routledge & Kegan Paul, 1981.

Hamaliam, Leo, ed. *Ladies on the Loose: Women Travellers of the 18th and 19th Centuries*. New York: Dodd, Mead, 1981.

Hammarstrom, David Lewis. *Big Top Boss: John Ringling North and the Circus*. Urbana: University of Illinois Press, 1992.

Haraway, Donna. *Primate Visions: Gender, Race, and Nature in the World of Modern Science*. New York: Routledge, 1989.

Harris, Michelle. "Whiteness as Property." *Harvard Law Review* 106 (June 1993): 1707–91.

Harris, Neil. "All the World a Melting Pot? Japan at American Fairs, 1876–1904." In *Mutual Images: Essays in American-Japanese Relations*, ed. Akira Iriye, 24–54. Cambridge: Harvard University Press, 1975.

———. *Humbug: The Art of P. T. Barnum*. Boston: Little, Brown, 1973.

Higginbotham, Evelyn Brooks. "African-American Women's History and the Metalanguage of Race." *Signs* 17 (Winter 1992): 251–74.

Hoage, R. J., and William A. Deiss. *New Worlds, New Animals: From Menagerie to Zoological Park in the Nineteenth Century*. Baltimore: Johns Hopkins University Press, 1996.

Hobsbawm, Eric. "Introduction: Inventing Traditions." In *The Invention of Tradition*, ed. Hobsbawm and Terence Ranger, 1–14. Cambridge: Cambridge University Press, 1983.

Hofstadter, Richard. *The Age of Reform: From Bryan to F. D. R.* New York: Vintage, 1955.

Hoganson, Kristin L. *Fighting for Manhood: How Gender Politics Provoked the Spanish-American and Philippine-American Wars*. New Haven: Yale University Press, 1998.

Horowitz, Helen L. "Seeing Ourselves through Bars." *Landscape* 25, no. 2 (1981): 12–19.

Horseman, Reginald. *Race and Manifest Destiny: The Origins of American Racial Anglo-Saxonism*. Cambridge: Harvard University Press, 1981.

Hoxie, Frederick E. *A Final Promise: The Campaign to Assimilate the Indians, 1880–1920*. Cambridge: Cambridge University Press, 1984.

Hughes, Langston, and Milton Meltzer. *Black Magic: A Pictorial History of the Negro in American Entertainment.* Englewood Cliffs, N.J.: Prentice-Hall, 1967.

Hunt, Michael H. *Ideology and U.S. Foreign Policy.* New Haven: Yale University Press, 1987.

Hunter, Jane. *The Gospel of Gentility: American Women Missionaries in Turn-of-the-Century China.* New Haven: Yale University Press, 1984.

Inden, Ronald. *Imagining India.* Cambridge, England: Basil Blackwell, 1990.

Irey, Elmer L., as told to William J. Slocum. *The Tax Dodgers: The Inside Story of the T-Men's War with America's Underworld Hoodlums.* New York: Greenberg, 1948.

Iriye, Akira. "Culture and Power: International Relations as Intercultural Relations." *Diplomatic History* 3 (Spring 1979): 115–28.

———. "Culture (A Round Table: Explaining the History of American Foreign Relations)." *Journal of American History* 77 (June 1990): 99–107.

Isaac, Rhys. *The Transformation of Virginia, 1740–1790.* New York: W. W. Norton, 1988 [1982].

Isenberg, Andrew C. *The Destruction of the Bison: An Environmental History, 1750–1920.* New York: Cambridge University Press, 2000.

Israel, Jerry. *Progressivism and the Open Door: America and China, 1905–1921.* Pittsburgh: University of Pittsburgh Press, 1971.

Jacobson, Matthew Frye. *Whiteness of a Different Color: European Immigrants and the Alchemy of Race.* Cambridge: Harvard University Press, 1998.

Jameson, Fredric. "Reification and Utopia in Mass Culture." Repr. in Jameson, *Signatures of the Visible,* 9–34. New York: Routledge, 1992.

Jay, Martin. *The Dialectical Imagination: The Frankfurt School and the Institute of Social Research.* Boston: Little, Brown, 1973.

Jewell, K. Sue. *From Mammy to Miss America and Beyond: Cultural Images and the Shaping of U.S. Social Policy.* New York: Routledge, 1993.

Johannsen, Robert W. *To the Halls of the Montezumas: The Mexican War in the American Imagination.* New York: Oxford University Press, 1985.

Johnson, Paul E. *A Shopkeeper's Millennium: Society and Revivals in Rochester, New York, 1815–1837.* New York: Hill & Wang, 1978.

Kammen, Michael G. *American Culture, American Tastes: Social Change and the Twentieth Century.* New York: Alfred A. Knopf, 1999.

Kanigel, Robert. *The One Best Way: Frederick Winslow Taylor and the Enigma of Efficiency.* New York: Viking, 1997.

Kantrowitz, Steven. *Ben Tillman and the Reconstruction of White Supremacy.* Chapel Hill: University of North Carolina Press, 2000.

Kaplan, Amy. "Black and Blue on San Juan Hill." In *Cultures of United States Imperialism,* ed. Kaplan and Donald E. Pease, 219–36. Durham: Duke University Press, 1993.

Kaplan, E. Ann. "Is the Gaze Male?" In *Powers of Desire: The Politics of Sexuality,* ed. Ann Snitow, Christine Stansell, and Sharon Thompson, 309–27. New York: Monthly Review, 1983.

Kasson, John F. *Amusing the Million: Coney Island at the Turn of the Century.* New York: Hill & Wang, 1978.

Kasson, Joy S. *Buffalo Bill's Wild West: Celebrity, Memory, and Popular History.* New York: Hill & Wang, 2000.

Kelley, Robin D. G. *Race Rebels: Culture, Politics and the Black Working Class.* New York: Free Press, 1994.

———. "Notes on Deconstructing 'The Folk' (AHR Forum)." *American Historical Review* 97 (Dec. 1992): 1400–1408.

Kerber, Linda K. *Women of the Republic: Intellect and Ideology in Revolutionary America.* New York: W. W. Norton, 1986.

Kevles, Daniel J. *In the Name of Eugenics: Genetics and the Uses of Human Heredity.* New York: Alfred A. Knopf, 1985.

Kibler, M. Alison. *Rank Ladies: Gender and Cultural Hierarchy in American Vaudeville.* Chapel Hill: University of North Carolina Press, 1999.

King, Orrin Copple. "Only Big Show Coming: Prodigal Profusion of Princely Paraphernalia." Vol. 1, chap. 2, pt. 1. *Bandwagon* 38 (Mar.–Apr. 1994): 33–41.

Korall, Burt. *Drummin' Men.* New York: Schirmer, 1990.

Kunhard, Philip B., Jr., Philip B. Kunhardt III, and Peter W. Kunhardt. *P. T. Barnum: America's Greatest Showman.* New York: Alfred A. Knopf, 1995.

LaFeber, Walter. *Michael Jordan and the New Global Capitalism.* New York: W. W. Norton, 1999.

———. *The American Age: United States Foreign Policy at Home and Abroad since 1750.* New York: W. W. Norton, 1989.

———. *The New Empire: An Interpretation of American Expansion, 1860–1898.* Ithaca, N.Y.: Cornell University Press, 1963.

Lamoreaux, Naomi R. *The Great Merger Movement in American Business, 1895–1904.* New York: Cambridge University Press, 1985.

Lasch, Christopher. "The Anti-Imperialists, the Philippines, and the Inequality of Man." *Journal of Southern History* 25 (Aug. 1958): 319–31.

Leach, William. *Land of Desire: Merchants, Power, and the Rise of a New American Culture.* New York: Vintage, 1993.

Lears, T. J. Jackson. "Making Fun of Popular Culture, (AHR Forum)." *American Historical Review* 97 (Dec. 1992): 1417–26.

———. "The Concept of Cultural Hegemony: Problems and Possibilities." *American Historical Review* 90 (June 1985): 567–93.

———. *No Place of Grace: Antimodernism and the Transformation of American Culture, 1880–1920.* New York: Pantheon, 1981.

Levine, Lawrence W. "The Folklore of Industrial Society (AHR Forum)." *American Historical Review* 97 (Dec. 1992): 1369–99.

———. *Highbrow/Lowbrow: The Emergence of Cultural Hierarchy in America.* Cambridge: Harvard University Press, 1988.

Lewis, David Levering. *W. E. B. Du Bois: Biography of a Race, 1868–1919.* New York: Henry Holt, 1993.

Lindfors, Bernth. "Hottentot, Bushman, Kaffir: Taxonomic Tendencies in Nineteenth-Century Racial Iconography." *Nordic Journal of African Studies* 5, no. 5 (1996): 1–30.

———. "Ethnological Show Business: Footlighting the Dark Continent." In *Freakery:*

Cultural Spectacles of the Extraordinary Body, ed. Rosemarie Garland Thomson, 207–18. New York: New York University Press, 1996.

———. "Circus Africans." *Journal of American Culture* 6, no. 2 (1983): 9–14.

Lipsitz, George. *Time Passages: Collective Memory and American Popular Culture.* Minneapolis: University of Minnesota Press, 1990.

Lopiano, Donna A. "Women's Sports: Coming of Age in the Third Millennium." In *The Olympics at the Millennium: Power, Politics, and the Games*, ed. Kay Schaffer and Sidonie Smith, 117–27. New Brunswick: Rutgers University Press, 2000.

Lott, Eric. *Love and Theft: Blackface Minstrelsy and the American Working Class.* New York: Oxford University Press, 1995.

Lutz, Catherine A., and Jane L. Collins. *Reading National Geographic.* Chicago: University of Chicago Press, 1993.

MacKenzie, John M., ed. *Imperialism and the Natural World.* Manchester: Manchester University Press, 1990.

Mannheim, Karl. *Ideology and Utopia: An Introduction to the Sociology of Knowledge.* New York: Harcourt, Brace & World, 1936.

May, Earl Chapin. *The Circus from Rome to Ringling.* New York: Dover, 1963 [1932].

McClintock, Anne. *Imperial Leather: Race, Gender and Sexuality in the Colonial Contest.* New York: Routledge, 1995.

McCormick, Thomas J. *The China Market.* Chicago: University of Chicago Press, 1967.

McEnaney, Laura. "He-Men and Christian Mothers: The America First Movement and the Gendered Meanings of Patriotism and Isolationism." *Diplomatic History* 18 (Winter 1994): 47–57.

Michaels, Walter Benn. "Anti-Imperial Americanism." In *Cultures of United States Imperialism*, ed. Amy Kaplan and Donald E. Pease, 365–91. Durham: Duke University Press, 1993.

Mihara, Aya, and Stuart Thayer. "Richard Risley Carlisle: Man in Motion." *Bandwagon* 12 (Jan.–Feb. 1997): 12–14.

Miller, Stuart Creighton. *"Benevolent Assimilation": The American Conquest of the Philippines, 1899–1903.* New Haven: Yale University Press, 1982.

Morgan, Edmund S. *American Slavery American Freedom: The Ordeal of Colonial Virginia.* New York: W. W. Norton, 1975.

Morris, Edmund. *The Rise of Theodore Roosevelt.* New York: Ballantine, 1979.

Moses, L. G. *Wild West Shows and the Images of American Indians, 1883–1933.* Albuquerque: University of New Mexico Press, 1996.

Mosse, George L. *The Image of Man: The Creation of Modern Masculinity.* Oxford: Oxford University Press, 1996.

———. *Toward the Final Solution: A History of European Racism.* New York: Howard Fertig, 1978.

Mukerji, Chandra, and Michael Schudson, eds. *Rethinking Popular Culture: Contemporary Perspectives in Cultural Studies.* Berkeley: University of California Press, 1991.

Mulvey, Laura. *Fetishism and Curiosity.* Bloomington: Indiana University Press, 1996.

———. "Visual Pleasure and Narrative Cinema." *Screen* 16 (Autumn 1975): 6–18.

Musser, Charles. *The Emergence of Cinema: The American Screen to 1907.* Berkeley: University of California Press, 1990.

Narayan, Kirin. *Storytellers, Saints, and Scoundrels: Folk Narrative in Hindu Religious Teaching.* Philadelphia: University of Pennsylvania Press, 1989.

Nasaw, David. *Going Out: The Rise and Fall of Public Amusements.* New York: Basic, 1993.

Nash, Gary B., et al., eds. *The American People: Creating a Nation and a Society.* 3d ed. New York: HarperCollins, 1994.

Noer, Thomas J. *Briton, Boer and Yankee: The U.S. and South Africa, 1870–1914.* Kent, Ohio: Kent State University Press, 1978.

Omi, Michael, and Howard Winant. *Racial Formation in the United States: from the 1960s to the 1980s.* New York: Routledge & Kegan Paul, 1987.

Oriard, Michael. *Reading Football: How the Popular Press Created an American Spectacle.* Chapel Hill: University of North Carolina Press, 1993.

Ownby, Ted. *Subduing Satan: Religion, Recreation, and Manhood in the Rural South, 1865–1920.* Chapel Hill: University of North Carolina Press, 1990.

Page, George. *Inside the Animal Mind.* New York: Doubleday, 1999.

Pal, Pratapaditya, and Vidya Dehejia. *From Merchants to Emperors: British Artists and India, 1757–1930.* Ithaca, N.Y.: Cornell University Press, 1986.

Peacock, Shane. "Farini the Great." *Bandwagon* 34 (Sept.–Oct. 1990): 13–20.

Peiss, Kathy. *Hope in a Jar: The Making of America's Beauty Culture.* New York: Owl, 1998.

———. *Cheap Amusements: Working Women and Leisure in Turn-of-the-Century New York.* Philadelphia: Temple University Press, 1986.

Pfening, Fred D., Jr. "The Big Show of the World." *Bandwagon* 8 (Jan.–Feb. 1964): 4–15.

Pfening, Fred D., III. "The Frontier and the Circus." *Bandwagon* 15 (Sept.–Oct. 1971): 16–20.

Plowden, Gene. *Merle Evans, Maestro of the Circus.* Miami: E. A. Seemann, 1971.

———. *Those Amazing Ringlings and Their Circus.* Caldwell, Idaho: Caxton, 1967.

Plummer, Brenda Gayle. *Rising Wind: Black Americans and U.S. Foreign Affairs, 1935–1960.* Chapel Hill: University of North Carolina Press, 1996.

———. *Haiti and the U.S.: The Psychological Moment.* Athens: University of Georgia Press, 1992.

Poole, Deborah A. "A One-Eyed Gaze: Gender in 19th Century Illustrations of Peru." *Dialectical Anthropology* 13 (1988): 333–64.

Post, Robert C. *1876, A Centennial Exhibition.* Washington: National Museum of History and Technology, Smithsonian Institution, 1976.

Pratt, Mary Louise. *Imperial Eyes: Travel Writing and Transculturation.* London: Routledge, 1992.

Price, Jennifer. *Flight Maps: Adventures with Nature in Modern America.* New York: Basic, 1999.

———. "Looking for Nature at the Mall: A Field Guide to the Nature Company." In *Uncommon Ground: Rethinking the Human Place in Nature,* ed. William Cronon, 186–203. New York: W. W. Norton, 1995.

Rabinovitz, Lauren. *For the Love of Pleasure: Women, Movies, and Culture in Turn-of-the-Century Chicago.* New Brunswick: Rutgers University Press, 1998.

Renda, Mary. *Taking Haiti: Military Occupation and the Culture of U.S. Imperialism, 1915–1940*. Chapel Hill: University of North Carolina Press, 2001.

Ritvo, Harriet. *The Animal Estate: The English and Other Creatures in the Victorian Age.* Cambridge: Harvard University Press, 1987.

Roberts, Randy, and James Olson. *Winning Is the Only Thing: Sports in America since 1945*. Baltimore: Johns Hopkins University Press, 1989.

Roediger, David. "White Looks: Hairy Apes, True Stories and Limbaugh's Laughs." *Minnesota Review* 47 (May 1997): 37–48.

———. *The Wages of Whiteness: Race and the Making of the American Working Class.* New York: Verso, 1991.

Rooks, Noliwe. *Hair Raising: Beauty, Culture, and African American Women.* New Brunswick: Rutgers University Press, 1996.

Rosen, Ruth. *The Lost Sisterhood: Prostitution in America, 1900–1918*. Baltimore: Johns Hopkins University Press, 1982.

Rosenberg, Emily S. "'Foreign Affairs' after World War II: Connecting Sexual and International Politics." *Diplomatic History* 18 (Winter 1994): 59–70.

———. "Gender (A Round Table: Explaining the History of American Foreign Relations)." *Journal of American History* 77 (June 1990): 116–24.

———. *Spreading the American Dream: American Economic and Cultural Expansion, 1890–1945*. New York: Hill & Wang, 1982.

Rosenzweig, Roy. *Eight Hours for What We Will: Workers and Leisure in an Industrial City, 1870–1920*. New York: Cambridge University Press, 1983.

Rothfels, Nigel. "Aztecs, Aborigines, and Ape-People: Science and Freaks in Germany, 1850–1900." In *Freakery: Cultural Spectacles of the Extraordinary Body*, ed. Rosemarie Garland Thomson, 158–72. New York: New York University Press, 1996.

Russett, Cynthia Eagle. *Sexual Science: The Victorian Construction of Womanhood.* Cambridge: Harvard University Press, 1989.

Russell, Don. *The Lives and Legends of Buffalo Bill.* Norman: University of Oklahoma Press, 1960.

Ryan, Mary P. *Cradle of the Middle Class: The Family in Oneida County, New York, 1790–1865*. Cambridge: Cambridge University Press, 1981.

Rydell, Robert W. *World of Fairs: The Century-of-Progress Expositions.* Chicago: University of Chicago Press, 1993.

———. *All the World's a Fair: Visions of Empire at American International Expositions, 1876–1916*. Chicago: University of Chicago Press, 1984.

Said, Edward. *Culture and Imperialism.* New York: Vintage, 1994.

———. *Orientalism.* New York: Pantheon, 1978.

St. Leon, Mark. "An Unbelievable Lady Bareback Rider May Wirth." *Bandwagon* 34 (May–June 1990): 4–13.

Saxon, A. H. "New Light on the Life of James A. Bailey." *Bandwagon* 40 (Nov.–Dec. 1996): 4–9.

———. *Barnumiana: A Select, Annotated Bibliography of Works by or Relating to P. T. Barnum*. Fairfield, Conn.: Jumbo's, 1995.

———. *P. T. Barnum, The Legend and the Man.* New York: Columbia University Press, 1989.

Saxton, Alexander. *The Rise and Fall of the White Republic: Class Politics and Mass Culture in Nineteenth-Century America*. New York: Verso, 1990.

Schivelbusch, Wolfgang. *Railway Journey: The Industrialization of Time and Space in the Nineteenth Century*. Berkeley: University of California Press, 1987.

Sell, Henry Blackman, and Victor Weybright. *Buffalo Bill and the Wild West*. New York: Oxford University Press, 1955.

Senelick, Laurence. "'Boys and Girls Together': Subcultural Origins of Glamour Drag and Male Impersonation on the Nineteenth-Century Stage." In *Crossing the Stage: Controversies on Cross-Dressing*, ed. Lesley Ferris, 80–95. New York: Routledge, 1993.

———. "Lady and the Tramp: Drag Differentials in the Progressive Era." In *Gender in Performance: The Presentation of Difference in the Performing Arts*, ed. Senelick, 26–45. Hanover, N.H.: University Press of New England, 1992.

Singal, Daniel Joseph. "Toward a Definition of American Modernism." *American Quarterly* 39 (Spring 1987): 5–26.

Singer, Milton. "Passage to More than India: A Sketch of Changing European and American Images." In *When a Great Tradition Modernizes: An Anthropological Approach to Indian Civilization*, ed. Singer, 11–37. New York: Praeger, 1970.

Slotkin, Richard. *Gunfighter Nation: The Myth of the Frontier in Twentieth-Century America*. New York: Harper Perennial, 1992.

Slout, William L. *A Royal Coupling: The Historic Marriage of Barnum and Bailey*. San Bernardino, Calif.: Borgo, 2000.

———. *Olympians of the Sawdust Circle: A Biographical Dictionary of the Nineteenth Century American Circus*. San Bernardino, Calif.: Borgo, 1998.

Smith-Rosenberg, Carol. "The New Woman as Androgyne: Social Order and Gender Crisis, 1870–1936." In *Disorderly Conduct: Visions of Gender in Victorian America*, ed. Smith-Rosenberg, 245–96. New York: Alfred A. Knopf, 1985.

Snyder, Robert W. *The Voice of the City: Vaudeville and Popular Culture in New York*. New York: Oxford University Press, 1989.

Speaight, George. *A History of the Circus*. London: Tantivy, 1980.

Springhall, John O. "'Up Guards and at Them!' British Imperialism and Popular Art, 1880–1914." In *Imperialism and Popular Culture*, ed. John MacKenzie, 49–72. Manchester: Manchester University Press, 1990.

Staiger, Janet. *Perverse Spectators: The Practices of Film Reception*. New York: New York University Press, 2000.

———. *Interpreting Films: Studies in the Historical Reception of American Cinema*. Princeton, N.J.: Princeton University Press, 1992.

Stallybrass, Peter, and Allon White. *The Politics and Poetics of Transgression*. Ithaca, N.Y.: Cornell University Press, 1986.

Strasser, Susan. *Waste and Want: A Social History of Trash*. New York: Metropolitan, 1999.

Summerfield, Penny. "Patriotism and Empire: Music-Hall Entertainment 1870–1914." In *Imperialism and Popular Culture*, ed. John M. MacKenzie, 17–48. Manchester: Manchester University Press, 1986.

Susman, Warren I. *Culture as History: The Transformation of American Society in the Twentieth Century.* New York: Pantheon, 1985.

Taper, Bernard. *Balanchine: A Biography.* New York: Times Books, 1984.

Taylor, Robert Lewis. *Center Ring: The People of the Circus.* Garden City, N.Y.: Doubleday, 1956.

Thayer, Stuart. *Traveling Showmen: The American Circus before the Civil War.* Detroit: Astley & Ricketts, 1997.

———. "The Circus Roots of Negro Minstrelsy." *Bandwagon* 43 (Nov.–Dec. 1996): 43–45.

———. *Annals of the American Circus, 1830–1847.* Vol. 2. Seattle: Peanut Butter, 1986.

———. "The Birth of the Blues: Early Circus Seating." *Bandwagon* 29 (Sept.–Oct. 1985): 24–26.

———. "Legislating the Shows: Vermont, 1824–1933." *Bandwagon* 25 (July–Aug. 1981): 20.

———. "Some Class Distinctions in the Early Circus Audience." *Bandwagon* 24 (July–Aug. 1980): 20–21.

———. "The Anti-Circus Laws in Connecticut, 1773–1840." *Bandwagon* 20 (Jan.–Feb. 1976): 18–20.

———. *Annals of the American Circus, 1793–1829.* Vol. 1. Manchester, Mich.: Rymack, 1976.

———. "One Sheet." *Bandwagon* 18 (Sept.–Oct. 1974): 23.

———. "The Oriental Influence on the American Circus." *Bandwagon* 17 (Sept.–Oct. 1973): 20–23.

———. "A Note on the Decline of the Circus." *Bandwagon* 16 (July–Aug. 1972): 17.

Thompson, E. P. "Time, Work-Discipline, and Industrial Capitalism." *Past & Present* 38 (Dec. 1967): 56–97.

Thomson, Rosemarie Garland. "Introduction: From Wonder to Error—A Genealogy of Freak Discourse in Modernity." In *Freakery: Cultural Spectacles of the Extraordinary Body,* ed. Thomson, 1–19. New York: New York University Press, 1996.

Todd, Jan. "Bring on the Amazons: An Evolutionary History." In *Picturing the Modern Amazon,* ed. Johanna Freuh, Judith Stein, and Laurie Fierstein, 48–61. New York: Rizzoli & New Museum of Contemporary Art, 1999.

———. *Physical Culture and the Body Beautiful: Purposive Exercise in the Lives of American Women, 1800–1870.* Macon, Ga.: Mercer University Press, 1998.

Trachtenberg, Alan. *Reading American Photographs: Images as History, Mathew Brady to Walker Evans.* New York: Hill & Wang, 1989.

———. *The Incorporation of America: Culture and Society in the Gilded Age.* New York: Hill & Wang, 1982.

Truzzi, Marcello, ed. "Circuses, Carnivals and Fairs." *Journal of Popular Culture* 6 (Spring 1973): 529–619.

———. "The Decline of the American Circus: The Shrinkage of an Institution." In *Sociology and Everyday Life,* ed. Truzzi, 314–22. Englewood Cliffs, N.J.: Prentice-Hall, 1968.

———. "The American Circus as a Source of Folklore: An Introduction." *Southern Folklore Quarterly* 30 (Dec. 1966): 289–300.

Turner, James. *Reckoning with the Beast: Animals, Pain, and Humanity in the Victorian Mind.* Baltimore: Johns Hopkins University Press, 1980.

Vertinsky, Patricia. *The Eternally Wounded Woman: Women, Doctors, and Exercise in the Late Nineteenth Century.* Manchester: University of Manchester Press, 1990.

Vickery, Jim dale. *Wilderness Visionaries.* Merrillville, Ind.: ICS, 1986.

Von Eschen, Penny M. *Race against Empire: Black Americans and Anticolonialism, 1937–1957.* Ithaca, N.Y.: Cornell University Press, 1997.

Weeks, David C. *Ringling: The Florida Years, 1911–1936.* Gainesville: University Press of Florida, 1993.

Wexler, Laura. *Tender Violence: Domestic Visions in an Age of U.S. Imperialism.* Chapel Hill: University of North Carolina Press, 2000.

Wiebe, Robert H. *The Search for Order, 1877–1920.* New York: Hill & Wang, 1967.

Williams, Raymond. *Marxism and Literature.* Oxford: Oxford University Press, 1977.

Williams, Walter L. "United States Indian Policy and the Debate over Philippine Annexation: Implications for the Origins of American Imperialism." *Journal of American History* 66 (Mar. 1980): 810–31.

Williams, William Appleman. *The Tragedy of American Diplomacy.* New ed. New York: W. W. Norton, 1972 [1959].

Wolpert, Stanley. *A New History of India.* 3rd ed. New York: Oxford University Press, 1989.

Woodward, C. Vann. *The Strange Career of Jim Crow.* New York: Oxford University Press, 1955.

UNPUBLISHED PAPERS AND THESES

Abrahams, Yvette. "A Khoisan Contribution to Western Science: Or; The Life and Legacy of Sara Bartman." Seminar paper, Queens University, Kingston, Ont., 1994.

Carmeli, Yoram S. "The Sight of Cruelty—The Case of Circus Animal Acts," article draft, 1996.

Carlyon, David. "Dan Rice's Aspirational Project: The Nineteenth-Century Circus Clown and Middle Class Formation." Ph.D. diss., Northwestern University, 1993.

Friedman, Andrea S. "Prurient Interests: Anti-Obscenity Campaigns in New York City, 1909–1945." Ph.D. diss., University of Wisconsin-Madison, 1995.

Grier, Katherine. "'Why Can't You Talk?': Representing the Relationships of Animals and Children in Popular Images, 1820 to the Present." Center for Twentieth Century Studies, University of Wisconsin-Milwaukee, Apr. 2000.

Hyson, Jeffrey. "Urban Jungles: Zoos and American Society." Ph.D. diss., Cornell University, 1999.

Mischler, Doug. "The Greatest Show on Earth: The Circus and the Development of Modern American Culture, 1860–1940." Ph.D. diss., University of Nevada-Reno, 1994.

Mizelle, David Brett. "The Downfall of Taste and Genius—Animal Exhibitions and the Struggle over Acceptable Leisure in the Early Nineteenth Century." Circus Historical Society Annual Meeting, Normal, Ill., Sept. 2000.

Narayan, Kirin. "From Self-Torturing Bodies to the Guru of Good Times:

Representing American Representations of Hindu Ascetics, 1833–1990." South Asian Studies Association Annual Meeting, Madison, Wis., Nov. 1989.

Neather, Andrew. "Race, Patriotism, and Labor Responses to Empire, 1890–1914." Organization of American Historians Annual Meeting, Atlanta, Apr. 1994.

Oberdeck, Kathryn. "Women, Vaudeville and Cultural Hierarchy in Turn-of-the-Century America." Ninth Berkshire Conference on the History of Women, Poughkeepsie, N.Y., June 1993.

Sappol, Michael. "Reading *Sammy Tubbs:* Anatomical Dissection, Minstrelsy, and the Technology of Self-Making in Postbellum America." Organization of American Historians Annual Meeting, Atlanta, Apr. 1994.

Solomon, Jeremy. "Professional Wrestling: An Exploration." Seminar paper, University of Pennsylvania, Philadelphia, Dec. 1991.

Strange, Carolyn, and Tina Loo. "Spectacular Justice: The Circus on Trial and the Trial as Circus, Picton, 1903." Annual Meeting of the Canadian Historical Association, Calgary, Alberta, June 1994.

Tchen, John Kuo Wei. "Barnum and the 'Siamese Twins' Reconsidered: Identity Formation and the Politics of Difference." Organization of American Historians Annual Meeting, Atlanta, April 1994.

INDEX

Anderson, Sherwood, 2, 8, 143, 144, 215

"Anglo-Saxonism," 215–16

Animal rights, contemporary, 1, 234, 240 (n. 1)

Animals: as human representations, xii, 27, 149–51; and sexuality, 96; and "marriage," 96, 156, 157; and males as stars, 148–49; and human evolution, 152; and labor, 154; escape from circus, 154–55, 162; and hunting, 155, 195–97; execution of, 161–63. *See also* Acts, big-top; Masculinity; Women; *names of specific species*

Animal trade, xiii, 13, 17; international, 194–97; and European imperialism, 195–97, 240 (n. 1), 275 (n. 40)

Animal training: by women, 101–3, 160–61; kindness method of, 159; by men, 159–63

Animal welfare movement, 69, 150–54, 159, 275 (n. 40), 276 (n. 52), 277 (n. 75)

Anticircus sentiment, early-twentieth-century, 138; and obscenity, 138–39; and authorities, 272 (n. 221)

Anti-imperialism, 221–24

Anti-Imperialist League, 221

Arlington & Beckman's Oklahoma Ranch Wild West show, 127

Astley, Philip, 16

Athleticism. *See* Physical-culture movement

Attendance: at circus, 3, 6–7, 229, 293 (n. 4)

Audience, circus: size of, 3, 5; as spectacle, 6–7, 28; interaction with performers, 27–28, 118–19, 127, 173, 184, 186; and crime, 29–30; and fighting, 31–32, 174, 188–90, 224; seating of, 32; diversity of, 32, 34–35; African Americans in, 32–34, 175; and racial segregation, 32–35, 133–34, 247, (nn. 82, 84), 271 (n. 199); Chinese in, 34; Native Americans in, 34, 184, 247 (n. 85); and children, 34–36, 140, 143, 190; in antebellum era, 35,

86–87; and voyeurism, 127, 140–41, 272 (n. 222), 273 (n. 230)

Australian aborigines, 198–99

Baba, Lotan ("rolling saint"), 234

Bailey, James A., 11, 19, 21, 32, 40, 52, 77, 118–19, 122, 164; as self-made man, 54–56; and treatment of employees, 55–56, 68, 70, 80

Bailey, Mollie, 26, 244 (n. 45)

Baker, Johnny, 57

Bakhtin, Mikhail, 28

Balanchine, George, 294 (n. 8)

Ballet girls: sleeping accommodations of, 64–65; work/conduct rules of, 105, 107; hiring of, 111–12, 266 (n. 123); and racial disguise, 112

Barber, Lottie ("Jolly Dolly Dimples"), 26–27

Barnum, P. T., 10, 18, 20–21, 40, 42, 98, 120, 155–56, 162–63, 180, 182, 197–99; as "children's friend," 35; early life of, 53; finances of, 53, 251 (n. 53); as self-made man, 53–54; and temperance/dry workplace, 54, 77; in contrast to Bailey, 55; compared to Cody, 56; on antebellum circus, 86; and purity reform, 139–40

Barnum & Bailey circus, 2, 3, 6, 11, 13, 22–24, 30–31, 33–35, 37–38, 41, 43–44, 46–48, 60, 64, 66–67, 69–70, 78, 80–82, 84, 93–99, 102–4, 107–10, 113, 119–22, 124, 126, 130–33, 139, 149, 158, 164–66, 171–73, 179, 183, 185, 189, 193, 195, 200–201, 208–9, 211, 216–18, 220; physical growth of, 21; merger of, 21, 40, 254 (n. 97)

Barnum and London circus, 118–19

Barnum, Bailey and Hutchinson's circus, 198

Barnum, Coup and Castello circus, 19–21

Barnum, Coup and Castello's Great Traveling World's Fair, 20

Barnum's Animals (crackers), 35

lum, 86; for Wild West shows, 101, 282 (n. 184); safety reasons for styling of, 104–5; and work rules, 105, 107; and gowning movement, 108–10; and physical-culture movement, 110; for ballet girls, 111–15; for snake charmers, 123–24; for clowns, 169–71, 173, 178; for minstrel shows, 172–73; for male members of ethnological congress, 185–86

Coup, W. C., 18, 20, 42, 128, 196

Coxy, W. D., 109–10

Crime: on Circus Day, 29–30, 31–32, 174, 188–90, 224, 258 (n. 180), 279 (n. 135)

Cross, Gary, 35

Crusades, 214

Cuba, 9, 193, 206, 207–8, 209, 211, 222, 223

Cultural hierarchy, 32, 246 (n. 75), 247 (n. 91)

Cunningham, R. A., 198

Curtis, Bill ("Cap"), 70

Custer, George Armstrong, 203; and Battle of Little Big Horn, 203–4

Custer, Elizabeth, 204

Curzon, George, Lord, 218

Curzon, Lady, 219

Dahlinger, Fred, 19, 42, 240 (n. 2)

Dan Castello's Circus and Menagerie, 20

Darwin, Charles, 152

Darwinism (evolutionary theory), 176, 183; social, 9, 52; and discourse of "vanishing races," 58, 184; and juxtaposition of human and animal, 128–30, 183–84, 152, 223; and "Darwin" the monkey, 153; and "Race of the Races," 153

Davis, Susan, 28, 31–32

Dawes Allotment Act, 204

Deloria, Philip, 146

DeMott & Ward circus, 93

De Tiers, Mauricia, 103, 164

Dewey, George, 212

Dewey, John, 35, 290 (n. 117)

Diamonds, 73–74

Diana (princess of Wales), 81

Dime museums, 34, 139, 169

Dime novels, 146, 190, 205, 286 (n. 33)

"Dog and pony show," 24, 34, 74

Dogs, 12, 71, 97, 103, 116, 126, 147–48, 149–51, 154, 161, 171

Drag: and reinforcement of traditional female gender roles, 115–16; in cooch show, 127–28; and circus culture, 168–69; and clowns, 178–79

Dress: change in women's, 87; and anti-corset movement, 91–92

Drinnon, Richard, 205, 221

Du Bois, W. E. B., 112–13, 211, 221

Dudziak, Mary, 293 (n. 154)

Dumbo, 229

Durbar of India, 218

Dutton, Nellie, 26

Eagleton, Terry, 28–29

Economy: depressions and panics, 8, 40, 52, 144, 176; and circus "drain" on, 29; women's changing roles in, 88

Ederle, Gertrude, 116

Edward VII (king of England), 218

Electricity: advent of, at circus, 65, 228, 251 (n. 43)

Elephants, xi, xii, 1, 3, 11, 27, 38, 45, 49, 64, 71, 72, 81, 101, 103, 138, 143, 150, 154, 195, 196, 216, 218, 225, 229, 230, 236, 237; Crowninshield, 17–18; and "to see the elephant," 17–18; as laborers, 37, 46, 47; and white elephant war, 53, 252 (n. 58); Jumbo, 148, 155–56, 197; "John L. Sullivan, the Boxing Elephant," 149, 175; execution of, 161–63; "Tipoo Sultan," 215

Ellis, Havelock, 90

Eltinge, Julian, 169

Empire: dimensions of, 9, 193; and race, 9–10, 92, 119, 128–36, 159–60, 182–86, 210, 215–16, 221–23, 226; and popular culture, 35–36, 92, 118, 128, 155, 199, 248 (n. 102), 286 (n. 31); and continental expansion, 148, 184, 186, 203–6; and

American exceptionalism, 194, 202–3, 219, 225–26; formal versus informal, 194, 207, 221; "empire of liberty," 206, 210–11; and Turner's frontier thesis, 212; critics of, 220–23. *See also* Circus; Imperialism; Race; Spectacles

Employees, circus: and "holdback" pay, 79; and resistance, 79–80

Entertainment, circus, 13–14

Ethnological congress: decline of, 22; and labor, 49, 50, 131–33, 185; women in, 83, 131–32; and racial ideology, 96, 157, 183, 185–86, 206, 223; origins of, 118; composition of, 119–20; and exhibiting families, 131; and masculinity, 185–86; and representations of developing world, 232

Evans, Merle, 135, 294 (n. 8)

Ewing, Ella, 122–23

Excursion cars, 24, 45–46

Excursion fares, 2, 22–24

Expansionism. *See* Empire

Fagen, David, 222

Farini, El Nino ("Lulu"), 168

Farini, G. A., 128, 168, 180

Farini, Krao ("Missing Link"), 67, 128–31, 180, 128–31

Feld, Kenneth, 231, 234, 236

Fiction: circus-boy, 142–43; boy hunter, 148, 155; dime novels, 286 (n. 33)

Fires: and dangers of smoking, 16; at American Museum, 251 (n. 53); at Hartford, Conn. (July 6, 1944), 254 (n. 97)

"Flying Clarkonians," 21

Ford, Henry, 8, 103

Forepaugh, Adam, 53, 162, 252 (n. 58)

Forepaugh and Sells Bros. circus, 11, 108, 179, 207

Foucault, Michel, 261 (n. 7)

Fox, Richard, 115

Franzen, Wayne, 235, 237

Freaks, 20; "made" and "born," 118, 179; marriage of, 120, 122–23, 268

(n. 150); circus role of, 223; and medical technology, 271 (n. 210). *See also* Sideshow

French Revolution, 202–3

Friedman, Andrea, 138

Fuston, Frederick, 222

Garland, Hamlin, 6, 13–14

Garner, R. L.: and Chiko, 156–57

Gaylord, J. B., 198

Ghost Dance Movement, 204

Gilman, Charlotte Perkins, 9, 90, 133

"Giraffe-Neck Women of Burma" (Padungs), 136–38, 271 (n. 210)

Gollmar Brothers circus, 24, 66–67, 68, 71, 77, 127

Goodman, Theodosia, 115

Great Circus Parade, 230, 233, 237, 294 (n. 12)

Green & Waring's Eagle circus, 172

Grier, Katherine, 159

Grimaldi, Stephanie (princess of Monaco), 235–36

Guinness World Records, 234

Gumpertz, Sam, 40

Hagenbeck, Carl, 118, 159, 160; and animal trade, 195–96; and South Asian performers, 198, 268 (n. 143)

Hagenbeck, Lorenz, 195, 196, 281 (n. 177)

Hagenbeck-Wallace circus, 105, 136–38, 225

Haiti, 207, 211

Hall, G. Stanley, 35, 145, 190

Hall, Stuart, 25

Hamilton, Tody, 193–94

"Happy Family" menagerie, 96–98, 161

Haraway, Donna, 96, 157, 197

Harris, Townsend, 197

Hay, John, 213

Hayworth, Cheryl, 232

"Hey rube," 72–73

Hodgini, Albert ("Original Miss Daisy"), 115–16; and drag culture, 169

Hoffa, Jimmy, 229

Willson, Dixie, 99–98, 247 (n. 82)

Wirth, May, 9, 94; marriage of, 99–101; endorsements by, 116; early childhood of, 263 (n. 54)

Witt, Katarina, 232

Wisconsin: as circus center, 18

Woman's Christian Temperance Union, 89–90

Women: changing work roles of, 9, 88–89, 132–33; "New Woman," 12, 82–83, 91, 262 (n. 37); and race, 12, 83, 85, 90–93, 93–96, 97–100, 104–13, 115–16, 118–38; and marriage, 74, 95–96; as performers, 82–85, 93–96, 98–107, 110–15, 119–38; and nudity, 83, 92–93; and animals, 83, 96–97, 101–3, 123–26, 128–31; and exclusion from antebellum circus, 86–87; as reformers, 89; and higher education, 90; and physical-culture movement, 90–91, 110; and anticorset movement, 91–92; numbers of, in circus, 93; from old circus families, 93–95; and domestic ideal, 93–96, 98– 102, 105; as animal trainers, 101–3; in automobile acts, 103–4; costumes for, 104–5, 111–12, 185–88; conduct codes for, 105, 107

Workingmen. *See* Roustabouts

World's Columbian Exposition, 131, 219

World's fairs, 13, 24, 34, 92, 119, 195, 200, 216, 219

"World's Most Shocking Medical Videos," 233–34

World War I, 78, 101, 182, 224–25

World War II, 226, 254 (n. 97)

Wovoka (Paiute holy man), 204

Yankee Robinson and Ringling Brothers Great Double Shows, 60

Yellow Hand, 203, 287 (n. 44)

Ziegfeld Follies, 111, 261 (n. 19)

Zoological garden: and relationship to ethnological congress, 118, 268 (nn. 142, 143)

Zora, Lucia, 101–2, 265 (n. 84)